CAMBRIDGE LIBRARY COLLECTION

Books of enduring scholarly value

Botany and Horticulture

Until the nineteenth century, the investigation of natural phenomena, plants and animals was considered either the preserve of elite scholars or a pastime for the leisured upper classes. As increasing academic rigour and systematisation was brought to the study of 'natural history', its subdisciplines were adopted into university curricula, and learned societies (such as the Royal Horticultural Society, founded in 1804) were established to support research in these areas. A related development was strong enthusiasm for exotic garden plants, which resulted in plant collecting expeditions to every corner of the globe, sometimes with tragic consequences. This series includes accounts of some of those expeditions, detailed reference works on the flora of different regions, and practical advice for amateur and professional gardeners.

Hortus Kewensis

Trained as a gardener in his native Scotland, William Aiton (1731–93) had worked in the Chelsea Physic Garden prior to coming to Kew in 1759. He met Joseph Banks in 1764, and the pair worked together to develop the scientific and horticultural status of the gardens. Aiton had become superintendent of the entire Kew estate by 1783. This important three-volume work, first published in 1789, took as its starting point the plant catalogue begun in 1773. In its compilation, Aiton was greatly assisted with the identification and scientific description of species, according to the Linnaean system, by the botanists Daniel Solander and Jonas Dryander (the latter contributed most of the third volume). Aiton added dates of introduction and horticultural information. An important historical resource, it covers some 5,600 species and features a selection of engravings. Volume 3 covers Diadelphia to Cryptogamia, and includes addenda and indexes of generic and English names.

Cambridge University Press has long been a pioneer in the reissuing of out-of-print titles from its own backlist, producing digital reprints of books that are still sought after by scholars and students but could not be reprinted economically using traditional technology. The Cambridge Library Collection extends this activity to a wider range of books which are still of importance to researchers and professionals, either for the source material they contain, or as landmarks in the history of their academic discipline.

Drawing from the world-renowned collections in the Cambridge University Library and other partner libraries, and guided by the advice of experts in each subject area, Cambridge University Press is using state-of-the-art scanning machines in its own Printing House to capture the content of each book selected for inclusion. The files are processed to give a consistently clear, crisp image, and the books finished to the high quality standard for which the Press is recognised around the world. The latest print-on-demand technology ensures that the books will remain available indefinitely, and that orders for single or multiple copies can quickly be supplied.

The Cambridge Library Collection brings back to life books of enduring scholarly value (including out-of-copyright works originally issued by other publishers) across a wide range of disciplines in the humanities and social sciences and in science and technology.

Hortus Kewensis

Or, a Catalogue of the Plants Cultivated in the Royal Botanic Garden at Kew

Volume 3:
Diadelphia to Cryptogamia

William Aiton

CAMBRIDGE
UNIVERSITY PRESS

University Printing House, Cambridge, CB2 8BS, United Kingdom

Published in the United States of America by Cambridge University Press, New York

Cambridge University Press is part of the University of Cambridge.
It furthers the University's mission by disseminating knowledge in the pursuit of education, learning and research at the highest international levels of excellence.

www.cambridge.org
Information on this title: www.cambridge.org/9781108069694

This edition first published 1789
This digitally printed version 2014

ISBN 978-1-108-06969-4 Paperback

HORTUS KEWENSIS;

OR, A

CATALOGUE

OF THE

PLANTS

CULTIVATED IN THE

ROYAL BOTANIC GARDEN AT KEW.

BY WILLIAM AITON,

GARDENER TO

HIS MAJESTY.

IN THREE VOLUMES.

VOL. III.

DIADELPHIA — CRYPTOGAMIA.

LONDON:

PRINTED FOR GEORGE NICOL, BOOKSELLER TO HIS MAJESTY, PALL MALL.

M.DCC.LXXXIX.

Classis XVII.

DIADELPHIA
HEXANDRIA.

FUMARIA. *Gen. pl.* 849.

Cal. diphyllus. *Cor.* ringens. Filamenta 2, membranacea, singula *Antheris* 3.

* *Corollis bicalcaratis.*

1. F. scapo nudo. *Sp. pl.* 983. *Cucullaria.*
Naked-stalk'd Fumitory.
Nat. of Virginia and Canada.
Cult. 1759, by Mr. Philip Miller. *Mill. dict. edit.* 7. *n.* 9.
Fl. June and July. H. ♃.

2. F. floribus basi bigibbis, siliquis linearibus ancipitibus corolla fungoso-inflata tectis, foliis scandentibus. *fungosa.*
Spongy-flower'd Fumitory.
Nat. of North America.
Introd. about 1778, by John Fothergill, M.D.
Fl. June——September. H. ⊙.

** *Corollis unicalcaratis.*

3. F. caulibus simplicibus, bracteis flore brevioribus indivisis. *Syst. veget.* 636. *Jacqu. hort.* 2. *p.* 53. *t.* 116. *nobilis.*
Great-flower'd Fumitory.
Nat. of Siberia.

Introd. 1783, by Mr. John Græfer.
Fl. May. H. ♃.

bulbosa. 4. F. caule simplici, bracteis longitudine florum. *Sp. pl.* 983.

cava. α Fumaria bulbosa radice cava major. *Bauh. pin.* 143.
Hollow-rooted Bulbous Fumitory.

solida. β Fumaria bulbosa radice non cava major. *Bauh. pin.* 144.
Solid-rooted Bulbous Fumitory.
Nat. of Europe.
Cult. 1596, by Mr. John Gerard. *Hort. Ger.*
Fl. February——April. H. ♃.

sempervirens. 5. F. siliquis linearibus paniculatis, caule erecto. *Sp. pl.* 984.
Glaucous Fumitory.
Nat. of North America.
Cult. 1683, by Mr. James Sutherland. *Sutherl. hort. edin.* 125. *n.* 3.
Fl. July and August. H. ☉.

lutea. 6. F. siliquis teretibus, caulibus diffusis: angulis obtusis. *Linn. mant.* 258.
Yellow Fumitory.
Nat. of Barbary.
Cult. 1596, by Mr. John Gerard. *Hort. Ger.*
Fl. April——October. H. ♃.

capnoides. 7. F. siliquis linearibus tetragonis, caulibus diffusis acutangulis. *Sp. pl.* 984.
White-flower'd Fumitory.
Nat. of the South of Europe.
Cult. 1596, by Mr. John Gerard. *Hort. Ger.*
Fl. May——October. H. ☉.

8. F.

8. F. pericarpiis monofpermis racemofis, caule diffufo. *Sp. pl.* 984. *Curtis lond.* *officinalis.*
Common Fumitory.
Nat. of Britain.
Fl. May——Auguft. H. ☉.

9. F. pericarpiis monofpermis racemofis, foliis fcandentibus fubcirrhofis. *Sp. pl.* 985. *capreolata.*
Ramping Fumitory.
Nat. of Britain.
Fl. May——Auguft. H. ☉.

10. F. pericarpiis monofpermis fpicatis, caule erecto, foliolis filiformibus. *Sp. pl.* 985. *fpicata.*
Narrow-leav'd Fumitory.
Nat. of the South of Europe.
Cult. 1714. *Philofoph. tranf. n.* 344. *p.* 279. *n.* 74.
Fl. July and Auguft. H. ☉.

11. F. filiquis linearibus, foliis cirrhiferis. *Sp. pl.* 985. *claviculata.*
Climbing Fumitory.
Nat. of Britain.
Fl. June and July. H. ☉.

12. F. filiquis globofis acutis inflatis, foliis cirrhiferis. *Sp. pl.* 985. *vefìcaria.*
Bladder'd Fumitory.
Nat. of the Cape of Good Hope.
Cult. 1696, in Chelfea Garden. *Pluk. alm.* 400. *t.* 335. *f.* 3.
Fl. July. G. H. ☉.

OCTANDRIA.

POLYGALA. *Gen. pl.* 851.

Cal. 5-phyllus: foliolis alæformibus, coloratis. *Legumen* obcordatum, biloculare.

* *Criſtatæ (flores appendice penicilliformi.)*

amara. 1. P. floribus criſtatis racemoſis, caulibus erectiuſculis, foliis radicalibus obovatis majoribus. *Sp. pl.* 987. *Jacqu. auſtr.* 5. *p.* 6. *t.* 412.
Bitter Milk-wort.
Nat. of Europe.
Introd. 1775, by the Doctors Pitcairn and Fothergill.
Fl. June. H. ♃.

vulgaris. 2. P. floribus criſtatis racemoſis, caulibus herbaceis ſimplicibus procumbentibus, foliis lineari-lanceolatis. *Sp. pl.* 986.
α floribus cæruleis.
Blue Milk-wort.
β floribus albis.
White Milk-wort.
γ floribus carneis.
Fleſh-colour'd Milk-wort.
Nat. of Britain.
Fl. May and June. H. ♃.

bracteolata. 3. P. floribus criſtatis racemoſis, bracteolis triphyllis, foliis lineari-lanceolatis, caule fruticoſo. *Syſt. veget.* 638.
Spear-leav'd Milk-wort.

Nat.

Nat. of the Cape of Good Hope.
Introd. 1787, by Mr. Francis Masson.
Fl. G. H. ♄.

4. P. floribus cristatis: carina lunulata, caule fruticoso, foliis lævibus oblongis obtusis. *Sp. pl.* 988. *myrtifolia.*
Myrtle-leav'd Milk-wort.
Nat. of the Cape of Good Hope.
Cult. 1707, by the Dutchess of Beaufort. *Br. Mus. H. S.* 133. *fol.* 2.
Fl. Most part of the Year. G. H. ♄.

5. P. floribus cristatis lateralibus, caule fruticoso, ramis spinosis, foliis sparsis ovali-oblongis, fructu drupaceo. *spinosa.*
α foliis ovalibus obtusis. ovalifolia.
Polygala spinosa. *Sp. pl.* 989.
Oval-leav'd prickly Milk-wort.
β foliis lanceolato-linearibus glabris acutis, ramis ramulisque foliigeris. angustata.
Ulex capensis. *Sp. pl.* 1046.
Narrow-leav'd prickly Milk-wort.
Nat. of the Cape of Good Hope.
Cult. 1759. *Mill. dict. edit.* 7. Ulex 2.
Fl. January and February. G. H. ♄.

** *Imberbes (flores absque penicillo carinali) frutescentes.*

6. P. floribus imberbibus sparsis: carina apice subrotundo, caule fruticoso, foliis lanceolatis. *Sp. pl.* 989. *Jacqu. austr.* 3. *p.* 19. *t.* 233. *Chamæbuxus.*
Box-leav'd Milk-wort.
Nat. of Austria and Switzerland.
Cult. 1658, in Oxford Garden. *Hort. oxon. edit.* 2. *p.* 34.
Fl. May and June. H. ♄.

Heisteria. 7. P. floribus imberbibus lateralibus, caule arborescente, foliis triquetris mucronato-spinosis. *Sp. pl.* 989.
Heath-leav'd Milk-wort.
Nat. of the Cape of Good Hope.
Introd. 1787, by Mr. Francis Masson.
Fl. G. H. ♄.

*** *Imberbes herbaceæ, caule simplicissimo.*

Senega. 8. P. floribus imberbibus spicatis, caule erecto herbaceo simplicissimo, foliis lato-lanceolatis. *Sp. pl.* 990.
Officinal Milk-wort, or Rattlesnake-root.
Nat. of North America.
Cult. 1759, by Mr. Ph. Miller. *Mill. dict. edit.* 7. *n.* 5.
Fl. July. H. ♃.

SECURIDACA. *Gen. pl.* 852.

Cal. 3-phyllus. *Cor.* papilionacea: vexillo diphyllo intra alas. *Legumen* ovatum, uniloculare, monospermum, desinens in alam ligulatam.

scandens. 1. S. caule scandente. *Sp. pl.* 992.
Climbing Securidaca.
Nat. of the West Indies.
Introd. 1786, by Mr. Alexander Anderson.
Fl. S. ♄.

DECANDRIA.

NISSOLIA. *Gen. pl.* 853.

Cal. 5-dentatus. *Caps.* monosperma, desinens in alam ligulatam.

fruticosa. 1. N. caule fruticoso volubili. *Sp. pl.* 992. *Jacqu. hort.* 2. *p.* 78. *t.* 167.

Shrubby

Shrubby Niſſolia.
Nat. of South America.
Cult. 1766, by Robert James Lord Petre.
Fl. July——November. S. ♄.

A B R U S.

Cal. obſolete quadrilobus: lobo ſuperiore latiore. *Filamenta* 9, baſi infima connata, dorſo hiantia. *Stigma* obtuſum. *Sem.* ſphærica.

1. Abrus. *Syſt. veget.* 641. *precatorius.*
Glycine Abrus. *Sp. pl.* 1025.
Jamaica Wild Liquorice.
Nat. of both Indies.
Cult. before 1680, by Biſhop Compton. *Moriſ. hiſt.* 2. *p.* 71.
Fl. S. ♄.

P T E R O C A R P U S. *Gen. pl.* 854.

Cal. 5-dentatus. *Capſ.* falcata, foliacea, varicoſa. *Sem.* aliquot ſolitaria.

1. P. foliis ſimplicibus aggregatis obovatis aveniis. *Syſt. veget.* 642. *buxifolius.*
Aſpalathus Ebenus. *Syſt. veget.* 647.
Box-leav'd Pterocarpus.
Nat. of the Weſt Indies.
Cult. 1739, by Mr. Philip Miller. *Rand. chel.* Spartium 3.
Fl. July and Auguſt. S. ♄.

ERYTHRINA. *Gen. pl.* 855.

Cal. 2-labiatus : $\frac{1}{1}$. *Cor.* vexillum longissimum, lanceolatum.

herbacea. 1. E. foliis ternatis, caulibus simplicissimis fruticoso-annuis. *Sp. pl.* 992.
Herbaceous Coral-tree.
Nat. of Carolina.
Introd. 1724, by Mr. Mark Catesby. *Mill. dict. edit.* 8.
Fl. September. G. H. ♃.

carnea. 2. E. foliis ternatis glabris, caule arboreo aculeato, calycibus campanulatis truncatis.
Erythrina americana. *Mill. dict.*
Flesh-colour'd Coral-tree.
Nat. of Vera Cruz.
Cult. 1759, by Mr. Ph. Miller. *Mill. dict. edit.* 7. *n.* 5.
Fl. May. S. ♄.

Corallodendrum. 3. E. foliis ternatis inermibus, caule arboreo aculeato, calycibus truncatis quinquedentatis.
Erythrina Corallodendrum. *Sp. pl.* 992.
Smooth-leav'd Coral-tree.
Nat. of the West Indies.
Cult. 1714, by the Dutchess of Beaufort. *Br. Mus. H. S.* 132. *fol.* 2.
Fl. May and June. S. ♄.

picta. 4. E. foliis ternatis aculeatis, caule arboreo aculeato. *Sp. pl.* 993.
Prickly-leav'd Coral-tree.
Nat. of India.

Cult.

Cult. 1768, by Mr. Philip Miller. *Mill. dict. edit.* 8.
Fl. S. ♄.

5. E. foliis ternatis: petiolis subaculeatis glandulosis, caule arboreo inermi. *Linn. mant.* 99. *crista galli.*
Cock's-comb Coral-tree.
Nat. of Brasil.
Introd. 1771, by Francis Bearsly, Esq.
Fl. S. ♄.

PISCIDIA. *Gen. pl.* 856.

Stigma acutum. *Legumen* quadrifariam alatum.

1. P. foliolis ovatis. *Sp. pl.* 993. *Erythrina.*
Jamaica Dogwood.
Nat. of the West Indies.
Cult. 1690, in the Royal Garden at Hampton-court. *Catal. mss.*
Fl. S. ♄.

BORBONIA. *Gen. pl.* 857.

Stigma emarginatum. *Cal.* acuminato-spinosus. *Legum.* mucronatum.

1. B. foliis lanceolatis multinerviis integerrimis. *Sp. pl.* 994. *lanceolata.*
Spear-leav'd Borbonia.
Nat. of the Cape of Good Hope.
Cult. 1748, by Mr. Philip Miller. *Mill. dict. edit.* 5. Genista 13.
Fl. July and August. G. H. ♄.

2. B. foliis cordatis multinerviis denticulatis. *Syst. veget.* 643. *crenata.*

Heart-

Heart-leav'd Borbonia.
Nat. of the Cape of Good Hope.
Introd. 1774, by Mr. Francis Maſſon.
Fl. Auguſt. G. H. ♄.

SPARTIUM. *Gen. pl.* 858.

Stigma longitudinale, ſupra villoſum. *Filam.* germini adhærentia. *Cal.* deorſum productus.

contaminatum. 1. S. ramis teretibus, foliis alternis filiformibus baſi contaminatis. *Linn. mant.* 268.
Narrow-leav'd Broom.
Nat. of the Cape of Good Hope.
Introd. 1787, by Mr. Francis Maſſon.
Fl. G. H. ♄.

junceum. 2. S. ramis oppoſitis teretibus apice floriferis, foliis lanceolatis. *Sp. pl.* 995.
ſimplex. α floribus ſimplicibus.
Common Spaniſh Broom.
plenum. β floribus plenis.
Double-flower'd Spaniſh Broom.
Nat. of the South of Europe.
Cult. 1562, by Lord Cobham. *Turn. herb. part* 2. *fol.* 144.
Fl. July——September. H. ♄.

monoſpermum. 3. S. ramis teretibus ſtriatis, racemis paucifloris: floribus ſubaggregatis, foliis lanceolatis ſericeis.
Spartium monoſpermum. *Sp. pl.* 995.
White-flower'd Single-ſeeded Broom.
Nat. of Spain and Portugal.
Introd. 1690, by Mr. Bentick. *Br. Muſ. Sloan. mſſ.* 3370.
Fl. June and July. G. H. ♄.

4. S. ramis teretibus ſtriatis, racemis multifloris: floribus remotis, foliis lanceolatis ſeſſilibus ſubtus pubeſcentibus. *ſphærocarpum.*

Spartium ſphærocarpum. *Linn. mant.* 571.

Yellow-flower'd Single-ſeeded Broom.

Nat. of the South of Europe.

Introd. 1778, by Monſ. Thouin.

Fl. June and July. G. H. ♄.

5. S. ramis teretibus ſtriatis, foliis lanceolato-oblongis ſericeis, calycibus infundibuliformibus bilabiatis hirtis, vexillo carinaque pubeſcentibus. *virgatum.*

Cytiſus tener. *Jacqu. ic. collect.* 1. *p.* 40.

Nat. of Madeira. Mr. *Francis Maſſon.*

Introd. 1777

Fl. March——June. G. H. ♄.

6. S. caule decumbente ramoſo, foliis ſolitariis ovatis, floribus longe petiolatis. *Hall. hiſt.* 355. *Durande bourg.* 1. *p.* 299. *decumbens.*

Geniſta pedunculata. *L'Herit. ſtirp. nov. tab.* 89.

Trailing Broom.

Nat. of France and Switzerland.

Introd. 1775, by the Doctors Pitcairn and Fothergill.

Fl. May and June. H. ♄.

7. S. ramis ſpinoſis patentibus, foliis ovatis. *Sp. pl.* 995. *Scorpius.*

Scorpion Broom.

Nat. of the South of Europe.

Cult. 1640, by Mr. John Parkinſon. *Park. theat.* 999. *n.* 3.

Fl. March and April. H. ♄.

8. S. foliis ternatis ſimplicibuſque ſericeis, virgultis ſtrictis ſtriatis undique floridis. *L'Herit. ſtirp. nov. tab.* 87. *multiflorum.*

Portugal

Portugal white Broom.
Nat. of Portugal.
Introd. about 1770, by Mr. James Gordon.
Fl. May. H. ♄.

scoparium. 9. S. foliis simplicibus ternatisque, ramis acutangulis, floribus axillaribus solitariis, leguminibus margine villosis.
Spartium scoparium. *Sp. pl.* 996. *Curtis lond.*
Common Broom.
Nat. of Britain.
Fl. April——June. H. ♄.

patens. 10. S. foliis ternatis, ramis virgatis, floribus lateralibus geminis cernuis. *Syst. veget.* 644. *L'Herit. stirp. nov. tab.* 86.
Cytisus patens. *Syst. veget.* 666.
Cytisus pendulinus. *Linn. suppl.* 328. *Syst. veget.* 667.
Woolly-podded Broom.
Nat. of Portugal.
Introd. about 1764, by Mr. James Gordon.
Fl. June and July. H. ♄.

sericeum. 11. S. inerme sericeum, foliis ternatis: foliolis linearibus, racemis terminalibus, ramis angulatis.
Silky Broom.
Nat. of the Cape of Good Hope. Mr. *Fr. Masson.*
Introd. 1774.
Fl. April. G. H. ♄.

cytisoides. 12. S. inerme sericeum, foliis ternatis: foliolis lanceolatis obtusiusculis, racemis terminalibus, ramis teretibus.
Spartium cytisoides. *Berg. cap.* 199. (exclusis synonymis) *Linn. suppl.* 320.

Cytisus-

Cytifus-leav'd Broom.
Nat. of the Cape of Good Hope.
Introd. 1774, by Mr. Francis Maffon.
Fl. April. G. H. ♄.

13. S. foliis ternatis lanceolatis pilofis petiolatis, floribus lateralibus fafciculatis, leguminibus glabris, ramis teretibus ftriatis. *nubigena.*
Spartium fupranulium. *Linn. fuppl.* 319.
Clufter-flower'd Spartium.
Nat. of the Pic of Teneriffe. Mr. *Francis Maffon.*
Introd. 1779.
Fl. G. H. ♄.

14. S. foliis ternatis linearibus feffilibus, petiolis perfiftentibus, ramis oppofitis angulatis. *Sp. pl.* 996. *radiatum.*
Starry Broom.
Nat. of Italy.
Cult. 1758, by Mr. Ph. Miller. *Mill. ic.* 173. *t.* 259. *f.* 1.
Fl. June and July. H. ♄.

15. S. foliis ternatis, ramis angulatis fpinofis. *Sp. pl.* 997. *fpinofum.*
Prickly Broom, or Cytifus.
Nat. of the South of Europe.
Cult. 1640, by Mr. John Parkinfon. *Park. theat.* 999. *f.* 4.
Fl. June and July. G. H. ♄.

GENISTA. *Gen. pl.* 859.

Cal. 2-labiatus: $\frac{2}{3}$. *Vexillum* oblongum, a piftillo ftaminibufque deorfum reflexum.

* *Inermes.*

1. G. foliis ternatis utrinque pubefcentibus, ramis angulatis. *Syft. veget.* 645. *canarienfis.*

Canary

Canary Genista, or Cytisus.
Nat. of Spain and the Canaries.
Cult. 1656, by Mr. John Tradescant, Jun. *Mus. Trad.* 107.
Fl. May——September. G. H. ♄.

candicans. 2. G. foliis ternatis subtus villosis, pedunculis lateralibus subquinquefloris foliatis, leguminibus hirsutis. *Sp. pl.* 997.
Hoary Genista, or Montpelier Cytisus.
Nat. of Spain, Italy, and France.
Cult. 1748, by Mr. Philip Miller. *Mill. dict. edit.* 5. Cytisus 9.
Fl. April——June. H. ♄.

linifolia. 3. G. foliis ternatis sessilibus linearibus subtus sericeis. *Sp. pl.* 997.
Flax-leav'd Genista, or Broom.
Nat. of Spain.
Introd. 1786, by Sir Francis Drake, Bart.
Fl. Most part of the Summer. G. H. ♄.

triquetra. 4. G. foliis ternatis: summis simplicibus, ramis triquetris procumbentibus. *L'Herit. stirp. nov. tab.* 88.
Triangular Genista, or Broom.
Nat. of Corsica.
Cult. 1770, by John Ord, Esq.
Fl. May and June. H. ♄.

sagittalis. 5. G. ramis ancipitibus membranaceis articulatis, foliis ovato-lanceolatis. *Syst. veget.* 645. *Jacqu. austr.* 3. *p.* 5. *t.* 209.
Jointed Genista, or Broom.
Nat. of Germany, France, and Italy.
Cult. 1758, by Mr. Ph. Miller. *Mill. ic.* 173. *t.* 259. *f.* 2.
Fl. May and June. H. ♄.

6. G.

6. G. foliis lanceolatis glabris, ramis ſtriatis teretibus erectis. *Sp. pl.* 998. *tinctoria.*
Common Dyer's Geniſta, or Broom.
Nat. of England.
Fl. June——Auguſt. H. ♄.

7. G. foliis lanceolatis ſericeis, ramis ſtriatis teretibus, racemis ſecundis. *Sp. pl.* 998. *florida.*
Spaniſh Dyer's Geniſta, or Broom.
Nat. of Spain.
Cult. 1768, by Mr. Philip Miller. *Mill. dict. edit.* 8.
Fl. June——Auguſt. H. ♄.

8. G. foliis lanceolatis obtuſis, caule tuberculato decumbente. *Sp. pl.* 999. *Jacqu. auſtr.* 3. *p.* 5. *t.* 208. *piloſa.*
Hairy Geniſta, or Broom.
Nat. of England.
Fl. May and June. H. ♄.

** *Spinoſæ.*

9. G. ſpinis ſimplicibus, ramis floriferis inermibus, foliis lanceolatis. *Sp. pl.* 999. *anglica.*
Petty-whin.
Nat. of Britain.
Fl. May and June. H. ♄.

10. G. ſpinis compoſitis, ramis floriferis inermibus, foliis lanceolatis. *Sp. pl.* 999. *germanica.*
German Geniſta, or Broom.
Nat. of Germany.
Introd. 1773, by John Earl of Bute.
Fl. June——Auguſt. H. ♄

11. G. ſpinis decompoſitis, ramis floriferis inermibus, foliis linearibus piloſis. *Sp. pl.* 999. *hiſpanica.*

§ Dwarf

Dwarf prickly Geniſta, or Broom.
Nat. of Spain.
Cult. 1759, by Mr. Ph. Miller. *Mill. dict. edit.* 7. *n.* 9.
Fl. June and July. H. ♄.

luſitanica. 12. G. caule aphyllo, ſpinis decuſſatis. *Sp. pl.* 999.
Portugal Geniſta, or Broom.
Nat. of Portugal.
Introd. 1771, by Mrs. Primmet.
Fl. March——May. H. ♄.

ASPALATHUS. *Gen. pl.* 860.

Cal. 5-fidus: lacinia ſuperiore majore. *Legum.* ovatum, muticum, ſubdiſpernum.

albens. 1. A. foliis faſciculatis ſubulatis ſericeis apice patulis, faſciculis floreis ſparſis. *Syſt. veget.* 646.
Silky Aſpalathus.
Nat. of the Cape of Good Hope,
Introd. 1774, by Mr. Francis Maſſon.
Fl. July. G. H. ♄.

pedunculata. 2. A. foliis faſciculatis ſubulatis glabris, pedunculis filiformibus folio duplo longioribus. *L'Herit. ſert. angl. tab.* 26.
Small-leav'd Aſpalathus.
Nat. of the Cape of Good Hope. Mr. *Fr. Maſſon.*
Introd. 1775.
Fl. Auguſt. G. H. ♄.

indica. 3. A. foliis quinatis ſeſſilibus, pedunculis unifloris. *Sp. pl.* 1001.
Small-flower'd Aſpalathus.
Nat. of the Eaſt Indies.

Cult.

Cult. 1759, by Mr. Ph. Miller. *Mill. dict. edit.* 7. *n.* 2.
Fl. S. ♄.

4. A. foliis trinis linearibus ſericeis, ſtipulis ſimplicibus mucronatis, floribus ſparſis tomentoſis. *Sp. pl.* 1002. *argentea.*
Silvery Aſpalathus.
Nat. of the Cape of Good Hope.
Cult. 1759, by Mr. Ph. Miller. *Mill. dict. edit.* 7. *n.* 3.
Fl. July and Auguſt. G. H. ♄.

5. A. foliis trinis faſciculatiſque filiformibus ſericeis, floribus ſublateralibus, vexillis nudis. *candicans.*
White Aſpalathus.
Nat. of the Cape of Good Hope. Mr. *Fr. Maſſon.*
Intord. 1774.
Fl. June and July. G. H. ♄.

U L E X. *Gen. pl.* 881.

Cal. 2-phyllus. *Legum.* vix calyce longius.

1. U. foliis villoſis acutis; ſpinis ſparſis. *Sp. pl.* 1045. *europæus.*
α Geniſta ſpinoſa major longioribus aculeis. *Bauh. pin.* 394. vulgaris.
Common Furze, or Whin.
β Geniſta ſpinoſa minor. *Park. theat.* 1003. nanus.
Dwarf Whin.
Nat. of Britain.
Fl. α. April and May; β. Auguſt. H. ♄.

A M O R P H A. *Gen. pl.* 861.

Corollæ vexillum ovatum, concavum. *Alæ* o. *Carina* o.

1. Amorpha. *Sp. pl.* 1003. *fruticoſa.*
Shrubby Baſtard Indigo.

Nat. of Carolina.
Introd. 1724, by Mr. Mark Catesby. *Mill. ic.* 18. *t.* 27.
Fl. June and July. H. ♄.

CROTALARIA. *Gen. pl.* 862.

Legumen turgidum, inflatum, pedicellatum. *Filam.* connata cum fissura dorsali.

* *Foliis simplicibus.*

perfoliata. 1. C. foliis perfoliatis cordato-ovatis. *Sp. pl.* 1003.
Perfoliate Crotalaria.
Nat. of Carolina.
Cult. 1732, by James Sherard, M.D. *Dill. elth.* 122. *t.* 102. *f.* 122.
Fl. August. G. H. ♃.

sagittalis. 2. C. foliis simplicibus lanceolatis, petiolis decurrentibus solitariis bidentatis. *Syst. veget.* 649.
Virginian Crotalaria.
Nat. of America.
Introd. 1734, by Mr. Robert Millar. *Mart. dec.* 5. *p.* 43.
Fl. June. S. ☉.

juncea. 3. C. foliis simplicibus lanceolatis petiolato-sessilibus, caule striato. *Sp. pl.* 1004.
Channel-stalk'd Crotalaria.
Nat. of the East Indies.
Cult. 1768, by Mr. Philip Miller. *Mill. dict. edit.* 8.
Fl. June and July. S. ☉.

retusa. 4. C. foliis simplicibus oblongis cuneiformibus retusis. *Sp. pl.* 1004.
Wedge-leav'd Crotalaria.

Nat.

Nat. of the East Indies.
Cult. 1731, by Mr. Philip Miller. *Mill. dict. edit.* 1. *n.* 2.
Fl. June and July. S. ☉.

5. C. foliis simplicibus ovatis sessilibus glabris, ramis angulatis, pedunculis ternis lateralibus unifloris. *Sp. pl.* 1004. *triflora.*
Three-flower'd Crotalaria.
Nat. of the Cape of Good Hope.
Introd. 1786, by Mr. Francis Masson.
Fl. June. G. H. ♂.

6. C. foliis simplicibus ovatis, stipulis lunatis declinatis, ramis tetragonis. *Sp. pl.* 1005. *verrucosa.*
Blue-flower'd Crotalaria.
Nat. of the East Indies.
Cult. 1731, by Mr. Philip Miller. *Mill. dict. edit.* 1. *n.* 1.
Fl. July and August. S. ☉.

** *Foliis compositis.*

7. C. foliis ternatis oblongis glabris, floribus lateralibus subracemosis, leguminibus sessilibus. *lotifolia.*
Crotalaria latifolia. *Sp. pl.* 1005.
Lotus-leav'd Crotalaria.
Nat. of Jamaica.
Cult. 1732, by James Sherard, M.D. *Dill. elth.* 121. *t.* 102. *f.* 121.
Fl. June and July. S. ☉.

8. C. foliis ternatis cuneiformibus, ramis pubescentibus, leguminibus pedicellatis glabris rugosis compressis carinatis. *floribunda.*
Small-flower'd Crotalaria.
Nat. of the Cape of Good Hope. Mr. *William Paterson.*

Introd. 1780, by the Counteſs of Strathmore.
Fl. July and Auguſt. G. H. ♄.

axillaris. 9. C. foliis ternatis ovato-ellipticis ſubtus piloſis, ſtipulis ſubulatis minutis, pedunculis axillaribus geminis unifloris.
Two-flower'd Crotalaria.
Nat. of Guinea. Mr. *William Braſs.*
Introd. 1781, by the Earl of Tankerville and Dr. Pitcairn.
Fl. July. S. ⊙.

incaneſcens. 10. C. foliis ternatis obovatis, ſtipulis foliiformibus petiolatis, racemis terminalibus, leguminibus pedicellatis.
Crotalaria incaneſcens. *Linn. ſuppl.* 323.
Crotalaria capenſis. *Jacqu. hort.* 3. *p.* 36. *t.* 64.
Spreading ſhrubby Crotalaria.
Nat. of the Cape of Good Hope.
Introd. 1774, by Mr. Francis Maſſon.
Fl. June——October. G. H. ♄.

incana. 11. C. foliis ternatis ovalibus ſubtus villoſis, racemis ſpiciformibus, carina margine tomentoſa, leguminibus ſeſſilibus hirſutis.
Crotalaria incana. *Sp. pl.* 1005.
Hoary Crotalaria.
Nat. of the Weſt Indies.
Cult. 1714, by the Dutcheſs of Beaufort. *Br. Muſ. H. S.* 134. *fol.* 6.
Fl. June and July. S. ⊙.

pallida. 12. C. foliis ternatis lanceolatis glabris, racemis terminalibus ſpiciformibus.
Pale-flower'd Crotalaria.
Nat. of Africa. *James Bruce*, Eſq.

Introd.

Introd. 1775.
Fl. June and July. S. ⊙.

ONONIS. *Gen. pl.* 863.

Cal. 5-partitus: laciniis linearibus. *Vexillum* ſtriatum. *Legumen* turgidum, ſeſſile. *Filamenta* connata abſque fiſſura.

* *Floribus ſubſeſſilibus.*

1. O. floribus axillaribus geminatis, foliis ternatis: ſuperioribus ſolitariis, ramiſque villoſis. *ſpinoſa.*
Ononis ſpinoſa β. *Sp. pl.* 1006.
α floribus purpureis.
Thorney purple Reſt-harrow, or Cammock.
β floribus albis.
Thorney white Reſt-harrow.
Nat. of Britain.
Fl. June and July. H. ♃.

2. O. floribus ſubſpicatis geminatis, foliis inferioribus ternatis; ſuperioribus ſolitariis villoſiuſculis, ramis piloſo-villoſis. *hircina.*
Ononis hircina. *Jacqu. hort.* 1. *p.* 40. *t.* 93.
Ononis fœtens. *Allion. pedem.* 1. *p.* 317. *t.* 41. *f.* 1.
Ononis altiſſima. *De Lamarck. encycl.* 1. *p.* 506.
Ononis arvenſis. *Retz. obſ. bot.* 2. *p.* 21. *n.* 67.
Tagglöſt Puktörne, *Retzius in act. lund.* 1. *p.* 130.
Anonis non ſpinoſa, flore thyrſoide carneo, polonica. *Ban. ic.* 1214.
Ononis mitior flore purpureo. *Beſl. eyſt. æſtiv.* 10. *t.* 2. *f.* 2.
Ononis mitior I. purpureo flore. *Cluſ. hiſt.* 1. *p.* 99.
Stinking Reſt-harrow.
Nat. of Sweden, Germany, Italy, and Hungary.
Introd. 1774, by Joſeph Nicholas de Jacquin, M.D.
Fl. May——Auguſt. H. ♃.

minutissima. 3. O. floribus subsessilibus lateralibus, foliis ternatis glabris, stipulis ensiformibus, calycibus scariosis corolla longioribus. *Syst. veget.* 651. *Jacqu. austr.* 3. *p.* 23. *t.* 240.
Small-flower'd Rest-harrow.
Nat. of the South of Europe.
Cult. 1739, by Mr. Philip Miller. *Mill. dict. vol.* 2. Anonis 22.
Fl. June and July. H. ♃.

mitissima. 4. O. floribus sessilibus spicatis, bracteis stipularibus ovatis ventricosis scariosis imbricatis. *Sp. pl.* 1007.
Cluster-flower'd annual Rest-harrow.
Nat. of Portugal and Spain.
Cult. 1732, by James Sherard, M. D. *Dill. elth.* 28. *t.* 24. *f.* 27.
Fl. June. H. ☉.

alopecuroides. 5. O. spicis foliosis, foliis simplicibus ovatis obtusis, stipulis dilatatis. *Sp. pl.* 1008.
Foxtail Rest-harrow.
Nat. of Sicily and Portugal.
Cult. 1696, by Mr. Jacob Bobart. *Br. Mus. Sloan. mss.* 3343.
Fl. July and August. H. ☉.

** *Floribus pedunculatis: pedunculis muticis.*

pubescens. 6. O. pedunculis muticis brevissimis, foliis superioribus simplicibus, stipulis ovato-lanceolatis integerrimis. *Linn. mant.* 267.
Downy Rest-harrow.
Nat. of the South of Europe.
Introd. 1783, by Mons. Thouin.
Fl. August. H. ☉.

7. O.

7. O. racemis ſtrictis, foliis cuneiformibus, leguminibus cernuis linearibus recurvatis. *Syſt. veget.* 652. *cernua.*
Hanging-podded Reſt-harrow.
Nat. of the Cape of Good Hope.
Introd. 1774, by Mr. Francis Maſſon.
Fl. July——September. G. H. ♄.

8. O. foliis ternatis obovatis, pedunculis lateralibus bifloris. *geminata.*
Two-flower'd Reſt-harrow.
Nat. of the Cape of Good Hope. Mr. *Fr. Maſſon.*
Introd. 1787.
Fl. July. G. H. ♃.

9. O. pedunculis muticis unifloris, foliis ternis cuneatis, ſtipulis ſerratis, caulibus proſtratis. *Linn. mant.* 267. *ceniſia.*
Narrow-leav'd trailing Reſt-harrow.
Nat. of Italy.
Introd. 1778, by Monſ. Thouin.
Fl. June——Auguſt. H. ♃.

*** *Pedunculis ariſtatis.*

10. O. pedunculis unifloris ariſtatis, foliis ternatis, ſtipulis ſerratis. *Syſt. veget.* 653. *Cherleri.*
Dwarf Reſt-harrow.
Nat. of the South of Europe.
Introd. 1771, by Monſ. Richard.
Fl. June and July. H. ♃.

11. O. pedunculis unifloris ariſtatis, foliis ſimplicibus: infimis ternatis. *Sp. pl.* 1009. *viſcoſa.*
Clammy Reſt-harrow.
Nat. of the South of France and Spain.

Cult. 1768, by Mr. Philip Miller. *Mill. dict. edit.* 8.
Fl. July. G. H. ⊙.

Natrix. 12. O. pedunculis unifloris ariſtatis, foliis ternatis viſcoſis, ſtipulis integerrimis, caule fruticoſo. *Syſt. veget.* 653.
Yellow-flower'd ſhrubby Reſt-harrow.
Nat. of the South of France and Spain.
Cult. 1683, by Mr. James Sutherland. *Sutherl. Hort. edin.* 29. *n.* 2.
Fl. May——September. G. H. ♄.

**** *Fruticoſæ.*

criſpa. 13. O. fruticoſa, foliis ternatis ſubrotundis undulatis dentatis viſcoſo-pubeſcentibus, pedunculis unifloris muticis. *Sp. pl.* 1010.
Curl'd-leav'd Reſt-harrow.
Nat. of Spain.
Cult. 1752, by Mr. Philip Miller. *Mill. dict. edit.* 6. Anonis 14.
Fl. June——Auguſt. G. H. ♄.

fruticoſa. 14. O. fruticoſa, foliis ſeſſilibus ternatis lanceolatis ſerratis, ſtipulis vaginalibus, pedunculis ſubtrifloris. *Syſt. veget.* 653.
α floribus purpuraſcentibus.
Purple-flower'd ſhrubby Reſt-harrow.
β floribus albis.
White-flower'd ſhrubby Reſt-harrow.
Nat. of the South of France.
Cult. 1748, by Mr. Philip Miller. *Mill. dict. edit.* 5. Anonis 6.
Fl. May and June. H. ♄.

15. O.

15. O. fruticosa, foliis ternatis ovatis dentatis, calycibus triphyllo-bracteatis, pedunculis subtrifloris. *Syst. veget.* 653. *Jacqu. austr.* 5. *p.* 55. *tab. app.* 49. *rotundifolia.*
Round-leav'd Rest-harrow.
Nat. of Switzerland.
Cult. 1570, by Mr. Hugh Morgan. *Lobel. adv.* 400.
Fl. May——July. H. ♄.

ANTHYLLIS. *Gen. pl.* 864.

Cal. ventricosus. *Legumen* subrotundum, tectum.

Herbaceæ.

1. A. herbacea, foliis quaterno-pinnatis, floribus lateralibus. *Sp. pl.* 1011. *tetraphylla.*
Four-leav'd Anthyllis, or Kidney-vetch.
Nat. of the South of Europe.
Cult. 1640, by Mr. John Parkinson. *Park. theat.* 1094. *n.* 3.
Fl. July. H. ⊙.

2. A. herbacea, foliis pinnatis inæqualibus, capitulo duplicato. *Sp. pl.* 1012. *Vulneraria.*
α floribus luteis.
Common yellow Anthyllis, or Kidney-vetch.
β floribus coccineis.
Scarlet Anthyllis, or Kidney-vetch.
Nat. of Britain.
Fl. May——July. H. ♃.

3. A. herbacea, foliis pinnatis æqualibus, capitulo terminali secundo, floribus obliquatis. *Sp. pl.* 1012. *Jacqu. austr.* 4. *p.* 17. *t.* 334. *montana.*
Mountain Anthyllis.
Nat. of the South of Europe.

Cult.

Cult. 1759, by Mr. Ph. Miller. *Mill. dict. edit.* 7. *n.* 4.
Fl. June and July. H. ♃.

** *Fruticosæ.*

Barba jovis. 4. A. fruticosa, foliis pinnatis æqualibus tomentosis, floribus capitatis. *Syst. veget.* 654.
Silvery Anthyllis, or Jupiter's Beard.
Nat. of Spain, Italy, and the Levant.
Cult. 1640. *Park. theat.* 1459.
Fl. March——May. G. H. ♄.

Cytisoides. 5. A. fruticosa, foliis ternatis inæqualibus, calycibus lanatis lateralibus. *Sp. pl.* 1013.
Downy-leav'd Anthyllis.
Nat. of Spain.
Cult. 1759, by Mr. Ph. Miller. *Mill. dict. edit.* 7. *n.* 7.
Fl. April——June. G. H. ♄.

hermaniæ. 6. A. fruticosa, foliis ternatis subpedunculatis, calycibus nudis. *Sp. pl.* 1014.
Lavender-leav'd Anthyllis.
Nat. of the Levant.
Cult. before 1739. *Mill. dict. edit.* 8.
Fl. April——July. G. H. ♄.

Erinacea. 7. A. fruticosa spinosa, foliis simplicibus. *Sp. pl.* 1014.
Prickly Anthyllis.
Nat. of Spain.
Cult. 1759, by Mr. Ph. Miller. *Mill. dict. edit.* 7. *n.* 8.
Fl. April and May. G. H. ♄.

ARACHIS.

ARACHIS. *Gen. pl.* 876.

Cal. 2-labiatus. *Cor.* supinata. *Filamenta* connexa. *Legum.* gibbum, torulosum, venosum, coriaceum.

1. ARACHIS. *Sp. pl.* 1040. *hypogæa.*
American Earth-nut.
Nat. of South America.
Cult. 1712, in Chelsea Garden. *Philosoph. transf. n.* 337. *p.* 62. *n.* 101.
Fl. May and June. S. ☉.

EBENUS. *Gen. pl.* 895.

Cal. dentes longitudine corollæ. *Cor.* alis subnullis. *Semen.* 1, hirtum.

1. E. foliis ternatis quinatis pinnatisve bijugis, spicis terminalibus. *cretica.*
Ebenus cretica. *Sp. pl.* 1076.
Cretan Ebony.
Nat. of Candia.
Cult. 1748, by Mr. Philip Miller. *Mill. dict. edit.* 5. Barba-jovis 9.
Fl. June and July. G. H. ♄.

2. E. foliis pinnatis quadrijugis, spicis axillaribus longissime pedunculatis. *pinnata.*
Pinnated Ebony.
Nat. of Barbary and the Levant.
Introd. 1786, by Monf. Thouin.
Fl. July. G. H. ♂.

LUPINUS, *Gen. pl.* 865.

Cal. 2-labiatus. *Antheræ* 5 oblongæ; 5 ſubrotundæ. *Legumen* coriaceum.

perennis. 1. L. calycibus alternis inappendiculatis: labio ſuperiore emarginato; inferiore integro. *Sp. pl.* 1014.
Perennial Lupine.
Nat. of Virginia.
Cult. 1658, in Oxford Garden. *Hort. oxon. edit.* 2. *p.* 98.
Fl. May——July. H. ♃.

albus. 2. L. calycibus alternis inappendiculatis: labio ſuperiore integro; inferiore tridentato. *Sp. pl.* 1015.
White Lupine.
Nat.
Cult. 1596, by Mr. John Gerard. *Hort. Ger.*
Fl. July and Auguſt. H. ☉.

varius. 3. L. calycibus ſemiverticillatis appendiculatis: labio ſuperiore bifido; inferiore ſubtridentato. *Sp. pl.* 1015.
Small blue Lupine.
Nat. of Spain, France, and Sicily.
Cult. 1596, by Mr. John Gerard. *Hort. Ger.*
Fl. July and Auguſt. H. ☉.

hirſutus. 4. L. calycibus alternis appendiculatis: labio ſuperiore bipartito; inferiore tridentato. *Sp. pl.* 1015.
Great blue Lupine.
Nat. of the South of Europe and the Levant.
Cult. 1629. *Park. parad.* 335. *n.* 2.
Fl. July and Auguſt. H. ☉.

piloſus. 5. L. calycibus verticillatis appendiculatis: labio ſuperiore bipartito; inferiore integro. *Syſt. veget.* 655.

Roſe

Rose Lupine.
Nat. of the South of Europe.
Cult. 1731, by Mr. Ph. Miller. *Mill. dict. edit.* 1. *n.* 5.
Fl. July and August. H. ☉.

6. L. calycibus alternis appendiculatis: labio superiore bipartito; inferiore integro. *Sp. pl.* 1015. *angustifolius.*
Narrow-leav'd blue Lupine.
Nat. of Spain and Sicily.
Cult. 1731, by Mr. Ph. Miller. *Mill. dict. edit.* 1. *n.* 2.
Fl. July and August. H. ☉.

7. L. calycibus verticillatis appendiculatis: labio superiore bipartito; inferiore tridentato. *Sp. pl.* 1015. *luteus.*
Yellow Lupine.
Nat. of Sicily.
Cult. 1596, by Mr. John Gerard. *Hort. Ger.*
Fl. July and August. H. ☉.

PHASEOLUS. *Gen. pl.* 866.

Carina cum staminibus styloque spiraliter tortis.

1. P. caule volubili, floribus racemosis geminis, bracteis calyce minoribus, leguminibus pendulis. *Syst. veget.* 656. *vulgaris.*
α Phaseolus vulgaris. *Lob. icon.* 2. *p.* 59.
Kidney-bean.
β Phaseolus puniceo flore. *Corn. canad.* 184. *t.* 185. coccineus.
Scarlet Kidney-bean.
Nat. of India.
Cult. 1597. *Ger. herb.* 1038.
Fl. June——September. H. ☉.

2. P. caule volubili, leguminibus acinaciformibus sublunatis lævibus. *Syst. veget.* 656. *lunatus.*

Phaseolus

Phaſeolus rufus. *Jacqu. hort.* 1. *p.* 13. *t.* 34. conf. *vol.* 3. *p.* 1.
Scymitar-podded Kidney-bean.
Nat. of the Eaſt Indies.
Introd. 1779, by Mr. William Roxburgh.
Fl. June and July. S. ⊙.

trilobus. 3. P. caule ſemivolubili decumbente ſubglabro, foliolis trilobis: lobis ovatis, ſtipulis ovatis, leguminibus cylindraceis.
Dolichos trilobus. *Sp. pl.* 1021.
Three-lobed Phaſeolus.
Nat. of the Eaſt Indies.
Introd. 1777, by Sir Joſeph Banks, Bart.
Fl. July. S. ⊙.

vexillatus. 4. P. caule volubili, pedunculis petiolo craſſioribus capitatis, alis ſubfalcatis difformibus, leguminibus linearibus ſtrictis. *Syſt. veget.* 656. *Jacqu. hort.* 2. *p.* 46. *t.* 102.
Sweet-ſcented Kidney-bean.
Nat. of the Weſt Indies.
Cult. 1732, by James Sherard, M.D. *Dill. elth.* 313. *t.* 234. *f.* 202.
Fl. July. S. ⊙.

ſemierectus. 5. P. caule ſemivolubili, floribus ſpicatis, calycibus ebracteatis, alis expanſis majoribus, foliolis ovatis. *Syſt. veget.* 656. *Jacqu. ic. vol.* 2. *collect.* 1. *p.* 134.
Dark-red-flower'd Kidney-bean.
Nat. of the Weſt Indies.
Introd. 1781, by Mr. Francis Maſſon.
Fl. July. S. ⊙.

Caracalla. 6. P. caule volubili, vexillis carinaque ſpiraliter convolutis. *Syſt. veget.* 656.

§ Twiſted-

Twiſted-flower'd Kidney-bean, or Snail-flower.
Nat. of India.
Cult. 1690, in the Royal Garden at Hampton-court. *Catal. mſſ.*
Fl. Auguſt and September. S. ♃.

7. P. caule erecto anguloſo hiſpido, leguminibus pendulis hirtis. *Sp. pl.* 1018. *Max.*
Hairy-podded Kidney-bean.
Nat. of India.
Cult. 1758, in Chelſea Garden.
Fl. June and July. S. ⊙.

DOLICHOS. *Gen. pl.* 867.

Vexilli baſis callis 2, parallelis, oblongis, alas ſubtus comprimentibus.

1. D. volubilis, leguminibus ovato-acinaciformibus, ſeminibus ovatis hilo arcuato verſus alteram extremitatem. *Sp. pl.* 1019. *Lablab.*
Black-ſeeded Dolichos.
Nat. of Egypt.
Cult. 1714, by the Dutcheſs of Beaufort. *Br. Muſ. H. S.* 132. *fol.* 1.
Fl. June and July. S. ⊙.

2. D. volubilis, pedunculis multifloris erectis, leguminibus pendulis cylindricis toruloſis. *Sp. pl.* 1018. *Jacqu. hort.* 3. *p.* 39. *t.* 71. *ſinenſis.*
Chineſe Dolichos.
Nat. of India.
Introd. 1776, by Monſ. Thouin.
Fl. July. S. ⊙.

3. D.

unguiculatus. 3. D. volubilis, leguminibus capitatis ſubcylindraceis: apice recurvo concavo. *Sp. pl.* 1019. *Jacqu. hort.* 1. *p.* 8. *t.* 23.
Bird's-foot Dolichos.
Nat. of the Weſt Indies.
Introd. 1780, by William Philp Perrin, Eſq.
Fl. June and July. S. ☉.

ſeſquipedalis. 4. D. volubilis, leguminibus ſubcylindricis lævibus longiſſimis. *Sp. pl.* 1019. *Jacqu. hort.* 1. *p.* 27. *t.* 67.
Long-podded Dolichos.
Nat. of the Weſt Indies.
Introd. 1781, by Mr. Francis Maſſon.
Fl. Auguſt. S. ☉.

pruriens. 5. D. volubilis, leguminibus racemoſis: valvulis ſubcarinatis hirtis, pedunculis ternis. *Syſt. veget.* 657.
Horſe-eye Bean, or Dolichos.
Nat. of the Weſt Indies.
Introd. 1781, by Mr. Gilbert Alexander.
Fl. S. ♄.

urens. 6. D. volubilis, leguminibus racemoſis hirtis tranſverſim lamellatis, ſeminibus hilo cinctis. *Sp. pl.* 1020.
Cow itch Dolichos.
Nat. of the Weſt Indies.
Cult. 1691, by the Dutcheſs of Beaufort. *Br. Muſ. Sloan. mſſ.* 3343.
Fl. June and July. S. ♄.

minimus. 7. D. volubilis, leguminibus racemoſis compreſſis tetraſpermis, foliis rhombeis. *Sp. pl.* 1020.
Small Dolichos.
Nat. of Jamaica.
Introd. 1776, by Monſ. Thouin.
Fl. July. S. ☉.

8. D. volubilis, foliis ovatis tomentosis, floribus solitariis, seminibus bicornibus. *Sp. pl.* 1020. *scarabæ-cides.*
Silvery-leav'd Dolichos.
Nat. of the East Indies.
Introd. 1773, by Joseph Nicholas de Jacquin, M. D.
Fl. June and July. S. ☉.

9. D. volubilis, foliis ovatis acutis rugosis reticulatis villosis, racemis paucifloris. *reticula-tus.*
Netted-leav'd Dolichos.
Nat. of New South Wales.
Introd. 1781, by Sir Joseph Banks, Bart.
Fl. S. ♄.

10. D. volubilis, foliis glabris multangulis dentatis. *Sp. pl.* 1021. *bulbosus.*
Bulbous Dolichos.
Nat. of the West Indies.
Introd. 1781, by Mr. Francis Masson.
Fl. S. ♃.

11. D. volubilis, caule perenni, pedunculis capitatis, leguminibus strictis linearibus. *Sp. pl.* 1022. *lignosus.*
Purple Dolichos.
Nat. of the East Indies.
Introd. 1776, by Monf. Thouin.
Fl. July and August. S. ♄.

12. D. caule suberecto, leguminibus acinaciformibus tricarinatis, seminibus arillatis. *Syst. veget.* 659. *ensifor-mis.*
Scymitar-podded Dolichos.
Nat. of the East Indies.
Introd. 1778, by Mr. William Roxburgh.
Fl. July and August. S. ☉.

biflorus. 13. D. caule perenni lævi, pedunculis bifloris, leguminibus erectis. *Sp. pl.* 1023.
Two-flower'd Dolichos.
Nat. of India.
Introd. 1776, by Monf. Thouin.
Fl. July and Auguft. S. ☉.

GLYCINE. *Gen. pl.* 868.

Cal. 2-labiatus. *Corollæ* carina apice vexillum reflectens.

monoica. 1. G. foliis ternatis nudiufculis, caule pilofo, racemis pendulis, floribus fructiferis apetalis. *Sp. pl.* 1023.
Pale-flower'd Glycine.
Nat. of North America.
Introd. 1781, by Mr. William Curtis.
Fl. September. H. ♃.

debilis. 2. G. foliis ternatis: foliolis ovalibus fubtus pilofis, leguminibus fubfolitariis linearibus polyfpermis, ftylo perfiftenti erecto.
Hairy Glycine.
Nat. of the Eaft Indies. *John Gerard Kœnig*, M.D.
Introd. 1778, by Sir Jofeph Banks, Bart.
Fl. June and July. S. ♂.

caribæa. 3. G. foliis ternatis fubvillofis: foliolis rhombeis, racemis patulis, leguminibus hirfutis, caule fruticofo volubili. *Jacqu. ic. collect.* 1. *p.* 66.
Trailing Glycine.
Nat. of the Weft Indies.
Cult. before 1742, by Robert James Lord Petre.
Fl. September and October. S. ♄.

4. G.

4. G. foliis ternatis oblongo-lanceolatis pubescentibus subtus reticulato-venosis, racemis axillaribus subsessilibus, leguminibus oblongis compressis. *Swartz prodr.* 105. *reticulata.*
Netted-leav'd Glycine.
Nat. of Jamaica.
Introd. 1779, by John Fothergill, M.D.
Fl. S. ♄.

5. G. foliis ternatis, floribus racemosis, leguminibus tumidis villosis. *Sp. pl.* 1024. *bituminosa.*
Clammy Glycine.
Nat. of the Cape of Good Hope.
Introd. 1774, by Mr. Francis Masson.
Fl. April——September. G. H. ♄.

6. G. foliis impari-pinnatis ovato-lanceolatis: foliolis septenis. *Syst. veget.* 660. *Apios.*
Tuberous-rooted Glycine.
Nat. of Virginia.
Cult. 1640. *Park. theat.* 1062. *f.* 6.
Fl. August and September. H. ♃.

7. G. foliis impari-pinnatis, caule perenni. *Sp. pl.* 1025. *frutescens.*
Shrubby Glycine, or Carolina Kidney-bean-tree.
Nat. of Carolina.
Introd. 1724, by Mr. Mark Catesby. *Hort. angl.* 55. *t.* 15. Phaseoloides.
Fl. June——September. H. ♄.

8. G. foliis simplicibus cordatis, caule pubescente triquetro. *Linn. mant.* 101. *monophylla.*
Simple-leav'd Glycine.
Nat. of the Cape of Good Hope.
Introd. 1787, by Mr. Francis Masson.
Fl. August. G. H. ♃.

CYLISTA.

Cal. maximus, 4-partitus: lacinia ſuprema apice bifida. *Cor.* perſiſtens.

villoſa. 1. CYLISTA.
Hairy Cyliſta.
Nat.
Introd. 1776.
Fl. April and May. S. ♄.

CLITORIA. *Gen. pl.* 869.

Cor. ſupinata: vexillo maximo patente alas obumbrante.

Ternatea. 1. C. foliis pinnatis. *Sp. pl.* 1025.
Winged-leav'd Clitoria.
Nat. of the Eaſt Indies.
Cult. 1739, by Mr. Ph. Miller. *Rand. chel.* Ternatea.
Fl. July and Auguſt. S. ♃.

virginiana. 2. C. foliis ternatis, calycibus geminis campanulatis. *Syſt. veget.* 660.
Small-flower'd Clitoria.
Nat. of Virginia and the Weſt Indies.
Cult. 1732, by James Sherard, M. D. *Dill. elth.* 90. *t.* 76. *f.* 87.
Fl. July. S. ☉.

PISUM. *Gen. pl.* 870.

Stylus triangularis, ſupra carinatus, pubeſcens. *Calycis* laciniæ ſuperiores 2 breviores.

ſativum. 1. P. petiolis teretibus, ſtipulis inferne rotundatis crenatis, pedunculis multifloris. *Sp. pl.* 1026.

α Piſum

α Pisum hortense majus. *Bauh. pin.* 342.
Common Marrowfat, or Garden Pea.
β Pisum arvense, fructu viridi. *Tournef. inst.* 394.
Green Rouncival Pea.
γ Pisum cortice eduli. *Tournef. inst.* 394.
Sugar Pea.
δ Pisum umbellatum. *Bauh. pin.* 342.
Rose, or Crown Pea.
Nat. of the South of Europe.
Fl. June——September. H. ⊙.

2. P. petiolis supra planiusculis, caule angulato, stipulis sagittatis, pedunculis multifloris. *Sp. pl.* 1027. *maritimum.*
Sea Pea.
Nat. of England.
Fl. July. H. ♃.

3. P. petiolis decurrentibus membranaceis diphyllis, pedunculis unifloris. *Sp. pl.* 1027. *Ochrus.*
Yellow-flower'd Pea.
Nat. of the South of Europe and the Levant.
Cult. 1633. *Ger. emac.* 1249. *f.* 1.
Fl. June and July. H. ⊙.

OROBUS. *Gen. pl.* 871.

Stylus linearis. *Cal.* basi obtusus: laciniis superioribus profundioribus, brevioribus.

1. O. foliis conjugatis subsessilibus, stipulis dentatis. *Sp. pl.* 1027. *Lathyroides.*
Upright Bitter-vetch.
Nat. of Siberia.
Cult. 1759, by Mr. Ph. Miller. *Mill. dict. edit.* 7. *n.* 7.
Fl. June. H. ♃.

luteus. 2. O. foliis pinnatis ovato-oblongis, ſtipulis rotundato-lunatis dentatis. *Sp. pl.* 1028.
Yellow Bitter-vetch.
Nat. of Siberia and Italy.
Cult. 1757, by Mr. Philip Miller. *Mill. ic.* 129. *t.* 193. *f.* 1.
Fl. June and July. H. ♃.

vernus. 3. O. foliis pinnatis ovatis, ſtipulis ſemiſagittatis integerrimis, caule ſimplici. *Sp. pl.* 1028.
Spring Bitter-vetch.
Nat. of Europe.
Cult. 1629. *Park. parad.* 337. *f.* 13.
Fl. March and April. H. ♃.

tuberoſus. 4. O. foliis pinnatis lanceolatis, ſtipulis ſemiſagittatis integerrimis, caule ſimplici. *Sp. pl.* 1028. *Curtis lond.*
Tuberous Bitter-vetch.
Nat. of Britain.
Fl. May and June. H. ♃.

anguſtifolius. 5. O. foliis bijugis enſiformibus, ſtipulis ſubulatis, caule ſimplici. *Sp. pl.* 1028.
Narrow-leav'd Bitter-vetch.
Nat. of Siberia.
Cult. 1766, in Oxford Garden.
Fl. May and June. H. ♃.

niger. 6. O. caule ramoſo, foliis ſexjugis ovato-oblongis. *Sp. pl.* 1028.
Black Bitter-vetch.
Nat. of Europe.
Cult. 1656, by Mr. John Tradeſcant, Jun. *Muſ. Trad.* 149.
Fl. June and July. H. ♃.

7. O. caule ramoso, foliis bijugis lanceolatis nervosis, stipulis subspinosis. *Sp. pl.* 1029. *pyrenaicus.*
Pyrenean Bitter-vetch.
Nat. of Spain.
Cult. 1699, by the Dutchess of Beaufort. *Br. Mus. Sloan. mss.* 525 and 3349.
Fl. May. H. ♃.

8. O. caule ramoso hirsuto decumbente, foliis subseptemjugis. *Sp. pl.* 1029. *sylvaticus.*
Wood-orobus, or Bitter-vetch.
Nat. of Britain.
Fl. June and July. H. ♃.

LATHYRUS. *Gen. pl.* 872.

Stylus planus, supra villosus, superne latior. *Cal.* laciniæ superiores 2 breviores.

* *Pedunculis unifloris.*

1. L. pedunculis unifloris, cirrhis aphyllis, stipulis sagittato-cordatis. *Sp. pl.* 1029. *Curtis lond.* *Aphaca.*
Yellow Lathyrus, or Vetchling.
Nat. of England.
Fl. June and July. H. ⊙.

2. L. pedunculis unifloris, foliis simplicibus, stipulis subulatis. *Sp. pl.* 1029. *Nissolia.*
Crimson Lathyrus, or Grass-vetch.
Nat. of England.
Fl. May. H. ⊙.

3. L. pedunculis unifloris calyce longioribus, cirrhis diphyllis simplicissimis. *Sp. pl.* 1029. *amphicarpos.*
Subterranean Lathyrus, or Earth-pea.

Nat. of the Levant.
Cult. 1680, in Oxford Garden. *Morif. hift.* 2. *p.* 51, *n.* 5.
Fl. June and July. H. ⊙.

Cicera. 4. L. pedunculis unifloris, cirrhis diphyllis, leguminibus ovatis compreffis dorfo canaliculatis. *Sp. pl.* 1030.
Flat-podded Lathyrus, or Dwarf Chickling-vetch.
Nat. of France and Spain.
Cult. 1633. *Ger. emac.* 1230. *f.* 3.
Fl. June and July. H. ⊙.

fativus. 5. L. pedunculis unifloris, cirrhis diphyllis tetraphyllifque, leguminibus ovatis compreffis dorfo bimarginatis. *Sp. pl.* 1030.
Common Lathyrus, or Blue Chickling-vetch.
Nat. of France and Spain.
Cult. 1739, in Chelfea Garden. *Rand. chel. n.* 16.
Fl. June and July. H. ⊙.

inconfpicuus. 6. L. pedunculis unifloris calyce brevioribus, cirrhis diphyllis fimplicibus: foliolis lanceolatis. *Syft. veget.* 662. *Jacqu. hort.* 1. *p.* 37. *t.* 86.
Small-flower'd Lathyrus.
Nat. of the Levant.
Cult. 1739, by Mr. Ph. Miller. *Rand. chel. n.* 11.
Fl. July. H. ⊙

fetifolius. 7. L. pedunculis unifloris, cirrhis diphyllis: foliolis fetaceo-linearibus. *Sp. pl.* 1031.
Narrow-leav'd Lathyrus.
Nat. of the South of France and Italy.
Cult. 1739, by Mr. Philip Miller. *Rand. chel. n.* 4.
Fl. June and July. H. ⊙.

8. L.

8. L. pedunculis subunifloris, cirrhis polyphyllis: foliolis alternis. *Sp. pl.* 1031. *articulatus.*

Jointed-podded Lathyrus.

Nat. of the South of Europe.

Cult. 1739, in Chelsea Garden. *Rand. chel.* Clymenum 2.

Fl. July. H. ☉.

** *Pedunculis bifloris.*

9. L. pedunculis bifloris, cirrhis diphyllis: foliis ovato-oblongis, leguminibus hirsutis. *Sp. pl.* 1032. *Curtis magaz.* 60. *odoratus.*

α Lathyrus siculus. *Rupp. jen.* 210. siculus.

Purple Lathyrus, or Sweet-pea.

β Lathyrus zeylanicus, odorato flore amœne ex albo & rubro vario. *Burm. zeyl.* 138. zeylanicus.

Variegated Lathyrus, or Painted Lady Sweet-pea.

Nat. α. of Sicily; β. of Zeylon.

Cult. 1700, by Dr. Uvedale. *Pluk. mant.* 114. conf. *Philosoph. trans. n.* 337. *p.* 210. *n.* 122.

Fl. June and July. H. ☉.

10. L. pedunculis bifloris, cirrhis diphyllis: foliolis ensiformibus, leguminibus glabris, stipulis bipartitis. *Sp. pl.* 1032. *annuus.*

Two-flower'd yellow annual Lathyrus.

Nat. of France and Spain.

Cult. 1640, by Mr. John Coys. *Park. theat.* 1064. *f.* 8.

Fl. July. H. ☉.

11. L. pedunculis bifloris, cirrhis diphyllis: foliolis alternis lanceolatis glabris, stipulis lunatis. *Sp. pl.* 1032. *Jacqu. hort.* 1. *p.* 18. *t.* 46. *tingitanus.*

Tangier Lathyrus, or Pea.

Nat.

Nat. of Barbary.
Cult. 1680, by Rob. Morison, M.D. *Moris. hist.* 2. *p.* 57. *n.* 1.
Fl. June and July. H. ⊙.

Clymenum. 12. L. pedunculis bifloris, cirrhis polyphyllis, stipulis dentatis. *Sp. pl.* 1032.
Various-flower'd Lathyrus.
Nat. of the Levant.
Introd. 1787, by Mons. Thouin.
Fl. July. H. ⊙.

*** *Pedunculis multifloris.*

hirsutus. 13. L. pedunculis subtrifloris, cirrhis diphyllis: foliolis lanceolatis, leguminibus hirsutis, serminibus scabris. *Syst. veget.* 663.
Hairy Lathyrus.
Nat. of England.
Fl. July. H. ⊙.

tuberosus. 14. L. pedunculis multifloris, cirrhis diphyllis: foliolis ovalibus, internodiis nudis. *Sp. pl.* 1033.
Tuberous Lathyrus.
Nat. of Holland and France.
Cult. 1640, by Mr. John Parkinson. *Park. theat.* 1061. *f.* 4.
Fl. July and August. H. ♃.

pratensis. 15. L. pedunculis multifloris, cirrhis diphyllis simplicissimis: foliolis lanceolatis. *Sp. pl.* 1033. *Curtis lond.*
Meadow Lathyrus.
Nat. of Britain.
Fl. June——August. H. ♃.

16. L.

16. L. pedunculis multifloris, cirrhis diphyllis: foliolis ensiformibus, internodiis membranaceis. *Sp. pl.* 1033. *ſylveſtris.*
Wild Lathyrus, or narrow-leav'd Everlaſting-pea.
Nat. of Britain.
Fl. July——September. H. ♃.

17. L. pedunculis multifloris, cirrhis diphyllis: foliolis lanceolatis, internodiis membranaceis. *Sp. pl.* 1033. *latifolius.*
Broad-leav'd Lathyrus, or Everlaſting-pea.
Nat. of England.
Fl. July——September. H. ♃.

18. L. pedunculis multifloris, cirrhis polyphyllis, ſtipulis lanceolatis. *Sp. pl.* 1034. *paluſtris.*
Marſh Lathyrus.
Nat. of Britain.
Fl. July and Auguſt. H. ♃.

19. L. pedunculis multifloris, cirrhis polyphyllis, ſtipulis ovatis foliolo latioribus. *Syſt. veget.* 663. *piſiformis.*
Siberian Lathyrus.
Nat. of Siberia.
Cult. 1759, by Mr. Philip Miller. *Mill. dict. edit.* 7. *n.* 17.
Fl. June. H. ♃.

V I C I A. *Gen. pl.* 873.

Stigma latere inferiore tranſverſe barbatum.

* *Pedunculis elongatis.*

1. V. pedunculis multifloris, petiolis polyphyllis: foliolis ovatis: infimis ſeſſilibus. *Sp. pl.* 1034. *Jacqu. auſtr.* 4. *p.* 33. *t.* 364. *piſiformis.*

Pale-

Pale-flower'd Vetch.
Nat. of Auſtria.
Cult. 1758, by Mr. Philip Miller.
Fl. July and Auguſt. H. ♃.

dumetorum. 2. V. pedunculis multifloris, foliolis reflexis ovatis mucronatis, ſtipulis ſubdentatis. *Sp. pl.* 1035.
Great Wood Vetch.
Nat. of France and Germany.
Cult. 1759, by Mr. Philip Miller.
Fl. May. H. ♃.

ſylvatica. 3. V. pedunculis multifloris, foliolis ovalibus, ſtipulis denticulatis. *Sp. pl.* 1035.
Common Wood Vetch.
Nat. of Britain.
Fl. July. H. ♃.

caſſubica. 4. V. pedunculis ſubſexfloris, foliolis denis ovatis acutis, ſtipulis integris. *Sp. pl.* 1035.
Vicia Gerardi. *Jacqu. auſtr.* 3. *p.* 16. *t.* 229.
Caſſubian Vetch.
Nat. of Germany.
Cult. 1711. *Philoſoph. tranſ. n.* 332. *p.* 389. *n.* 43.
Fl. June and July. H. ♃.

Cracca. 5. V. pedunculis multifloris, floribus imbricatis, foliolis lanceolatis pubeſcentibus, ſtipulis integris. *Sp. pl.* 1035. *Curtis lond.*
Tufted Vetch.
Nat. of Britain.
Fl. June and July. H. ♃.

niſſoliana. 6. V. pedunculis multifloris, foliolis oblongis, ſtipulis integris, leguminibus villoſis ovato-oblongis. *Sp. pl.* 1036.

Red-

Red-flower'd Vetch.
Nat. of the Levant.
Introd. 1773, by John Earl of Bute.
Fl. June and July. H. ⊙.

7. V. pedunculis multifloris, petiolis ſulcatis ſubdodecaphyllis: foliolis lanceolatis glabris. *Sp. pl.* 1036. *biennis.*
Biennial Vetch.
Nat. of Siberia.
Cult. 1759, by Mr. Ph. Miller. *Mill. dict. edit.* 7. *n.* 4.
Fl. July——September. H. ♂.

** *Floribus axillaribus, ſubſeſſilibus.*

8. V. leguminibus ſeſſilibus ſubbinatis erectis, foliis retuſis, ſtipulis notatis. *Sp. pl.* 1037. *ſativa.*
α Vicia ſativa vulgaris, ſemine nigro. *Bauh. pin.* 344.
Common Vetch, or Tare.
β Vicia ſativa alba. *Bauh. pin.* 344.
White-ſeeded common Vetch, or Tare.
Nat. of Britain.
Fl. May and June. H. ⊙.

9. V. leguminibus ſeſſilibus ſolitariis erectis glabris, foliolis ſenis: inferioribus obcordatis. *Sp. pl.* 1037. *lathyroides.*
Dwarf Vetch.
Nat. of Britain.
Fl. April——June. H. ⊙.

10. V. leguminibus ſeſſilibus reflexis piloſis ſolitariis pentaſpermis, corollæ vexillis glabris. *Sp. pl.* 1037. *lutea.*
Smooth-flower'd yellow Vetch.
Nat. of England.
Fl. June and July. H. ⊙.

hybrida. 11. V. leguminibus sessilibus reflexis pilosis pentaspermis, corollæ vexillis villosis. *Sp. pl.* 1037. *Jacqu. hort.* 2. *p.* 68. *t.* 146.
Hairy-flower'd yellow Vetch.
Nat. of England.
Fl. June——August. H. ⊙.

peregrina. 12. V. leguminibus subsessilibus pendulis glabris tetraspermis, foliolis linearibus emarginatis. *Sp. pl.* 1038.
Broad-podded Vetch.
Nat. of France.
Introd. 1779, by Mons. Thouin.
Fl. July. H. ⊙.

sepium. 13. V. leguminibus pedicellatis subquaternis erectis, foliolis ovatis integerrimis: exterioribus decrescentibus. *Sp. pl.* 1038.
Bush Vetch.
Nat. of Britain.
Fl. May. H. ♃.

narbonensis. 14. V. leguminibus subsessilibus subternatis erectis, foliolis senis subovatis, stipulisque denticulatis. *Sp. pl.* 1038.
Broad-leav'd Vetch.
Nat. of France.
Cult. 1597, by Mr. John Gerard. *Ger. herb.* 1036. *f.* 2.
Fl. June. H. ⊙.

Faba. 15. V. caule erecto, petiolis absque cirrhis. *Sp. pl.* 1039.
α Faba. *Bauh. pin.* 338.
Common Garden-bean.

β Faba

β Faba minor ſive equina. *Bauh. pin.* 338.
Horſe-bean.
Nat. of Egypt.
Fl. June and July. H. ☉.

E R V U M. *Gen. pl.* 874.

Cal. 5-partitus, longitudine corollæ.

1. E. pedunculis ſubbifloris, ſeminibus compreſſis convexis. *Sp. pl.* 1039. *Lens.*
Flat-ſeeded Tare, or Common Lentil.
Nat. of France.
Cult. 1596, by Mr. John Gerard. *Hort. Ger.*
Fl. May. H. ☉.

2. E. pedunculis ſubbifloris, ſeminibus globoſis quaternis. *Sp. pl.* 1039. *Curtis lond.* *tetraſpermum.*
Smooth Tare.
Nat. of Britain.
Fl. June. H. ☉.

3. E. pedunculis multifloris, ſeminibus globoſis binis. *Sp. pl.* 1039. *Curtis lond.* *hirſutum.*
Hairy Tare.
Nat. of Britain.
Fl. June. H. ☉.

4. E. pedunculis ſubbifloris ariſtatis, petiolis acuminatis, foliolis obtuſis. *Sp. pl.* 1040. *ſoloniense.*
Spring Tare.
Nat. of France.
Cult. 1739, in Chelſea Garden. *Rand. chel.* Vicia 10.
Fl. April and May. H. ☉.

5. E.

monanthos. 5. E. pedunculis unifloris. *Sp. pl.* 1040.
One-flower'd Tare, or Lentil.
Nat. of Russia.
Cult. 1731, by Mr. Ph. Miller. *Mill. dict. edit.* 1. Lens 3.
Fl. May and June. H. ⊙.

Ervilia. 6. E. germinibus undato-plicatis, foliis impari pinnatis. *Sp. pl.* 1040.
Officinal Tare.
Nat. of France, Italy, and the Levant.
Cult. 1597, by Mr. John Gerard. *Ger. herb.* 1051.
Fl. June. H. ⊙.

C I C E R. *Gen. pl.* 875.

Cal. 5-partitus, longitudine corollæ: laciniis quatuor superioribus vexillo incumbentibus. *Legumen* rhombeum, turgidum, dispermum.

arietinum. 1. C. foliolis serratis. *Sp. pl.* 1040.
Chick Pea.
Nat. of the South of Europe and the Levant.
Cult. 1551, at Kew by William Turner, M. D. *Turn. herb. sign. Kij.*
Fl. July and August. H. ⊙.

L I P A R I A. *Linn. mant.* 156.

Cal. 5-fidus: lacinia infima elongata. *Cor.* alæ inferius bilobæ. *Staminis* majoris dentes tres breviores. *Legum.* ovatum.

villosa. 1. L. floribus fasciculatis, foliis ovatis villoso-tomentosis. *Syst. veget.* 666.
Borbonia tomentosa. *Sp. pl.* 994.

Woolly

Woolly Liparia.
Nat. of the Cape of Good Hope.
Introd. 1774, by Mr. Francis Maſſon.
Fl. G. H. ♄.

CYTISUS. *Gen. pl.* 877.

Cal. 2-labiatus: $\frac{2}{3}$. *Legumen* baſi attenuatum.

1. C. racemis ſimplicibus pendulis, foliolis ovato-oblongis. *Sp. pl.* 1041. *Jacqu. auſtr.* 4. *p.* 3. *t.* 306. *Laburnum.*
α foliis ſubtus incanis, leguminibus ſericeis.
Common Laburnum.
β foliis leguminibuſque glabris.
Cytiſus alpinus. *Mill. dict.*
Nat. of Auſtria and Switzerland.
Cult. 1596, by Mr. John Gerard. *Hort. Ger.*
Fl. May and June. H. ♄.

2. C. racemis terminalibus erectis, calycibus piloſis: denticulis minutis, foliolis ellipticis ſubtus piloſis. *nigricans.*
Cytiſus nigricans. *Sp. pl.* 1041. *Jacqu. auſtr.* 4. *p.* 45. *t.* 387.
Black Cytiſus.
Nat. of Auſtria, Hungary, and Italy.
Cult. 1731, by Mr. Ph. Miller. *Mill. dict. edit.* 1. *n.* 4.
Fl. June and July. H. ♄.

3. C. racemis terminalibus erectis, calycibus villoſis: laciniis falcatis, foliolis obovato-oblongis. *folioloſus.*
Leafy Cytiſus.
Nat. of the Canary Iſlands. Mr. *Francis Maſſon.*
Introd. 1779.
Fl. July and Auguſt. G. H. ♄.

divaricatus. 4. C. racemis terminalibus erectis, calycibus leguminibuſque ramentaceo-viſcidis, foliolis oblongis.
Cytiſus divaricatus. *L'Herit. ſtirp. nov. tab.* 91.
Spartium complicatum. *Sp. pl.* 996.
Clammy Cytiſus.
Nat. of Spain, the South of France, and Madeira.
Cult. 1763, by Mr. James Gordon.
Fl. July and Auguſt. H. ♄.

wolgaricus. 5. C. canus, racemis ſimplicibus erectis, floribus ſecundis, foliis pinnatis: foliolis ſubrotundis, ſtipulis ſubulatis. *Linn. ſuppl.* 327.
Cytiſus pinnatus. *Pallas roſſ.* 1. *p.* 73. *tab.* 47.
Winged-leav'd Cytiſus.
Nat. of Siberia.
Introd. 1786, by William Pitcairn, M.D.
Fl. H. ♄.

ſeſſilifolius. 6. C. racemis erectis, calycibus bractea triplici, foliis floralibus ſeſſilibus. *Sp. pl.* 1041.
Common Cytiſus.
Nat. of Italy.
Cult. 1713. *Philoſoph. tranſ. n.* 337. *p.* 51. *n.* 62.
Fl. May and June. H. ♄.

Cajan. 7. C. racemis axillaribus erectis, foliolis ſublanceolatis tomentoſis: intermedio longius petiolato. *Sp. pl.* 1041.
Pigeon Cytiſus, or Pea.
Nat. of the Eaſt Indies.
Cult. 1694, by the Dutcheſs of Beaufort. *Br. Muſ. Sloan. mſſ.* 3343.
Fl. July and Auguſt. S. ♄.

hirſutus. 8. C. pedunculis ſimplicibus lateralibus, calycibus hirſutis trifidis obtuſis ventricoſo-oblongis. *Syſt. veget.* 566.

Hairy

Hairy Cytiſus.
Nat. of Siberia and the South of Europe.
Cult. 1739, by Mr. Philip Miller. *Rand. chel. n.* 5.
Fl. June. H. ♄.

9. C. floribus capitatis, ramis erectis ſtrictis teretibus villoſis, foliolis ovato-ellipticis villoſis, bractea lineari ſubcalycina. *capitatus.*
Cytiſus capitatus. *Jacqu. auſtr.* 1. *p.* 22. *t.* 33.
Cluſter-flower'd Cytiſus.
Nat. of Auſtria.
Introd. 1774, by Joſeph Nicholas de Jacquin, M.D.
Fl. June and July. H. ♄.

10. C. floribus umbellatis terminalibus, caulibus erectis, foliolis lanceolatis. *Sp. pl.* 1042. *Jacqu. auſtr.* 1. *p.* 16. *t.* 21. *auſtriacus.*
Siberian Cytiſus.
Nat. of Siberia and Auſtria.
Cult. 1756, by Mr. Ph. Miller. *Mill. ic.* 78. *t.* 117.
Fl. June——September. H. ♄.

11. C. floribus pedunculatis ſubbinatis, caulibus decumbentibus fruteſcentibus. *Syſt. veget.* 667. *Jacqu. auſtr.* 1. *p.* 15. *t.* 20. *ſupinus.*
Trailing Cytiſus.
Nat. of Siberia and the South of Europe.
Cult. 1759, by Mr. Philip Miller. *Mill. dict. edit.* 7. *n.* 8.
Fl. May——Auguſt. H. ♄.

12. C. floribus umbellatis lateralibus, caulibus erectis, foliis ellipticis acutis ſubtus ſericeis, calycibus lanatis. *proliferus.*
Cytiſus proliferus. *Linn. ſuppl.* 328.

Silky Cytiſus.
Nat. of the Canary Iſlands. Mr. *Francis Maſſon.*
Introd. 1779.
Fl. April and May. G. H. ♄.

argenteus. 13. C. floribus ſubbinatis ſubſeſſilibus, foliis tomentoſis, caulibus decumbentibus, ſtipulis minutis. *Syſt. veget.* 667.
Silvery Cytiſus.
Nat. of the South of France.
Cult. 1739, by Mr. Philip Miller. *Rand. chel. n.* 9.
Fl. Auguſt. G. H. ♄.

biflorus. 14. C. pedunculis ſubbinatis lateralibus, calycibus oblongis villoſis bilabiatis: labio ſuperiore bifido, foliolis oblongo-ellipticis.
Cytiſus biflorus. *L'Herit. ſtirp. nov. tab.* 92.
Smooth Cytiſus.
Nat.
Cult. 1760, by Mr. James Gordon.
Fl. May and June. H. ♄.

GEOFFROYA. *Gen. pl.* 878.

Cal. 5-fidus. *Drupa* ovata. *Nucleus* compreſſus.

inermis. 1. G. inermis, foliolis lanceolatis. *Swartz prodr.* 106.
Geoffræa jamaicenſis inermis. *Wright in philoſ. tranſact.* 1777. *p.* 512. *tab.* 10.
Smooth Geoffroya, or Baſtard Cabbage-tree.
Nat. of Jamaica.
Introd. 1778, by Meſſrs. Kennedy and Lee.
Fl. S. ♄.

ROBINIA. *Gen. pl.* 879.

Cal. 4-fidus. *Legumen* gibbum, elongatum.

1. R. racemis pedicellis unifloris, foliis impari pinnatis, ftipulis fpinofis. *Sp. pl.* 1043. *Pfeud-Acacia.*
False, or Common Acacia.
Nat. of North America.
Cult. 1640, by Mr. John Tradefcant, Sen. *Park. theat.* 1550. *f.* 2.
Fl. May and June. H. ♄.

2. R. racemis pedicellis unifloris, foliis impari pinnatis, caule inermi. *Syft. veget.* 668. *violacea.*
Afh-leav'd Robinia.
Nat. of the Weft Indies.
Cult. before 1759, by Mr. Philip Miller. *Mill. dict. edit.* 7. *n.* 8.
Fl. S. ♄.

3. R. racemis axillaribus, foliis impari pinnatis, caule inermi hifpido. *Linn. mant.* 101. *hifpida.*
Rofe-acacia, or Robinia.
Nat. of Carolina.
Cult. 1758, by Mr. Ph. Miller. *Mill. ic.* 163. *t.* 244.
Fl. May——September. H. ♄.

4. R. pedunculis fimplicibus plurimis, foliis abrupte pinnatis fubquadrijugis, petiolis inermibus, leguminibus cylindricis. *L'Herit. ftirp. nov. p.* 160. *Caragana.*
Robinia Caragana. *Sp. pl.* 1044.
Nat. of Siberia.
Cult. 1768, by Mr. Ph. Miller. *Mill. dict. edit.* 8.
Fl. April and May. H. ♄.

ſpinoſa. 5. R. pedunculis ſimplicibus ſubſolitariis, foliis abrupte pinnatis multijugis, petiolis ſpineſcentibus. *L'Herit. ſtirp. nov. p.* 160.
Robinia ſpinoſa. *Linn. mant.* 269.
Robinia ferox. *Pallas roſſ.* 1. *p.* 70. *tab.* 44.
Thorny Robinia.
Nat. of Siberia.
Introd. 1775, by Hugh Duke of Northumberland.
Fl. April and May. H. ♄.

Chamlagu. 6. R. pedunculis ſimplicibus ſubſolitariis, foliis abrupte pinnatis bijugis, petiolis ſtipuliſque ſpineſcentibus. *L'Herit. ſtirp. nov. p.* 161. *tab.* 77.
Shining Robinia.
Nat.
Introd. 1773, by Monſ. Richard.
Fl. May and June. H. ♄.

Halodendron. 7. R. pedunculis trifloris, foliis abrupte pinnatis bijugis ſericeis, petiolis ſpineſcentibus perſiſtentibus, leguminibus veſicariis. *L'Herit. ſtirp. nov. p.* 162.
Robinia Halodendron. *Linn. ſuppl.* 330. *Pallas roſſ.* 1. *p.* 72. *tab.* 46.
Salt-tree Robinia.
Nat. of Siberia.
Introd. 1779, by William Pitcairn, M. D.
Fl. H. ♄.

fruteſcens. 8. R. pedunculis ſimplicibus, foliis quaternatis ſubpetiolatis terminatis ſpina inermi. *Syſt. veg.* 668. *Pallas roſſ.* 1. *p.* 69. *tab.* 43.
Shrubby Robinia.
Nat. of Siberia.
Cult. 1759, by Mr. Philip Miller. *Mill. dict. edit.* 7. *n.* 10.
Fl. April and May. H. ♄.

9. R.

9. R. pedunculis ſimpliciſſimis, foliis quaternatis ſeſſilibus. *Sp. pl.* 1044. *Pallas roſſ.* 1. *p.* 71. *tab.* 45. *pygmæa.*
Dwarf Robinia.
Nat. of Siberia.
Cult. 1768, by Mr. Philip Miller. *Mill. dict. edit.* 8.
Fl. April and May. H. ♄.

COLUTEA. *Gen. pl.* 880.

Cal. 5-fidus. *Legumen* inflatum, baſi ſuperiore dehiſcens.

1. C. foliolis ovali-obcordatis, vexilli gibbis abbreviatis. *arboreſcens.*
Colutea arboreſcens. *Sp. pl.* 1045.
Common Bladder-ſenna.
Nat. of France and Italy.
Cult. 1570. *Lobel. adv.* 406.
Fl. June—August. H. ♄.

2. C. fruticoſa, foliolis cuneiformi-obcordatis, vexilli gibbis obtuſis minimis. *cruenta.*
Colutea cruenta. *L'Herit. ſtirp. nov. tom.* 2. *tab.* 41.
Colutea humilis. *Scop. inſubr.* 2. *p.* 23. *tab.* 12.
Colutea orientalis. *Mill. dict. De Lamarck encycl.* 1. *p.* 353.
Oriental Bladder-ſenna.
Nat. of the Levant.
Cult. 1731, by Mr. Ph. Miller. *Mill. dict. edit.* 1. *n.* 3.
Fl. June and July. H. ♄.

3. C. fruticoſa, foliolis ovatis, vexilli gibbis elongatis adſcendentibus. *Pocockii.*
Colutea procumbens. *L'Herit. ſtirp. nov. tom.* 2. *tab.* 42.
Colutea Iſtria. *Mill. dict.*
Colutea halepica. *De Lamarck encycl.* 1. *p.* 353.
Pocock's Bladder-ſenna.

Nat. of the Levant.
Cult. 1752, by Mr. Ph. Miller. *Mill. dict. edit.* 6. *n.* 6.
Fl. May——October. H. ♄.

frutescens. 4. C. fruticosa, foliolis ovato-oblongis. *Sp. pl.* 1045.
Scarlet Bladder-senna.
Nat. of the Cape of Good Hope.
Cult. 1683, by Mr. James Sutherland. *Sutherl. hort. edin.* 91. *n.* 5.
Fl. June and July. G. H. ♄.

perennans. 5. C. herbacea, foliolis ovato-oblongis pubescentibus.
Colutea perennans. *Jacqu. hort.* 3. *p.* 5. *t.* 3. *Murray in nov. comm. gotting.* 5. *p.* 38. *t.* 7.
Perennial Bladder-senna.
Nat. of Africa.
Introd. 1776, by Joseph Nicholas de Jacquin, M.D.
Fl. August. G. H. ♃.

herbacea. 6. C. herbacea, foliolis linearibus glabris. *Syst. veget.* 669.
Annual Bladder-senna.
Nat. of the Cape of Good Hope.
Cult. 1731, by Mr. Ph. Miller. *Mill. dict. edit.* 1. *n.* 5.
Fl. June and July. G. H. ⊙.

GLYCYRRHIZA. *Gen. pl.* 882.

Cal. 2-labiatus: $\frac{1}{3}$. *Legumen* ovatum, compressum.

echinata. 1. G. leguminibus echinatis, foliis stipulatis: foliolo impari sessili. *Syst. veget.* 669. *Jacqu. hort.* 1. *p.* 41. *t.* 95.
Prickly-headed Liquorice.
Nat. of Italy.

Cult.

Cult. 1596, by Mr. John Gerard. *Hort. Ger.*
Fl. June——September. H. ♃.

2. G. leguminibus glabris, ſtipulis nullis, foliolo impari petiolato. *Syſt. veget.* 669. *glabra.*
Common Liquorice.
Nat. of the South of Europe.
Cult. 1562. *Turn. herb. part* 2. *fol.* 12.
Fl. July. H. ♃.

3. G. leguminibus hirſutis, foliolo impari petiolato. *Sp. pl.* 1046. *hirſuta.*
Hairy Liquorice.
Nat. of the Levant.
Cult. 1739, by Mr. Philip Miller. *Rand. chel. n.* 3.
Fl. H. ♃.

CORONILLA. *Gen. pl.* 883.

Cal. 2-labiatus : $\frac{2}{3}$: dentibus ſuperioribus connatis. *Vexillum* vix alis longius. *Legumen* iſthmis interceptum.

1. C. fruticoſa, pedunculis ſubtrifloris, corollarum unguibus calyce triplo longioribus, caule angulato. *Sp. pl.* 1046. *Emerus.*
Scorpion Senna.
Nat. of France and Germany.
Cult. 1596, by Mr. John Gerard. *Hort. Ger.*
Fl. April——June. H. ♄.

2. C. fruticoſa, foliis quinatis ternatiſque lineari-lanceolatis ſubcarnoſis obtuſis. *Sp. pl.* 1047. *juncea.*
Linear-leav'd Coronilla.
Nat. of the South of France.

Cult.

Cult. 1768, by Mr. Philip Miller. *Mill. dict. edit.* 8.
Fl. June and July. G. H. ♄.

valentina. 3. C. fruticosa, foliis subnovenis, stipulis suborbiculatis. *Sp. pl.* 1047.
Small shrubby Coronilla.
Nat. of Spain and Italy.
Cult. 1656, by Mr. John Tradescant, Jun. *Mus. Trad.* 155.
Fl. November——March. H. ♄.

glauca. 4. C. fruticosa, foliolis septenis obtusissimis, stipulis lanceolatis. *Syst. veget.* 669. *Curtis magaz.* 13.
Great shrubby Coronilla.
Nat. of the South of France.
Cult. 1739, in Chelsea Garden. *Rand. chel. n.* 2.
Fl. September——May. G. H. ♄.

coronata. 5. C. fruticosa, foliis novenis obovatis: foliolis intimis cauli aproximatis, stipulis oppositifoliis bipartitis. *Sp. pl.* 1047. *Jacqu. austr.* 1. *p.* 59. *t.* 95.
Crown'd Coronilla.
Nat. of the South of Europe.
Introd. 1776, by Joseph Nicholas de Jacquin, M. D.
Fl. July. H. ♄.

x *minima.* 6. C. suffruticosa procumbens, foliolis novenis ovatis, stipula oppositifolia emarginata, leguminibus angulatis nodosis. *Syst. veget.* 670. *Jacqu. austr.* 3. *p.* 39. *t.* 271.
Least Coronilla.
Nat. of the South of Europe.
Cult. 1683, by Mr. James Sutherland. *Sutherl. hort. edin.* 117. *n.* 5.
Fl. July. H. ♃.

7. C.

7. C. herbacea, leguminibus falcato-gladiatis, foliolis plurimis. *Sp. pl.* 1048. *Securidaca.*
Hatchet Coronilla, or Vetch.
Nat. of Spain.
Cult. 1596, by Mr. John Gerard. *Hort. Ger.*
Fl. July and August. H. ⊙.

8. C. herbacea, leguminibus erectis teretibus torosis numerosis, foliolis plurimis glabris. *Sp. pl.* 1048. *varia.*
Purple Coronilla.
Nat. of France, Germany, and Denmark.
Cult. 1640. *Park. theat.* 1088. *n.* 3.
Fl. June——November. H. ♃.

9. C. herbacea, leguminibus quinis erectis teretibus articulatis, foliolis undenis. *Sp. pl.* 1048. *Jacqu. hort.* 1. *p.* 9. *t.* 25. *cretica.*
Cretan Coronilla.
Nat. of Candia.
Cult. 1731, by Mr. Ph. Miller. *Mill. dict. edit.* 1. *n.* 4.
Fl. June and July. H. ⊙.

ORNITHOPUS. *Gen. pl.* 884.

Legumen articulatum, teres, arcuatum.

1. O. foliis pinnatis, leguminibus incurvatis. *Syst. veget.* 670. *perpusillus.*
Common Bird's-foot.
Nat. of Britain.
Fl. May——August. H. ⊙.

2. O. foliis pinnatis, leguminibus recurvatis compressis rugosis, bractea pinnata. *Syst. veget.* 670. *compressus.*
Hairy Bird's-foot.

Nat.

Nat. of the South of Europe.
Cult. 1731, by Mr. Philip Miller. *Mill. dict. edit.* 1. Ornithopodium 3.
Fl. June and July. H. ☉.

ſcorpioides. 3. O. foliis ternatis ſubſeſſilibus: impari maximo. *Sp. pl.* 1049.
Purſlane-leav'd Bird's-foot.
Nat. of the South of Europe.
Cult. 1656, by Mr. John Tradeſcant, Jun. *Muſ. Trad.* 166.
Fl. June and July. H. ☉.

HIPPOCREPIS. *Gen. pl.* 885.

Legumen compreſſum, altera ſutura pluries emarginatum, curvum.

uniſiliquoſa. 1. H. leguminibus ſeſſilibus ſolitariis rectis. *Syſt. veget.* 670.
Single-podded Horſe-ſhoe Vetch,
Nat. of Italy.
Cult. 1570. *Lobel. adv.* 403.
Fl. June and July. H. ☉.

multiſiliquoſa. 2. H. leguminibus pedunculatis confertis circularibus glabris margine exteriore lobatis, foliis calycibuſque glabris.
Hippocrepis multiſiliquoſa. *Sp. pl.* 1050.
Many-podded Horſe-ſhoe Vetch.
Nat. of the South of Europe.
Cult. 1739, by Mr. Philip Miller. *Rand. chel.* Ferrum equinum 3.
Fl. July and Auguſt. H. ☉.

3. H.

3. H. leguminibus pedunculatis confertis glabris margine exteriore lobatis, foliis calycibusque pilosiusculis, caulibus ancipitibus. *balearica.*

Hippocrepis balearica. *Jacqu. ic. miscell.* 2. *p.* 305.

Shrubby Horse-shoe Vetch.

Nat. of Minorca.

Introd. 1776, by Monsf. Thouin.

Fl. May and June. G. H. ♄.

4. H. leguminibus pedunculatis confertis arcuatis scabris utroque margine sinuatis. *comosa.*

Hippocrepis comosa. *Sp. pl.* 1050. *Jacqu. austr.* 5. *p.* 14. *t.* 431.

Tufted Horse-shoe Vetch.

Nat. of England.

Fl. April——June. H. ♃.

SCORPIURUS. *Gen. pl.* 886.

Legumen isthmis interceptum, revolutum, teres.

1. S. pedunculis unifloris, leguminibus tectis undique squamis obtusis. *Sp. pl.* 1050. *vermiculata.*

Common Caterpillar.

Nat. of the South of Europe.

Cult. 1621, by Mr. John Goodyer. *Ger. emac.* 1627. *n.* 2.

Fl. June and July. H. ⊙.

2. S. pedunculis bifloris, leguminibus extrorsum obtuse aculeatis. *Sp. pl.* 1050. *muricata.*

Two-flower'd Caterpillar.

Nat. of the South of Europe.

Cult. 1640. *Park. theat.* 1117. *n.* 1.

Fl. June and July. H. ⊙.

3. S.

sulcata. 3. S. pedunculis subbifloris, leguminibus extrorsum spinis distinctis acutis. *Syst. veget.* 671.
Furrowed Caterpillar.
Nat. of the South of Europe.
Cult. 1596, by Mr. John Gerard. *Hort. Ger.*
Fl. June and July. H. ☉.

ÆSCHYNOMENE. *Gen. pl.* 888.

Cal. bilabiatus. *Legum.* articulis truncatis, monospermis.

grandiflora. 1. Æ. caule arboreo, floribus maximis, leguminibus filiformibus. *Sp. pl.* 1060.
Great-flower'd Æschynomene.
Nat. of the East Indies.
Cult. 1768, by Mr. Philip Miller. *Mill. dict. edit.* 8.
Fl. S. ♄.

americana. 2. Æ. caule herbaceo hispido, foliolis acuminatis, leguminum articulis semicordatis, bracteis ciliatis. *Sp. pl.* 1061.
Hairy Æschynomene.
Nat. of Jamaica.
Cult. 1739, by Mr. Philip Miller. *Rand. chel.* Hedysarum 5.
Fl. July and August. S. ☉.

Sesban. 3. Æ. caule herbaceo lævi, foliolis obtusis, leguminibus cylindricis æqualibus. *Sp. pl.* 1061.
Egyptian Æschynomene.
Nat. of Egypt.
Cult. 1680, in Oxford Garden. *Moris. hist.* 2. *p.* 79. *n.* 9.
Fl. July and August. S. ♂.

HEDYSA-

HEDYSARUM. *Gen. pl.* 887.

Cor. carina transverse obtusa. *Legumen* articulis 1-spermis.

1. H. foliis simplicibus lanceolatis obtusis, caule fruticoso spinoso. *Sp. pl.* 1051. *Alhagi.*
Prickly Hedysarum.
Nat. of the Levant.
Cult. 1714, by the Dutchess of Beaufort. *Br. Mus.* *H. S.* 13. *fol.* 79.
Fl. G. H. ♄.

2. H. foliis simplicibus cuneiformibus. *Sp. pl.* 1051. *nummularifolium.*
Money-wort-leav'd Hedysarum.
Nat. of India.
Introd. 1777, by Daniel Charles Solander, LL. D.
Fl. July——September. S. ⊙.

3. H. foliis simplicibus ovatis acutis stipulatis. *Syst. veget.* 672. *gangeticum.*
Oval-leav'd Hedysarum.
Nat. of the East Indies.
Cult. 1768, by Mr. Philip Miller. *Mill. dict. edit.* 8.
Fl. July and August. S. ⊙.

4. H. foliis simplicibus ovatis obtusis. *Sp. pl.* 1051. *maculatum.*
Spotted Hedysarum.
Nat. of India.
Cult. 1732, by James Sherard, M. D. *Dill. elth.* 170. *t.* 141. *f.* 168.
Fl. August. S. ⊙.

5. H. foliis simplicibus ternatisque: foliolis intermediis bilobis: lobis lanceolatis divaricatis, leguminibus plicatis. *Vespertilionis.*

Hedysarum

Hedysarum Vespertilionis. *Linn. suppl.* 331.
Bat-wing'd Hedysarum.
Nat. of Cochinchina. *Father Joannes de Loureyro.*
Introd. 1780, by Sir Joseph Banks, Bart.
Fl. July and August. S. ♂.

canadense. 6. H. foliis simplicibus ternatisque, caule lævi, floribus racemosis. *Syst. veget.* 673.
Canadian Hedysarum.
Nat. of Virginia and Canada.
Cult. 1640. *Park. theat.* 1090. *f.* 10.
Fl. July and August. H. ♃.

gyrans. 7. H. foliis ternatis ovali-lanceolatis obtusis: lateralibus minutis.
Hedysarum gyrans. *Linn. suppl.* 332.
Sensitive Hedysarum.
Nat. of the East Indies.
Introd. 1775, by Patrick Russell, M.D.
Fl. July and August. S. ♂.

canescens. 8. H. foliis ternatis subtus scabris, caule hispido, floribus racemosis conjugatis. *Sp. pl.* 1054.
Rough-leav'd Hedysarum.
Nat. of the West Indies.
Introd. before 1733, by William Houstoun, M.D. *Mill. dict. edit.* 8.
Fl. July——September. S. ♃.

tortuosum. 9. H. foliis ternatis ovali-oblongis obtusis glabriusculis, racemis erectis axillaribus, leguminibus tortuosis compressis pubescentibus. *Swartz prodr.* 107.
Hedysarum triphyllum fruticosum, flore purpureo, siliqua varie distorta. *Sloan. jam.* 1. *p.* 184. *t.* 116. *f.* 2.
Twisted-podded Hedysarum.

Nat.

Nat. of Vera Cruz in America.
Cult. 1731, by Mr. Philip Miller. *R. S. n.* 472.
Fl. July and Auguſt. S. ♄.

10. H. foliis ternatis acutiuſculis, caule erecto, racemis longiſſimis erectis. *Sp. pl.* 1055. *viridiflorum.*
Green-flower'd Hedyſarum.
Nat. of North America.
Introd. 1787, by Thomas Walter, Eſq.
Fl. H. ♃.

11. H. foliis ternatis lanceolatis, leguminibus uniarticulatis rhombeis, pedunculis ſubumbellatis lateralibus. *Sp. pl.* 1053. *junceum.*
Slender-branch'd Hedyſarum.
Nat. of India.
Introd. 1776, by Monſ. Thouin.
Fl. July and Auguſt. S. ♃.

12. H. foliis ternatis ovatis, floribus geminatis, leguminibus nudis venoſis articulatis rhombeis. *Syſt. veget.* 674. *violaceum.*
Violet-flower'd Hedyſarum.
Nat. of North America.
Introd. 1787, by Thomas Walter, Eſq.
Fl. H. ♃.

13. H. foliis ternatis lineari-lanceolatis, floribus paniculatis, leguminibus rhombeis. *Syſt. veget.* 674. *paniculatum.*
Panicl'd Hedyſarum.
Nat. of Virginia.
Introd. 1781, by Monſ. Thouin.
Fl. S. ♄.

alpinum. 14. H. foliis pinnatis, leguminibus articulatis glabris pendulis, caule erecto. *Sp. pl.* 1057.
Alpine Hedyſarum.
Nat. of Switzerland.
Cult. 1640. *Park. theat.* 1083. *n.* 8.
Fl. July and Auguſt. H. ♃.

coronarium. 15. H. foliis pinnatis, leguminibus articulatis aculeatis nudis rectis, caule diffuſo. *Sp. pl.* 1058.
Common Hedyſarum, or French Honeyſuckle.
Nat. of Italy.
Cult. 1596, by Mr. John Gerard. *Hort. Ger.*
Fl. June and July. H. ♂.

flexuoſum. 16. H. foliis pinnatis, leguminibus articulatis aculeatis flexuoſis, caule diffuſo. *Sp. pl.* 1058.
Wav'd-podded Hedyſarum.
Nat. of Aſia.
Cult. 1739, by Mr. Philip Miller. *Mill. dict. vol.* 2. *n.* 1.
Fl. July and Auguſt. S. ⊙.

humile. 17. H. foliis pinnatis, leguminibus articulatis aſperis, corollæ alis obſoletis, ſpicis hirſutis, caulibus depreſſis. *Sp. pl.* 1058.
Dwarf Hedyſarum.
Nat. of Spain and the South of France.
Cult. 1640. *Park. theat.* 1082. *n.* 7.
Fl. July and Auguſt. H. ⊙.

Onobrychis. 18. H. foliis pinnatis, leguminibus monoſpermis aculeatis, corollarum alis calycem æquantibus, caule elongato. *Syſt. veget.* 676. *Jacqu. auſtr.* 4. *p.* 27. *t.* 352.

Cultivated

Cultivated Hedyſarum, or St. Foin.
Nat. of Britain.
Fl. July. H. ♃.

19. H. foliis pinnatis, leguminibus monoſpermis: criſtæ dentibus ſubulatis, alis breviſſimis, caule diffuſo. *Syſt. veget.* 676. *Caput galli.*
Cock's-head Hedyſarum.
Nat. of France.
Cult. 1748, by Mr. Philip Miller. *Mill. dict. edit.* 5. Onobrychis 3.
Fl. July and Auguſt. H. ○.

20. H. foliis pinnatis, racemis oblongis, leguminibus inflexis, caule fruticoſo. *Linn. mant.* 102. *crinitum.*
Crooked-podded Hedyſarum.
Nat. of the Eaſt Indies.
Introd. 1780, by Sir Joſeph Banks, Bart.
Fl. S. ♄.

INDIGOFERA. *Gen. pl.* 889.

Cal. patens. *Corollæ* carina utrinque calcari ſubulato patulo! *Legumen* lineare.

1. I. foliis ternatis lanceolatis, racemis longiſſimis, leguminibus cernuis. *Syſt. veget.* 677. *pſoraloides.*
Cytiſus pſoraloides. *Sp. pl.* 1043.
Long-ſpik'd Indigo.
Nat. of the Cape of Good Hope.
Cult. 1758, by Mr. Philip Miller.
Fl. July——September. G. H. ♄.

2. I. foliis ternatis lanceolato-linearibus ſubtus ſericeis, ſpicis pedunculatis paucifloris, leguminibus cylindraceis rectis. *candicans.*

White Indigo.
Nat. of the Cape of Good Hope. Mr. *Fr. Maſſon.*
Introd. 1774.
Fl. July——September. G. H. ♄.

amœna. 3. I. foliis ternatis ovalibus piloſiuſculis, ramis teretibus, ſpicis pedunculatis, ſtipulis ſetaceis, calycibus laxis, caule fruteſcente.
Scarlet-flower'd Indigo.
Nat. of the Cape of Good Hope. Mr. *Fr. Maſſon.*
Introd. 1774.
Fl. March and April. G. H. ♄.

ſarmentoſa. 4. I. foliis ternatis ovatis ſubſeſſilibus, pedunculis axillaribus ſubbifloris, caule proſtrato filiformi. *Syſt. veget.* 677.
Dwarf Indigo.
Nat. of the Cape of Good Hope.
Introd. 1786, by Mr. Francis Maſſon.
Fl. June. G. H. ♃.

coriacea. 5. I. foliis quinatis obovatis mucronatis piloſis, ſtipulis ſubulatis, leguminibus rectis glabris.
Ononis mauritanica. *Linn. mant.* 267.
Lotus mauritanicus. *Sp. pl.* 1091.
Lotus fruticoſus. *Berg. cap.* 226.
Leathery-leav'd Indigo.
Nat. of the Cape of Good Hope.
Introd. 1774, by Mr. Francis Maſſon.
Fl. July and Auguſt. G. H. ♄.

cytiſoides. 6. I. foliis quinato-pinnatis ternatiſque, racemis axillaribus, caule fruticoſo. *Syſt. veget.* 678.
Pſoralea cytiſoides. *Sp. pl.* 1076.
Angular-ſtalk'd Indigo.

Nat.

Nat. of the Cape of Good Hope.
Introd. 1774, by Mr. Francis Maſſon.
Fl. July. G. H. ♄.

7. I. foliis pinnatis cuneatis ſeptenis, caulibus proſtratis, ſpicis lateralibus. *Syſt. veget.* 678. *ennea-phylla.*
Trailing Indigo.
Nat. of the Eaſt Indies.
Introd. 1776, by Monſ. Thouin.
Fl. July and Auguſt. S. ⊙.

8. I. foliis pinnatis linearibus, racemis elongatis, caule fruticoſo. *Syſt. veget.* 678. *anguſti-folia.*
Narrow-leav'd Indigo.
Nat. of the Cape of Good Hope.
Introd. 1774, by Mr. Francis Maſſon.
Fl. June——October. G. H. ♄.

9. I. foliis pinnatis obovatis, racemis brevibus, caule ſuffruticoſo. *Syſt. veget.* 678. *tinctoria.*
Dyer's Indigo.
Nat. of India.
Cult. 1731, by Mr. Philip Miller. *Mill. dict. edit.* 1. Anil 1.
Fl. July and Auguſt. S. ♄.

10. I. foliis pinnatis tomentoſis obovatis, caule fruticoſo. *Syſt. veget.* 678. *argentea.*
Silvery-leav'd Indigo.
Nat. of the Weſt Indies.
Introd. 1776, by Mr. Gilbert Alexander.
Fl. July. S. ♄.

GALEGA. *Gen. pl.* 890.

Cal. dentibus ſubulatis, ſubæqualibus. *Legumen* ſtriis obliquis, ſeminibus interjectis.

officinalis. 1. G. leguminibus ſtrictis erectis, foliolis lanceolatis ſtriatis nudis. *Syſt. veget.* 679.

α floribus cæruleis.
Officinal blue Galega, or Goat's-rue.
β floribus albis.
Officinal white Galega, or Goat's-rue.
Nat. of Spain and Italy.
Cult. 1596, by Mr. John Gerard. *Hort. Ger.*
Fl. June——September. H. ♃.

grandiflora. 2. G. leguminibus patentibus, ſtipulis ovato-lanceolatis, foliolis oblongis nudiuſculis ariſtatis. *L'Herit. ſtirp. nov. tom.* 2. *tab.* 44.
Roſe-colour'd Galega.
Nat. of the Cape of Good Hope. Mr. *Fr. Maſſon.*
Introd. 1774.
Fl. May——September. G. H. ♄.

pallens. 3. G. leguminibus ſtrictis patentibus ciliatis, ſtipulis ſubulatis, foliolis (9-11) oblongis acutis ſubtus pubeſcentibus.
Pale-colour'd Galega.
Nat. of the Cape of Good Hope. Mr. *Fr. Maſſon.*
Introd. 1787.
Fl. July. G. H. ♄.

ſtricta. 4. G. leguminibus rectis villoſis, racemis oppoſitifoliis, ſtipulis ſubulatis, foliis villoſis oblongis, vexillis ſupra ſericeis.
Upright Galega.

§ *Nat.*

Nat. of the Cape of Good Hope. Mr. *Fr. Maſſon.*
Introd. 1774.
Fl. May and June. G. H. ♄.

5. G. leguminibus ſtrictis adſcendentibus ſubvilloſis, ſtipulis ſubulatis, foliolis (11-13) oblongis obtuſis ſubtus piloſiuſculis, pedunculis ancipitibus. *piſcatoria.*
Woolly Galega.
Nat. of India and the South Sea Iſlands.
Introd. 1778, by Patrick Ruſſell, M. D.
Fl. June and July. S. ♂.

6. G. leguminibus ſtrictis adſcendentibus glabris racemoſis terminalibus, ſtipulis ſubulatis, foliolis oblongis glabris. *Sp. pl.* 1063. *purpurea.*
Purple Galega.
Nat. of the Eaſt Indies.
Cult. 1768, by Mr. Philip Miller. *Mill. dict. edit.* 8.
Fl. July and Auguſt. S. ♃.

P H A C A. *Gen. pl.* 891.

Legumen ſemibiloculare.

1. P. cauleſcens erecta piloſa, leguminibus tereti-cymbiformibus. *Sp. pl.* 1064. *bœtica.*
Hairy Phaca, or Baſtard Vetch.
Nat. of Spain and Portugal.
Cult. 1640. *Park. theat.* 1084. *f.* 1.
Fl. July. H. ♃.

2. P. cauleſcens erecta glabra, leguminibus oblongis inflatis ſubpiloſis. *Sp. pl.* 1064. *Jacqu. ic. miſcell.* 2. *p.* 93. *alpina.*
Smooth Phaca, or Baſtard Vetch.

Nat. of Siberia, Lapland, and Auſtria.
Cult. 1759, by Mr. Ph. Miller. *Mill. dict. edit.* 7. *n.* 2.
Fl. July. H. ♃.

auſtralis. 3. P. caule ramoſo proſtrato, foliolis lanceolatis, florum alis ſemibifidis. *Linn. mant.* 103.
Trailing Phaca, or Baſtard Vetch.
Nat. of the South of Europe.
Introd. 1779, by Antony Chamier, Eſq.
Fl. May and June. H. ♃.

ASTRAGALUS. *Gen. pl.* 892.

Legumen biloculare, gibbum.

* *Caulibus folioſis erectis, nec proſtratis.*

alopecuroides. 1. A. cauleſcens, ſpicis cylindricis ſubſeſſilibus, calycibus leguminibuſque lanatis. *Sp. pl.* 1064.
Fox-tail Milk Vetch.
Nat. of Spain and Siberia.
Cult. 1739, by Mr. Philip Miller. *Rand. chel. n.* 4.
Fl. June and July. H. ♃.

piloſus. 2. A. cauleſcens erectus piloſus, floribus ſpicatis, leguminibus ſubulatis piloſis. *Sp. pl.* 1065. *Jacqu. auſtr.* 1. *p.* 32. *t.* 51.
Pale-flower'd Milk Vetch.
Nat. of Siberia and Germany.
Cult. 1732, by Mr. Philip Miller. *R. S. n.* 507.
Fl. June——Auguſt. H. ♃.

ſulcatus. 3. A. cauleſcens erectus glaber ſtriatus ſtrictus, foliolis lineari-lanceolatis acutis, leguminibus triquetris. *Syſt. veget.* 681. *Jacqu. hort.* 3. *p.* 23. *t.* 40.
Furrowed Milk Vetch.

Nat.

Nat. of Siberia.
Introd. 1785, by William Pitcairn, M. D.
Fl. July. H. ♃.

4. A. cauleſcens erectus, ſpicis pedunculatis, vexillo alis duplo longiore, foliolis linearibus. *tenuifolius.*
Aſtragalus tenuifolius. *Sp. pl.* 1065.
Aſtragalus Onobrychis β. *Syſt. veget.* 681.
Upright Milk Vetch.
Nat. of Siberia.
Introd. 1780, by Peter Simon Pallas, M. D.
Fl. July and Auguſt. H. ♃.

5. A. cauleſcens ſtrictus glaber, floribus racemoſis pendulis, leguminibus triquetris utrinque mucronatis. *Sp. pl.* 1066. *galegiformis.*
Goat's-rue-leav'd Milk Vetch.
Nat. of Siberia and the Levant.
Cult. 1739, by Mr. Philip Miller. *Rand. chel. n.* 8.
Fl. June——Auguſt. H. ♃.

6. A. cauleſcens erectiuſculus, floribus ſpicatis, leguminibus erectiuſculis nudis tumidis tereti-depreſſis: mucrone reflexo. *Sp. pl.* 1066. *uliginoſus.*
Violet-colour'd Milk Vetch.
Nat. of Siberia.
Introd. 1775, by Monſ. Thouin.
Fl. June——Auguſt. H. ♃.

7. A. cauleſcens erectus, leguminibus recurvatis, pedunculis multifloris folio longioribus, foliolis lanceolatis acutis. *vireſcens.*
Green-flower'd Milk Vetch.
Nat. of Siberia.
Introd. 1780, by Peter Simon Pallas, M. D.
Fl. June. H. ♃.

** *Caulibus*

** *Caulibus folioſis, diffuſis.*

canadenſis. 8. A. cauleſcens diffuſus, leguminibus ſubcylindricis mucronatis, foliolis nudiuſculis. *Syſt. veget.* 682.
Woolly Milk Vetch.
Nat. of Virginia and Canada.
Cult. 1732, by James Sherard, M. D. *Dill. elth.* 46.
Fl. June and July. H. ♃.

Cicer. 9. A. cauleſcens proſtratus, leguminibus ſubgloboſis inflatis mucronatis piloſis. *Sp. pl.* 1067. *Jacqu. auſtr.* 3. *p.* 29. *t.* 251.
Bladder'd Milk Vetch.
Nat. of Italy, Switzerland, and Auſtria.
Cult. 1739, by Mr. Philip Miller. *Rand. chel. n.* 13.
Fl. June and July. H. ♃.

microphyllus. 10. A. cauleſcens erecto-patulus: foliolis ovalibus, calycibus tumidiuſculis, leguminibus ſubrotundis. *Sp. pl.* 1067.
Small round-podded Milk Vetch.
Nat. of Siberia and Germany.
Introd. 1773, by Joſeph Nicholas de Jacquin, M. D.
Fl. June and July. H. ♃.

glycyphyllos. 11. A. cauleſcens proſtratus, leguminibus ſubtriquetris arcuatis, foliis ovalibus pedunculo longioribus. *Sp. pl.* 1067.
Liquorice Milk Vetch.
Nat. of Britain.
Fl. June and July. H. ♃.

hamoſus. 12. A. cauleſcens procumbens, leguminibus ſubulatis recurvatis glabris, foliolis obcordatis ſubtus villoſis. *Syſt. veget.* 682.
Dwarf yellow-flower'd Milk Vetch.

Nat.

Nat. of France and Sicily.
Cult. 1640. *Park. theat.* 1088. *f.* 2.
Fl. June and July. H. ⊙.

13. A. caulefcens procumbens, leguminibus contortuplicatis canaliculatis villoſis. *Sp. pl.* 1068. *contortuplicatus.*
Wave-podded Milk Vetch.
Nat. of Siberia.
Introd. 1783, by Monſ. Thouin.
Fl. July and Auguſt. H. ⊙.

14. A. caulefcens procumbens, ſpicis pedunculatis, leguminibus priſmaticis rectis triquetris apice uncinatis. *Sp. pl.* 1068. *bœticus.*
Triangular-podded Milk Vetch.
Nat. of Sicily, Spain, and Portugal.
Cult. 1759, by Mr. Philip Miller. *Mill. dict. edit.* 7. *n.* 7.
Fl. June and July. H. ⊙.

15. A. caulefcens diffuſus, capitulis ſubſeſſilibus lateralibus, leguminibus erectis ſubulatis acumine reflexis. *Sp. pl.* 1068. *ſeſameus.*
Starry Milk Vetch.
Nat. of France and Italy.
Cult. 1616, by Mr. John Parkinſon. *Ger. emac.* 1627. *n.* 7.
Fl. June and July. H. ⊙.

16. A. caulefcens procumbens, leguminibus capitatis ſeſſilibus cernuis cordatis mucronatis replicatis nudis. *Linn. mant.* 274. *epiglottis.*
Heart-podded Milk Vetch.
Nat. of the South of France, Spain, and the Levant.
Cult. 1768, by Mr. Philip Miller. *Mill. dict. edit.* 8.
Fl. June and July. H. ⊙.

17. A.

hypoglottis. 17. A. caulefcens proſtratus, leguminibus capitatis ovatis replicatis compreſſis piloſis acumine reflexo. *Linn. mant.* 274.
Aſtragalus arenarius. *Hudſ. angl.* 323.
Purple mountain Milk Vetch.
Nat. of Britain.
Fl. June and July. H. ♃.

Glaux. 18. A. caulefcens diffuſus, capitulis pedunculatis imbricatis ovatis, floribus erectis, leguminibus ovatis calloſis inflatis. *Syſt. veget.* 683.
Small Milk Vetch.
Nat. of Spain.
Cult. 1658, in Oxford Garden. *Hort. oxon. edit.* 2. *p.* 71.
Fl. June and July. H. ☉.

Onobrychis. 19. A. caulefcens procumbens diffuſus, ſpicis pedunculatis, vexillo alis duplo longiore, foliolis linearibus.
Aſtragalus Onobrychis. *Sp. pl.* 1070. *Jacqu. auſtr.* 1. *p.* 25. *t.* 38.
Purple-ſpik'd Milk Vetch.
Nat. of Auſtria.
Cult. 1640. *Park. theat.* 1082. *n.* 2.
Fl. June and July. H. ♃.

alpinus. 20. A. caulefcens procumbens, floribus pendulis racemoſis, leguminibus utrinque acutis piloſis. *Sp. pl.* 1070.
Alpine Milk Vetch.
Nat. of the Alps of Lapland and Switzerland.
Introd. about 1771.
Fl. June and July. H. ♃.

21. A.

21. A. ſubcauleſcens, ſcapis ſubbifloris, leguminibus hamatis ſubulatis bicarinatis. *Syſt. veget.* 684. *Jacqu. hort.* 2. *p.* 81. *t.* 174. *trimeſtris.*

Egyptian Milk Vetch.

Nat. of Egypt.

Introd. 1777, by Anthony Gouan, M. D.

Fl. June and July. H. ⊙.

*** *Scapo nudo, abſque caule folioſo.*

22. A. acaulis, ſcapo erecto foliis longiore, leguminibus ſubulatis inflatis villoſis erectis. *Sp. pl.* 1071. *Jacqu. ic. miſcell.* 1. *p.* 150. *uralenſis.*

Silky Milk Vetch.

Nat. of Scotland.

Fl. May—Auguſt. H. ♃.

23. A. acaulis, ſcapis declinatis longitudine foliorum, leguminibus ſubulatis teretibus ſubarcuatis glabris. *Sp. pl.* 1072. *monſpeſſulanus.*

Montpelier Milk Vetch.

Nat. of the South of France.

Introd. 1776, by William Pitcairn, M. D.

Fl. July. H. ♃.

24. A. acaulis, calycibus leguminibuſque villoſis, foliolis lanceolatis acutis, ſcapo decumbente. *Sp. pl.* 1072. *campeſtris.*

Field Milk Vetch.

Nat. of Switzerland and Germany.

Introd. 1778, by Monſ. Thouin.

Fl. June and July. H. ♃.

25. A. acaulis, ſcapis folio brevioribus, leguminibus cernuis, foliolis ſubemarginatis nudis. *Sp. pl.* 1073. *depreſſus.*

Dwarf white-flower'd Milk Vetch.

Nat. of Europe.

Cult.

Cult. 1772, in Oxford Garden.
Fl. May and June. H. ♂.

exſcapus. 26. A. acaulis exſcapus, leguminibus lanatis, foliis villoſis. *Linn. mant.* 275. *Jacqu. ic. vol.* 2.
Hairy-podded Milk Vetch.
Nat. of Hungary.
Introd. 1787, by Joſeph Nicholas de Jacquin, M.D.
Fl. H. ♃.

**** *Caule lignoſo.*

Tragacantha. 27. A. caudice arboreſcente, petiolis ſpineſcentibus. *Sp. pl.* 1073.
Goat's-thorn Milk Vetch.
Nat. of the South of Europe.
Cult. 1640. *Park. theat.* 996. *f.* 1.
Fl. May——July. H. ♄.

BISERRULA. *Gen. pl.* 893.

Legumen biloculare, planum: diſſepimento contrario.

Pelecinus. 1. Biserrula. *Sp. pl.* 1073.
Baſtard Hatchet Vetch.
Nat. of the South of Europe.
Cult. 1640. *Park. theat.* 1089. *f.* 5.
Fl. July and Auguſt. H. ☉.

PSORALEA. *Gen. pl.* 894.

Cal. punctis calloſis adſperſus, longitudine *Leguminis* 1-ſpermi.

pinnata. 1. P. foliis pinnatis linearibus, floribus axillaribus. *Sp. pl.* 1074.
Winged-leav'd Pſoralea.
Nat. of the Cape of Good Hope.

Cult.

Cult. 1690, in the Royal Garden at Hampton-court. *Catal. mſſ.*

Fl. May——July. G. H. ♄.

2. P. foliis ternatis: foliolis cuneiformibus recurvo-mucronatis, floribus axillaribus ſolitariis approximatis. *aculeata.*

Pſoralea aculeata. *Sp. pl.* 1074.

Prickly Pſoralea.

Nat. of the Cape of Good Hope.

Introd. 1774, by Mr. Francis Maſſon.

Fl. June and July. G. H. ♄.

3. P. foliis ternatis obovatis recurvato-mucronatis, ſpicis ovatis. *Linn. mant.* 264. *bracteata.*

Oval-ſpik'd Pſoralea.

Nat. of the Cape of Good Hope.

Cult. 1731, by Mr. Philip Miller. *Mill. dict. edit.* 1. Trifolium 9.

Fl. June and July. G. H. ♄.

4. P. foliis ternatis oblongis obtuſis, ſpicis cylindricis. *Linn. mant.* 264. *ſpicata.*

Long-ſpik'd Pſoralea.

Nat. of the Cape of Good Hope.

Introd. 1774, by Mr. Francis Maſſon.

Fl. July. G. H. ♄.

5. P. foliis ternatis linearibus, pedunculis axillaribus ſolitariis triniſve paucifloris. *L'Herit. ſtirp. nov. tom.* 2. *tab.* 45. *anguſtiſſima.*

Narrow-leav'd Pſoralea.

Nat. of the Cape of Good Hope. Mr. *Fr. Maſſon.*

Introd. 1774.

Fl. May——Auguſt. G. H. ♄.

6. P.

hirta. 6. P. foliis ternatis, foliolis obovatis recurvato-mucronatis, floribus ternis ſpicatis, calycibus tomentoſis.
Pſoralea hirta. *Sp. pl.* 1074.
Hairy Pſoralea.
Nat. of the Cape of Good Hope.
Cult. 1713, by the Dutcheſs of Beaufort. *Br. Muſ.* *H. S.* 142. *fol.* 42.
Fl. Moſt part of the Summer. G. H. ♄.

decumbens. 7. P. folis ternatis: foliolis cuneiformi-lanceolatis mucrone recurvato, floribus axillaribus.
Trailing Pſoralea.
Nat. of the Cape of Good Hope. Mr. *Fr. Maſſon.*
Introd. 1774.
Fl. April and May. G. H. ♄.

repens. 8. P. foliis ternatis obovatis emarginatis, caule repente, floribus ſubumbellatis. *Linn. mant.* 265.
Creeping Pſoralea.
Nat. of the Cape of Good Hope.
Introd. 1774, by Mr. Francis Maſſon.
Fl. July and Auguſt. G. H. ♃.

bituminoſa. 9. P. foliis omnibus ternatis: foliolis lanceolatis, petiolis lævibus, floribus capitatis. *Syſt. veget.* 686.
Bituminous Pſoralea.
Nat. of Italy and the South of France.
Cult. 1570. *Lobel. adv.* 380.
Fl. Moſt part of the Summer. G. H. ♄.

glanduloſa. 10. P. foliis omnibus ternatis: foliolis lanceolatis; petiolis ſcabris, floribus ſpicatis. *Syſt. veget.* 686.
Strip'd-flower'd Pſoralea.
Nat. of Peru.
Introd. about 1770.
Fl. May——Auguſt. G. H. ♄.

11. P.

11. P. foliis omnibus ternatis: foliolis ovatis, petiolis pubeſcentibus, floribus capitatis. *Syſt. veget.* 686. *Jacqu. hort.* 2. *p.* 86. *t.* 184. *palæſtina.*
Herbaceus Pſoralea.
Nat. of the Levant.
Introd. 1771, by Joſeph Nicholas de Jacquin, M.D.
Fl. Moſt part of the Summer. G. H. ♃.

12. P. foliis ternatis: foliolis ovatis dentato-angulatis, ſpicis lateralibus. *Syſt. veget.* 686. *americana.*
American Pſoralea.
Nat. of Madeira.
Cult. 1640. *Park. theat.* 717. *f.* 4.
Fl. July and Auguſt. G. H. ♄.

13. P. foliis ſimplicibus ovatis ſubdentatis, ſpicis ovatis. *Syſt. veget.* 686. *corylifolia.*
Nut-leav'd Pſoralea.
Nat. of India.
Cult. 1752, by Mr. Philip Miller. *Mill. dict. edit.* 6. Dorycnium 2.
Fl. June and July. S. ☉.

14. P. foliis pinnatis, ſpicis axillaribus. *Sp. pl.* 1076. *enneaphylla.*
Nine-leav'd Pſoralea.
Nat. of the Weſt Indies.
Introd. 1772, by Monſ. Richard.
Fl. July and Auguſt. S. ☉.

15. P. foliis pinnatis oblongo-linearibus numeroſiſſimis, ſpicis ebracteatis villoſis lanceolatis. *leporina.*
Downy-ſpik'd Pſoralea.
Nat.
Introd. 1780, by Benjamin Bewick, Eſq.
Fl. October and November. S. ☉.

DESCR. *Caulis* glaber, striatus. *Folia* sparsa, impari pinnata: *Foliola* numerosa, sæpe 30 parium, brevissime pedicellata, lineari-oblonga, integerrima, obtusiuscula cum brevi accumine, glabra, subtus punctata, vix semuncialia. *Spicæ* axillares, solitariæ, cylindricæ, vix biunciales: *pedunculi* digitales, sæpe foliis minutis pinnatis instructi. *Calyx* villis longis hirtus, trilinearis. *Corolla* cærulea, basi albida? *Vexillum* ovatum, integrum; *Alæ* ovatæ, vexillo paulo longiores; *Carina* longitudine alarum; *Ungues* omnium petalorum filiformes, calyce paulo breviores, subconnatæ. *Filamenta* connata in vaginam superne fissam, calyce longiorem. *Germen* ovatum, parum compressum, superne villis longis barbatum. *Stylus* ex altero latere germinis, erectus, filiformis, staminibus paulo longior. *Legumen* turbinato-subrotundum, compressum, membranaceum, pellucidum, diametro lineari, pilosiusculum, uniloculare. *Semen* unicum, reniforme, glabrum, cinereo-fuscum.

foliolosa. 16. P. foliis pinnatis oblongis numerosis, spicis terminalibus bracteatis globoso-ovatis, calycibus compressis.

Leafy Psoralea.

Nat.

Introd. 1780, by Benjamin Bewick, Esq.

Fl. October and November. S. ☉.

DESCR. *Caulis* teres, glaber, glandulis ferrugineis adspersus, ramosus. *Folia* sparsa, impari pinnata. *Foliola* numerosa, 10-14 parium, brevissime petiolata, oblonga, acuta, glabra, bilinearia. *Spicæ* ramorum terminales, ovato-globosæ, multifloræ, vix semunciales. *Bractea* sub singulo flore, ovata, concava, longitudine calycis. *Calyx* urceolatus, compressus, villosiusculus, bilinearis. *Corolla* purpurascens, basi pallida:

pallida: *Vexillum* ſubrotundum, acutum; *Alæ* oblongæ, vexillo longiores; *Carina* longitudine alarum. *Filamenta* connata in vaginam ſuperne fiſſam. *Germen* ovatum, compreſſum, pubeſcens. *Stylus* filiformis, ſtaminibus longior. *Legumen* ſubrotundum, oblique acuminatum, compreſſum, pubeſcens, uniloculare. *Semen* unicum, compreſſum.

TRIFOLIUM. *Gen. pl.* 896.

Flores ſubcapitati. *Legumen* vix calyce longius, non dehiſcens, deciduum.

* Meliloti, *leguminibus nudis polyſpermis.*

1. T. racemis ovatis, leguminibus ſeminudis mucronatis, caule erecto. *Syſt. veget.* 687. — *M. cærulea.*
Blue Melilot Trefoil.
Nat. of Germany.
Cult. before 1562, by William Turner, M.D. *Turn. herb. part* 2. *fol.* 158.
Fl. Auguſt and September. H. ⊙.

2. T. leguminibus racemoſis nudis monoſpermis, caule erecto. *Sp. pl.* 1077. — *M. indica.*
Indian Melilot Trefoil.
Nat. of India.
Cult. 1739, by Mr. Philip Miller. *Rand. chel.* Melilotus 5.
Fl. June——Auguſt. H. ⊙.

3. T. leguminibus racemoſis nudis diſpermis lanceolatis, caule erecto. *Sp. pl.* 1078. — *M. polonica.*
Polonian Melilot Trefoil.
Nat. of Poland.

Introd. 1778, by Monſ. Thouin.
Fl. June——Auguſt. H. ☉.

M. officinalis. 4. T. leguminibus racemoſis nudis diſpermis rugoſis acutis, caule erecto. *Sp. pl.* 1078.
α Melilotus officinarum germaniæ. *Bauh. pin.* 331.
Common Melilot Trefoil.
β Melilotus officinarum germaniæ, flore albo. *Tournef. inſt.* 407.
White Melilot Trefoil.
Nat. of Britain.
Fl. Auguſt and September. H. ♂.

M. italica. 5. T. leguminibus racemoſis nudis diſpermis rugoſis obtuſis, caule erecto, foliolis integris. *Sp. pl.* 1078.
Italian Melilot Trefoil.
Nat. of Italy.
Cult. 1596, by Mr. John Gerard. *Hort. Ger.*
Fl. June——Auguſt. H. ☉.

M. cretica. 6. T. leguminibus racemoſis nudis diſpermis membranaceis ovalibus, caule erectiuſculo. *Sp. pl.* 1078.
Cretan Melilot Trefoil.
Nat. of Candia.
Cult. 1713. *Philoſoph. tranſ. n.* 337. *p.* 209. *n.* 116.
Fl. June——Auguſt. H. ☉.

M. ornithopodioides. 7. T. leguminibus nudis octoſpermis ſubternis calyce duplo longioribus, caulibus declinatis. *Sp. pl.* 1078. *Curtis lond.*
Bird's-foot Melilot Trefoil.
Nat. of Britain.
Fl. June. H. ☉.

** Lotoidea,

** Lotoidea, *leguminibus tectis polyspermis.*

8. T. capitulis dimidiatis, foliis quinatis sessilibus, leguminibus polyspermis. *Sp. pl.* 1079. *Lupinaster.*
Bastard Lupine, or Trefoil.
Nat. of Siberia.
Cult. 1763, by Mr. James Gordon.
Fl. July and August. H. ♃.

9. T. capitulis umbellaribus, leguminibus tetraspermis, caule repente. *Sp. pl.* 1080. *Curtis lond.* *repens.*
Creeping white Trefoil, or Dutch Clover.
Nat. of Britain.
Fl. July and August. H. ♃.

10. T. capitulis umbellaribus, scapo nudo, leguminibus dispermis pendulis, foliis lineari-lanceolatis. *Syst. veget.* 688. *alpinum.*
Alpine Trefoil.
Nat. of Italy, Switzerland, and the Pyrenées.
Introd. 1775, by the Doctors Pitcairn and Fothergill.
Fl. June——August. H. ♃.

*** Lagopoda, *calycibus villosis.*

11. T. capitulis villosis quinquefloris, coma centrali reflexa rigida fructum obvolvente. *Syst. veget.* 688. *Curtis lond.* *subterraneum.*
Subterranean Trefoil.
Nat. of England.
Fl. May. H. ☉.

12. T. capitulis villosis globosis terminalibus solitariis, calycibus omnibus fertilibus, caulibus procumbentibus, foliis obcordatis. *Syst. veget.* 688. *cherleri.*
Hairy Trefoil.
Nat. of Montpelier.

Introd. 1768, by Monſ. Richard.
Fl. May and June. H. ♃.

rubens. 13. T. ſpicis villoſis longis, corollis monopetalis, caule erecto, foliis ſerrulatis. *Sp. pl.* 1081. *Jacqu. auſtr.* 4. *p.* 44. *t.* 385.
Long-ſpik'd Trefoil.
Nat. of the South of Europe.
Cult. 1633. *Ger. emac.* 1192. *f.* 2.
Fl. June——September. H. ♃.

pratenſe. 14. T. ſpicis globoſis ſubvilloſis, cinctis ſtipulis oppoſitis membranaceis, corollis monopetalis. *Syſt. veget.* 688.
Purple Trefoil, or Honeyſuckle Clover.
Nat. of Britain.
Fl. May——July. H. ♃.

alpeſtre. 15. T. ſpicis ſubgloboſis villoſis terminalibus, caule erecto, foliis lanceolatis ſerrulatis. *Syſt. veget.* 688. *Jacqu. auſtr.* 5. *p.* 15. *t.* 433.
Oval-ſpik'd Trefoil, or Clover.
Nat. of Scotland.
Fl. July. H. ♃.

pannonicum. 16. T. ſpicis villoſis longis, corollis monopetalis, foliis integerrimis cauleque erecto villoſiſſimis. *Linn. mant.* 276.
Pannonian Trefoil.
Nat. of Hungary.
Introd. 1775, by Joſeph Nicholas de Jacquin, M.D.
Fl. June and July. H. ♃.

ſquarroſum. 17. T. ſpicis oblongis ſubpiloſis, calycum infimo dente longiſſimo reflexo, caule herbaceo erecto. *Syſt. veget.* 689.

Round-

Round-leav'd Trefoil.
Nat. of Spain.
Introd. 1778, by Monf. Thouin.
Fl. July. H. ⊙.

18. T. fpicis villofis oblongis obtufis aphyllis, foliolis fubrotundis crenatis. *Syft. veget.* 689. *incarnatum.*
Flefh-colour'd Trefoil.
Nat. of Italy.
Cult. 1640. *Park. theat.* 1106. *f.* 1.
Fl. July. H. ⊙.

19. T. fpicis villofis, caule erecto pubefcente, foliolis infimis obcordatis. *Syft. veget.* 689. *Jacqu. auftr.* 1. *p.* 26. *t.* 40. *ochroleucum.*
Pallid Trefoil.
Nat. of England.
Fl. May——July. H. ♂.

20. T. fpicis villofis conico-oblongis, dentibus calycinis fetaceis fubæqualibus, foliolis linearibus. *Sp. pl.* 1083. *anguftifolium.*
Narrow-leav'd Trefoil.
Nat. of the South of France, Italy, and Madeira.
Cult. 1739, by Mr. Ph. Miller. *Mill. dict. vol.* 2. *n.* 5.
Fl. June——Auguft. H. ⊙.

21. T. fpicis villofis ovalibus, dentibus calycinis fetaceis villofis æqualibus. *Syft. veget.* 689. *arvenfe.*
Hare's-foot Trefoil.
Nat. of Britain.
Fl. July and Auguft. H. ⊙.

22. T. fpicis pilofis ovatis, calycibus patentibus, caule diffufo, foliolis obcordatis. *Syft. veget.* 689. *ftellatum.*

Star Trefoil.
Nat. of England.
Fl. July. H. ♃.

clypeatum. 23. T. spicis ovatis, calycibus patulis: lacinia infima maxima lanceolata, foliolis ovatis. *Sp. pl.* 1084.
Oriental Trefoil.
Nat. of the Levant.
Cult. 1711. *Philosoph. trans. n.* 332. *p.* 388. *n.* 39.
Fl. July and August. H. ⊙.

scabrum. 24. T. capitulis sessilibus lateralibus ovatis, calycibus laciniis inæqualibus rigidis recurvis. *Syst. veget.* 689.
Rough Trefoil.
Nat. of Britain.
Fl. May and June. H. ⊙.

glomeratum. 25. T. capitulis sessilibus hemisphæricis rigidis, calycibus striatis patulis æqualibus. *Syst. veget.* 689. *Curtis lond.*
Round-headed Trefoil.
Nat. of England.
Fl. June. H. ⊙.

striatum. 26. T. capitulis sessilibus sublateralibus ovatis, calycibus striatis rotundatis. *Sp. pl.* 1085.
Striated, or knotted Trefoil.
Nat. of Britain.
Fl. June. H. ⊙

**** Vesicaria, *calycibus inflatis ventricosis.*

spumosum. 27. T. spicis ovatis, calycibus inflatis glabris quinquedentatis, involucris universalibus pentaphyllis. *Sp. pl.* 1085.
Bladder'd Trefoil.

Nat.

Nat. of France and Italy.
Introd. 1771, by Monf. Richard.
Fl. June and July. H. ⊙.

28. T. fpicis feffilibus globofis tomentofis, calycibus inflatis obtufis. *Syft. veget.* 690. *tomentofum.*
Woolly Trefoil.
Nat. of the South of Europe.
Cult. 1640, by Mr. John Parkinfon. *Park. theat.* 1109. *n.* 6.
Fl. June and July. H. ⊙.

29. T. fpicis fubrotundis, calycibus inflatis bidentatis reflexis, caulibus repentibus. *Syft. veget.* 690. *Curtis lond.* *fragiferum.*
Strawberry Trefoil.
Nat. of England.
Fl. Auguft. H. ♃.

***** Lupulina, *vexillis corollæ inflexis.*

30. T. fpicis fubimbricatis fubtribus, vexillis fubulatis emarcefcentibus, calycibus nudis, caule erecto. *Sp. pl.* 1087. *montanum.*
Mountain Trefoil.
Nat. of Europe.
Introd. 1786, by William Pitcairn, M.D.
Fl. July. H. ♃.

31. T. fpicis ovalibus imbricatis, vexillis deflexis perfiftentibus, calycibus nudis, caule erecto. *Sp. pl.* 1087. *Curtis lond.* *agrarium.*
Hop Trefoil.
Nat. of Britain.
Fl. June. H. ⊙.

spadiceum. 32. T. spicis ovalibus imbricatis, vexillis deflexis persistentibus, calycibus pilosis, caule erecto. *Sp. pl.* 1087.
Pale-flower'd Trefoil.
Nat. of Europe.
Introd. 1778, by Mons. Thouin.
Fl. June——August. H. ♃.

procumbens. 33 T. spicis ovalibus imbricatis, vexillis deflexis persistentibus, caulibus procumbentibus. *Sp. pl.* 1088. *Curtis lond.*
Procumbent Trefoil.
Nat. of Britain.
Fl. May——September. H. ♃.

filiforme. 34. T. spicis subimbricatis, vexillis deflexis persistentibus, calycibus pedicellatis, caulibus procumbentibus. *Sp. pl.* 1088.
Small Trefoil.
Nat. of Britain.
Fl. May and June. H. ⊙.

L O T U S. *Gen. pl.* 879.

Legumen cylindricum, strictum. *Alæ* sursum longitudinaliter conniventes. *Cal.* tubulosus.

* *Leguminibus rarioribus, nec capitulum constituentibus.*

maritimus. 1. L. leguminibus solitariis membranaceo-quadrangulis, foliis glabris, bracteis lanceolatis. *Sp. pl.* 1089.
Sea Bird's-foot Trefoil.
Nat. of Europe.
Cult. 1759, by Mr. Philip Miller. *Mill. dict. edit.* 7. *n.* 12.
Fl. May——October. H. ♃.

2. L.

2. L. leguminibus ſolitariis membranaceo-quadrangulis, caulibus procumbentibus, foliis ſubtus pubeſcentibus. *Sp. pl.* 1089. *Jacqu. auſtr.* 4. *p.* 32. *t.* 361. *ſiliquoſus.*
Square-podded Bird's-foot Trefoil.
Nat. of the South of Europe.
Cult. 1683, by Mr. James Sutherland. *Sutherl. hort. edin.* 204. *n.* 5.
Fl. July and Auguſt. H. ♃.

3. L. leguminibus ſolitariis membranaceo-quadrangulis, bracteis ovatis. *Sp. pl.* 1089. *tetragonolobus.*
Winged Pea, or Bird's-foot Trefoil.
Nat. of Sicily.
Cult. 1596, by Mr. John Gerard. *Hort. Ger.*
Fl. July and Auguſt. H. ☉.

4. L. leguminibus conjugatis membranaceo-quadrangulis, bracteis oblongo-ovatis. *Sp. pl.* 1089. *conjugatus.*
Twin-podded Bird's-foot Trefoil.
Nat. of Montpelier.
Cult. 1759, by Mr. Ph. Miller. *Mill. dict. edit.* 7. *n.* 13.
Fl. July. H. ☉.

5. L. leguminibus ſubſolitariis gibbis incurvis. *Sp. pl.* 1090. *edulis.*
Eſculent Trefoil.
Nat. of Italy and Candia.
Cult. 1759, by Mr. Ph. Miller. *Mill. dict. edit.* 7. *n.* 11.
Fl. July and Auguſt. H. ☉.

6. L. leguminibus ſubbinatis linearibus compreſſis nutantibus. *Sp. pl.* 1090. *peregrinus.*
Flat-podded Bird's-foot Trefoil.
Nat. of the South of Europe.

Cult.

Cult. 1713. *Philosoph. transf. n.* 337. *p.* 208. *n.* 112.
Fl. July. H. ♃.

glaucus. 7. L. leguminibus fubbinatis cylindraceis glabris, foliolis fubcuneiformibus carnofis incanis, ftipulis foliiformibus.
Glaucous Bird's-foot Trefoil.
Nat. of Madeira. Mr. *Francis Maffon.*
Introd. 1777.
Fl. June——Auguft. G. H. ☿.

arabicus. 8. L. leguminibus cylindricis ariftatis, caulibus proftratis, pedunculis trifloris, bracteis monophyllis. *Linn. mant.* 104. *Jacqu. hort.* 2. *p.* 72. *t.* 155.
Red-flower'd Bird's-foot Trefoil.
Nat. of Arabia.
Introd. 1773, by Chevalier Murray.
Fl. July——November. G. H. ⊙.

ornithopodioides. 9. L. leguminibus fubternatis arcuatis compreffis, caulibus diffufis. *Sp. pl.* 1091.
Claw-podded Bird's-foot Trefoil.
Nat. of Sicily.
Cult. 1683, by Mr. James Sutherland. *Sutherl. hort. edin.* 206. *n.* 4.
Fl. June——Auguft. H. ⊙.

jacobæus. 10. L. leguminibus fubternatis, caule herbaceo erecto, foliolis linearibus. *Sp. pl.* 1091.
Dark-flower'd Bird's-foot Trefoil.
Nat. of the Cape-verd Iflands.
Cult. 1714, by the Dutchefs of Beaufort. *Br. Muf. H. S.* 134. *fol.* 41.
Fl. Moft part of the Summer. G. H. ♄.

11. L.

11. L. leguminibus ſubternatis, caule ſuffruticoſo, foliis ſericeis nitidis. *Sp. pl.* 1091. *creticus.*
Silvery Bird's-foot Trefoil.
Nat. of Spain and the Levant.
Cult. 1696. *Br. Muſ. Sloan. mſſ.* 3343.
Fl. June——September. G. H. ♄.

** *Pedunculis multifloris in capitulum.*

12. L. capitulis ſubrotundis, caule erecto hirto, leguminibus ovatis. *Syſt. veget.* 691. *hirſutus.*
Hairy Bird's-foot Trefoil.
Nat. of the South of Europe.
Cult. 1683, by Mr. James Sutherland. *Sutherl. hort. edin.* 206. *n.* 1.
Fl. June——Auguſt. H. ♄.

13. L. capitulis ſubgloboſis, caule erecto lævi, leguminibus rectis glabris. *Syſt. veget.* 691. *rectus.*
Upright Bird's-foot Trefoil.
Nat. of the South of Europe.
Cult. 1683, by Mr. James Sutherland. *Sutherl. hort. edin.* 206. *n.* 3.
Fl. June——Auguſt. H. ♃.

14. L. capitulis depreſſis, caulibus decumbentibus, leguminibus cylindricis patentibus. *Sp. pl.* 1092. *Curtis lond.* *corniculatus.*
Common Bird's-foot Trefoil.
Nat. of Britain.
Fl. June——Auguſt. H. ♃.

15. L. capitulis dimidiatis, caule diffuſo ramoſiſſimo, foliis tomentoſis. *Sp. pl.* 1092. *cytiſoides.*
Downy Bird's-foot Trefoil.
Nat. of the South of Europe.

Cult.

Cult. 1752, by Mr. Ph. Miller. *Mill. dict. edit.* 6. *n.* 9.
Fl. July. H. ♃.

Dorycnium. 16. L. capitulis aphyllis, foliis sessilibus quinatis. *Sp. pl.* 1093.
Shrubby Bird's-foot Trefoil.
Nat. of the South of Europe.
Cult. before 1640, by Mr. John Parkinson. *Park. theat.* 360. *n.* 1.
Fl. July——September. H. ♄.

TRIGONELLA. *Gen. pl.* 898.

Vexillum et *Alæ* subæquales, patentes, forma corollæ 3-petalæ.

ruthenica. 1. T. leguminibus pedunculatis congestis pendulis linearibus rectis, foliolis sublanceolatis. *Sp. pl.* 1093.
Small Fenugreek.
Nat. of Siberia.
Cult. 1759, by Mr. Ph. Miller. *Mill. dict. edit.* 7. *n.* 5.
Fl. June and July. H. ⊙.

platycarpos. 2. T. leguminibus pedunculatis congestis pendulis ovalibus compressis, caule diffuso, foliolis subrotundis. *Sp. pl.* 1093.
Round-leav'd Fenugreek.
Nat. of Siberia.
Cult. 1759, by Mr. Ph. Miller. *Mill. dict. edit.* 7. *n.* 4.
Fl. June——September. H. ♂.

polycerata. 3. T. leguminibus subsessilibus congestis erectis subrectis longis linearibus, pedunculis muticis. *Syst. veget.* 692.

Broad-leav'd, or Spanish Fenugreek.
Nat. of France, Italy, and Spain.
Cult. 1759, by Mr. Ph. Miller. *Mill. dict. edit.* 7. *n.* 3.
Fl. July. H. ⊙.

4. T. leguminibus pedunculatis racemosis declinatis hamosis teretibus, pedunculis spinosis folio longioribus. *Syst. veget.* 692. *hamosa.*
Egyptian Fenugreek.
Nat. of Egypt.
Cult. 1640. *Park. theat.* 720. *n.* 7.
Fl. July. H. ⊙.

5. T. leguminibus subpedunculatis congestis declinatis falcatis compressis, pedunculis spinosis brevissimis. *Syst. veget.* 692. *spinosa.*
Thorny Fenugreek.
Nat. of the Island of Candia.
Cult. 1739, by Mr. Philip Miller. *Rand. chel.* Fœnum græcum 5.
Fl. July and August. H. ⊙.

6. T. leguminibus pedunculatis congestis declinatis subfalcatis, pedunculo longo subspinoso, caule erecto. *Syst. veget.* 692. *corniculata.*
Horse-shoe Fenugreek.
Nat. of the South of Europe.
Cult. 1597, by Mr. John Gerard. *Ger. herb.* 1033. *f.* 1.
Fl. June and July. H. ⊙.

7. T. leguminibus sessilibus congestis arcuatis divaricatis inclinatis brevibus, pedunculo mucronato inermi. *Syst. veget.* 692. *monspeliaca.*
Trailing Fenugreek.

Nat.

Nat. of Montpelier.
Introd. 1771, by Monſ. Richard.
Fl. June and July. H. ⊙

Fœnum græcum. 8. T. leguminibus ſeſſilibus ſtrictis erectiuſculis ſubfalcatis acuminatis, caule erecto. *Sp. pl.* 1095.
Common Fenugreek.
Nat. of Montpelier.
Cult. 1597. *Ger. herb.* 1026.
Fl. June——Auguſt. H. ⊙.

MEDICAGO. *Gen. pl.* 899.

Legumen compreſſum, cochleatum. *Carina* corollæ a vexillo deflectens.

arborea. 1. M. leguminibus lunatis margine integerrimis, caule arboreo. *Sp. pl.* 1096.
Tree Medick, or Moon Trefoil.
Nat. of Italy.
Cult. 1596, by Mr. John Gerard. *Hort. Ger.*
Fl. May——November. G. H. ♄.

circinnata. 2. M. leguminibus reniformibus margine dentatis, foliis pinnatis. *Sp. pl.* 1096.
Kidney-podded Medick.
Nat. of Spain and Italy.
Introd. 1777, by Abbé Nolin.
Fl. July and Auguſt. H. ⊙.

ſativa. 3. M. pedunculis racemoſis, leguminibus contortis, caule erecto glabro. *Sp. pl.* 1096.
Cultivated Medick, or Lucern.
Nat. of England.
Fl. June and July. H. ♃.

4. M.

4. M. pedunculis racemosis, leguminibus lunatis, caule prostrato. *Sp. pl.* 1096. *falcata.*
Yellow Medick.
Nat. of England.
Fl. July. H. ♃.

5. M. spicis ovalibus, leguminibus reniformibus monospermis, caule procumbente. *Sp. pl.* 1097. *Curtis lond.* *lupulina.*
Black Medick.
Nat. of Britain.
Fl. May——August. H. ♂.

6. M. pedunculis racemosis, leguminibus cochleatis spinosis, caule procumbente tomentoso. *Sp. pl.* 1097. *marina.*
Sea Medick.
Nat. of the Coasts on the Mediterranean.
Cult. 1597, by Mr. John Gerard. *Hort. Ger.*
Fl. June and July. H. ♃.

7. M. leguminibus cochleatis, stipulis dentatis, caule diffuso. *Sp. pl.* 1097. *polymorpha.*
α Medicago leguminibus solitariis cochleatis compressis planis, stipulis ciliatis, caule diffuso. *Sauv. monsp.* 186. orbicularis.
Flat-podded Medick.
β Medica cochleata major dicarpos, capsula rotunda globosa scutellata. *Moris. hist.* 2. *p.* 152. *s.* 2. *t.* 15. *f.* 4. scutellata.
S nailMedick.
γ Medica tornata major & minor lenis. *Park. theat.* 1116. tornata.
Smooth-podded Medick.
δ Medicago fructu turbinato. *Sauv. monsp.* 187. turbinata.
Turban Medick.
ε Medica cochleata spinosa major dicarpos, capsula seu spinis intertexta.

ſpinis longioribus ſurſum & deorſum tendentibus. *Moriſ. hiſt.* 2. *p.* 153. *ſ.* 2. *t.* 15. *f.* 8, 9, 7.
Hedge-hog Medick.

muricata. ζ Medica cochleata dicarpos capſula ſpinoſa rotunda minore. *Moriſ. hiſt.* 2. *p.* 153. *ſ.* 2. *t.* 15. *f.* 11.
Prickly Medick.

arabica. η Medica cochleata minor polycarpos annua capſula majore alba, folio cordato macula fuſca. *Moriſ. hiſt.* 2. *p.* 154. *ſ.* 2. *t.* 15. *f.* 17. *Curtis lond.*
Heart Medick.

rigidula. θ Medicago triphylla, leguminibus cochleatis ſpinoſis, foliolis inferioribus cuneiformibus retuſis, ſuperioribus ſubrotundis. *Dalib. pariſ.* 230.
Thorny-podded Medick.

laciniata. ι Medicago fructu echinato, foliis linearibus dentatis. *Sauv. monſp.* 187.
Cut-leav'd Medick.
Nat. of Europe.
Fl. June—Auguſt. H. ⊙.

Claſſis

Classis XVIII.

POLYADELPHIA

PENTANDRIA.

THEOBROMA. *Gen. pl.* 900.

Cal. 3-phyllus. *Petala* 5, fornicata, bicornia. *Nectarium* 5 phyllum, regulare. *Stam.* nectario innata, ſingulo antheris 5.

1. T. foliis integerrimis. *Sp. pl.* 1100. *Cacao.*
Chocolate Nut-tree.
Nat. of South America.
Cult. 1739, by Mr. Philip Miller. *Mill. dict. vol.* 2. Cacao.
Fl. S. ♄.

2. T. foliis ſerratis. *Sp. pl.* 1100. *Guazuma.*
Elm-leav'd Theobroma, or Baſtard Cedar.
Nat. of Jamaica.
Cult. 1739, by Mr. Ph. Miller. *Rand chel.* Guazuma.
Fl. Auguſt and September. S. ♄.

AMBROMA. *Linn. ſuppl.* 54.

Pentagyna. Capſ. 5-locularis, 1-valvis, apice dehiſcens. *Sem.* reniformia.

1. AMBROMA. *Linn. ſuppl.* 341. *augusta.*
Abroma faſtuoſum. *Jacqu. hort.* 3. *p.* 3. *t.* 1.

Theobroma augusta. *Syst. nat. vol.* 3. *p.* 233.
Althæa Luzonis peregrina altera. *Camel. luz.* 12. *n.* 23. *Pet. gaz. tab.* 102. *f.* 8.
Maple-leav'd Ambroma.
Nat. of New South Wales and the Philippine Islands.
Introd. about 1770.
Fl. August. S. ♄.

DODECANDRIA.

MONSONIA. *Linn. mant.* 14.

Cal. 5-phyllus. *Cor.* 5-petala. *Stam.* 15, connata in 5 filamenta. *Stylus* 5-fidus. *Caps.* 5-cocca.

speciosa. 1. M. foliis quinatis: foliolis bipinnatis. *Syst. veget.* 697. *Curtis magaz.* 73.
Fine-leav'd Monsonia.
Nat. of the Cape of Good Hope.
Introd. 1774, by Mr. Francis Masson.
Fl. April and May. G. H. ♃.

lobata. 2. M. foliis cordatis lobatis dentatis.
Monsonia lobata. *Montin in act. gothob.* 2. *p.* 1. *tab.* 1. *Syst. veget.* 697.
Monsonia filia. *Linn. suppl.* 341. *Syst. veget.* 696. *Cavanill. diss.* 3. *p.* 180. *tab.* 74. *fig.* 2.
Broad-leav'd Monsonia.
Nat. of the Cape of Good Hope. Mr. *Fr. Masson.*
Introd. 1774.
Fl. April and May. G. H. ♃.

3. M. foliis oblongis subcordatis crenatis undulatis. *ovata.*
Monsonia ovata. *Cavanill. diss.* 4. *p.* 193. *tab.* 113. *f.* 1.
Geranium emarginatum. *Linn. suppl.* 306.
Undulated Monsonia.
Nat. of the Cape of Good Hope. Mr. *Fr. Masson.*
Introd. 1774.
Fl. August. G. H. ♂.

ICOSANDRIA.

CITRUS. *Gen. pl.* 901.

Cal. 5-fidus. *Petala* 5, oblonga. *Antheræ* 20, filamentis connatis in varia corpora. *Bacca* 9-locularis.

1. C. petiolis linearibus. *Sp. pl.* 1100. *Medica.*
α Malus medica. *Bauh. pin.* 435.
Lemon-tree.
β Malus Limonia acida. *Bauh. pin.* 436. Limon.
Lime-tree.
Nat. of Asia.
Cult. 1648, in Oxford Garden. *Hort. oxon. edit.* 1. *p.* 33.
Fl. Most part of the Summer. G. H. ♄.

2. C. petiolis alatis, foliis acuminatis. *Syst. veget.* 697. *Aurantium.*
α Malus Arantia major. *Bauh. pin.* 436.
Seville Orange-tree.
β Malus Arantia, cortice dulci eduli. *Bauh. pin.* 436. sinensis.
China Orange-tree.
Nat. of India.
Cult. 1629. *Park. parad.* 584.
Fl. Most part of the Summer. G. H. ♄.

decumanus. 3. C. petiolis alatis, foliis obtufis emarginatis. *Syft. veget.* 697.
Shaddock-tree.
Nat. of India.
Cult. 1739. *Mill. dict. edit.* 1. Aurantium 11.
Fl. Moft part of the Summer. G. H. ♄.

POLYANDRIA.

MELALEUCA. *Linn. mant.* 14.

Cal. 5-partitus, fuperus. *Cor.* 5-petala. *Filam.* multa, connata in 5 corpora. *Styl.* 1. *Capf.* femiveftita calyce baccato, 3-valvis, 3-locularis.

Leucadendron. 1. M. polyadelpha, foliis alternis lanceolatis fubfalcatis quinquenerviis, fpica elongata;
angustifolia. foliis anguftioribus oblongis vix falcatis brevioribus obtufis glaucis. *Linn. fuppl.* 342.
Aromatic Melaleuca.
Nat. of New Caledonia.
Introd. 1775, by John Reinhold Forfter, LL.D.
Fl. S. ♄.

HYPERICUM. *Gen. pl.* 902.

Cal. 5-phyllus. *Petala* 5. *Nect.* 0. *Capfula.*

* *Pentagyna.*

balearicum. 1. H. floribus pentagynis, caule fruticofo, foliis ramifque cicatrifatis. *Sp. pl.* 1101.
Warted St. John's-wort.
Nat. of Majorca.
Cult. 1714. *Philofoph. tranf. n.* 344. *p.* 277. *n.* 68.
Fl. March——September. G. H. ♄.

2. H.

2. H. floribus pentagynis solitariis, caule suffruticoso ramoso, calycibus obovatis obtusissimis, foliis distichis oblongis. *calycinum.*

Hypericum calycinum. *Linn. mant.* 106.

Hypericum Ascyron. *Mill. dict.*

Androsæmum constantinopolitanum flore maximo. *Wheler's journey into Greece, p.* 205. *cum fig.*

Androsæmum flore et theca quinquecapsulari omnium maximis. *Moris. hist.* 2. *p.* 472. conf. *Whel. loc. cit. p.* 206.

Great-flower'd St. John's-wort, or Tutsan.

Nat. of the Country near Constantinople.

Introd. 1676, by Sir George Wheler, Bart. *Whel. loc. cit.*

Fl. June——September. H. ♄.

3. H. floribus pentagynis subpaniculatis, caule subtetragono herbaceo ramoso, calycibus ovatis acutis. *pyramidatum.*

Pyramidal St. John's-wort.

Nat.

Cult. 1764, by Mr. James Gordon.

Fl. July and August. H. ♃.

* * *Trigyna.*

4. H. floribus trigynis, pericarpiis baccatis, caule fruticoso ancipiti. *Sp. pl.* 1102. *Curtis lond.* *Androsæmum.*

Common Tutsan, or St. John's-wort.

Nat. of Britain.

Fl. July——September. H. ♄.

5. H. floribus trigynis, calycibus acutis, staminibus corolla brevioribus, caule fruticoso. *Sp. pl.* 1102. *Olympicum.*

Olympian St. John's-wort.

Nat. of the Levant.

Cult. 1706, in Chelsea Garden. *Br. Mus. Sloan. mss.* 3370. *part* 3.
Fl. July——September. H. ♄.

foliosum. 6. H. floribus trigynis, staminibus longitudine petalorum, calycibus lanceolatis acutis, foliis ovali-oblongis sessilibus glabris.
Shining St. John's-wort.
Nat. of the Azores. Mr. *Francis Masson.*
Introd. 1778.
Fl. August. G. H. ♄.

floribundum. 7. H. floribus trigynis, calycibus ovatis acutis subciliatis, staminibus corolla brevioribus, foliis lanceolato-ellipticis, caule fruticoso.
Hypericum frutescens canariense multiflorum. *Comm. hort.* 2. *p.* 135. *t.* 68.
Hypericum s. Androsæmum magnum canariense ramosum, copiosis floribus fruticosum. *Pluk. alm.* 189. *t.* 302. *f.* 1.
Many-flower'd St. John's-wort.
Nat. of Madeira.
Introd. 1779, by Mr. Francis Masson.
Fl. August. G. H. ♄.

canariense. 8. H. floribus trigynis, calycibus obtusis, staminibus corolla brevioribus, caule fruticoso. *Syst. veget.* 700.
Canary St. John's-wort.
Nat. of the Canary Islands.
Cult. 1699, by the Dutchess of Beaufort. *Br. Mus. Sloan. mss.* 525 & 3343.
Fl. July——September. G. H. ♄.

elatum. 9. H. floribus trigynis, calycibus lanceolato-ovatis acutis, staminibus corolla longioribus, caule fruticoso, foliis ovato-oblongis.

Tall

Tall St. John's-wort.
Nat. of North America.
Cult. 1762, by Mr. James Gordon.
Fl. July and Auguſt. H. ♄.

10. H. floribus trigynis, ſtaminibus corolla longioribus, calycibus lanceolatis acutis, foliis oblongis, caule ſuffruticoſo. *hircinum.*
Hypericum hircinum. *Sp. pl.* 1103.
α Hypericum fœtidum fruteſcens majus. *Dill. elth.* 182. *t.* 151. *f.* 182. majus.
Common ſtinking ſhrubby St. John's-wort.
β Hypericum fœtidum fruteſcens minus. *Dill. elth.* 182. *t.* 151. *f.* 181. minus.
Small ſtinking ſhrubby St. John's-wort.
Nat. of the South of Europe.
Cult. 1640. *Park. theat.* 576. *f.* 4.
Fl. July——September. H. ♄.

11. H. floribus trigynis, nectariis petalorum lanceolatis, caulibus ſuffruticoſis compreſſis. *Sp. pl.* 1103. *ægyptiacum.*
Egyptian St. John's-wort.
Nat. of Egypt.
Introd. 1787, by Monſ. Thouin.
Fl. G. H. ♄.

12. H. floribus trigynis: primordialibus ſeſſilibus, caule ancipiti fruticoſo, foliis lanceolato-linearibus. *Syſt. veget.* 701. *prolificum.*
Proliferous St. John's-wort.
Nat. of North America.
Cult. 1758, by Mr. Philip Miller.
Fl. June——Auguſt. H. ♄.

13. H. floribus trigynis, foliis lineari-lanceolatis, caule quadrangulo, pericarpiis coloratis. *Sp. pl.* 1104. *canadenſe.*
Canadian

Canadian St. John's-wort.
Nat. of North America.
Introd. 1770, by Samuel Martin, M. D.
Fl. July——September. H. ♃.

læviga-tum. 14. H. floribus trigynis, foliis ovatis ſubamplexicaulibus, foliolis calycinis ovatis acutis, panicula trichotoma: flore intermedio ſeſſili.
Smooth St. John's-wort.
Nat. of North America. *Samuel Martin,* M. D.
Introd. 1772.
Fl. July. H. ♃.

reflexum. 15. H. floribus trigynis, foliis ſeſſilibus lanceolatis approximatis reflexis, ramis tomentoſis, panicula terminali. *Linn. ſuppl.* 346.
Reflex-leav'd St. John's-wort.
Nat. of the Iſland of Teneriffe.
Introd. 1778, by Mr. Francis Maſſon.
Fl. Moſt part of the Summer. G. H. ♄.

quadran-gulum. 16. H. floribus trigynis, caule quadrato herbaceo. *Sp. pl.* 1104. *Curtis lond.*
Square-ſtalk'd St. John's-wort, or St. Peter's-wort.
Nat. of Britain.
Fl. July. H. ♃.

perfora-tum. 17. H. floribus trigynis, caule ancipiti, foliis obtuſis pellucido-punctatis. *Sp. pl.* 1105. *Curtis lond.*
Perforated St. John's-wort.
Nat. of Britain.
Fl. July. H. ♃.

humifu-ſum. 18. H. floribus trigynis axillaribus ſolitariis, caulibus ancipitibus proſtratis filiformibus, foliis glabris. *Sp. pl.* 1105. *Curtis lond.*

Trailing St. John's-wort.
Nat. of Britain.
Fl. July. H. ♃.

19. H. floribus trigynis, calycibus ſerrato-glandulosis, caule tereti erecto, foliis ovatis glabris. *Syſt. veget.* 701. *montanum.*
Mountain St. John's-wort.
Nat. of Britain.
Fl. July. H. ♃.

20. H. floribus trigynis, calycibus acutis ſerrato-glandulosis, foliis oblongo-lanceolatis pellucido-punctatis margine glandulosis, caule fruticoſo. *glanduloſum.*
Glandulous St. John's-wort.
Nat. of Madeira. Mr. *Francis Maſſon.*
Introd. 1777.
Fl. May——Auguſt. G. H. ♄.

21. H. floribus trigynis, calycibus ſerrato-glandulosis, caule tereti erecto, foliis ovatis ſubpubeſcentibus. *Sp. pl.* 1105. *Curtis lond.* *hirſutum.*
Hairy St. John's-wort.
Nat. of Britain.
Fl. July. H. ♃.

22. H. floribus trigynis, calycibus ſerrato-glandulosis, foliis ſemiamplexicaulibus flexuosis tomentosis, caulibus proſtratis. *Sp. pl.* 1106. *tomentoſum.*
Woolly St. John's-wort.
Nat. of the South of Europe.
Introd. 1772, by Monſ. Richard.
Fl. July——September. G. H. ♃.

23. H. floribus trigynis, caule ſubancipiti, foliis amplexicaulibus ovatis, cyma floribus ſeſſilibus. *Syſt. veget.* 702. *perfoliatum.*

Perfoliate

Perfoliate St. John's-wort.
Nat. of Italy.
Introd. 1785, by Mr. John Græfer.
Fl. May and June. G. H. ♃.

elodes. 24. H. floribus trigynis, caule tereti repente foliisque villosis subrotundis. *Sp. pl.* 1106.
Marsh St. John's-wort, or St. Peter's-wort.
Nat. of Britain.
Fl. July. H. ♃.

pulchrum. 25. H. floribus trigynis, calycibus serrato-glandulosis, caule tereti, foliis amplexicaulibus cordatis glabris. *Syst. veget.* 702. *Curtis lond.*
Upright St. John's-wort.
Nat. of Britain.
Fl. July. H. ♃.

Coris. 26. H. floribus trigynis, calycibus serrato-glandulosis, foliis subverticillatis. *Sp. pl.* 1107.
Heath-leav'd St. John's-wort.
Nat. of the South of Europe and the Levant.
Cult. 1640. *Park. theat.* 570. *f.* 1.
Fl. G. H. ♄.

*** *Digyna.*

setosum. 27. H. floribus digynis, foliis linearibus. *Sp. pl.* 1107.
Bristly St. John's-wort.
Nat. of Virginia and Carolina.
Introd. 1787, by Thomas Walter, Esq.
Fl. H. ♃.

**** *Monogyna.*

monogynum. 28. H. floribus monogynis, staminibus corolla longioribus, calycibus coloratis, caule fruticoso. *Sp. pl.* 1107.

Chinese

Chinese St. John's-wort.
Nat. of China.
Introd. 1753, by Hugh Duke of Northumberland.
Fl. March——September. G. H. ♄.

ASCYRUM. *Gen. pl.* 903.

Cal. 4-phyllus. *Petala* 4. *Filamenta* multa, in 4 phalanges digesta.

1. A. foliis ovatis, caule tereti, panicula dichotoma. *Sp. pl.* 1107. *Crux andreae.*
Common Ascyrum, or St. Andrew's Cross.
Nat. of North America.
Cult. 1759, by Mr. Ph. Miller. *Mill. dict. edit.* 7. *n.* 1.
Fl. July and August. G. H. ♄.

Classis

Classis XIX.

SYNGENESIA

POLYGAMIA ÆQUALIS.

GEROPOGON. *Gen. pl.* 904.

Receptaculum setoso-paleaceum. *Cal.* simplex. *Sem.* disci pappo plumoso; Radii 5-aristato.

glabrum. 1. G. foliis glabris. *Sp. pl.* 1109. *Jacqu. hort.* 1. *p.* 12. *t.* 33.
Smooth Geropogon, or Old Man's-beard.
Nat. of Italy.
Cult. 1759, by Mr. Philip Miller. *Mill. dict. edit.* 7. Tragopogon 4.
Fl. July and August. H. ☉.

calyculatum. 2. G. calycibus calyculatis. *Syst. veget.* 709.
Tragopogon calyculatus. *Jacqu. hort.* 2. *p.* 48. *t.* 106.
Perennial Geropogon.
Nat. of Italy.
Introd. 1774, by Joseph Nicholas de Jacquin, M. D.
Fl. June. H. ♃.

TRAGOPOGON. *Gen. pl.* 905.

Recept. nudum. *Cal.* simplex. *Pappus* plumosus.

pratense. 1. T. calycibus corollæ radium æquantibus, foliis integris strictis. *Sp. pl.* 1109.
Yellow Goat's-beard.

Nat.

Nat. of Britain.

Fl. May and June. H. ♂.

2. T. calycibus corollæ radio longioribus, foliis integris ſtrictis, pedunculis ſuperne incraſſatis, corollulis ad apicem rotundatis. *Syſt. veget.* 710. *Jacqu. auſtr.* 1. *p.* 19. *t.* 29. *majus.*

Great Goat's-beard.

Nat. of Auſtria and Switzerland.

Introd. 1788, by Edmund Davall, Eſq.

Fl. H. ♂.

3. T. calycibus corollæ radio longioribus, foliis integris ſtrictis, pedunculis ſuperne incraſſatis, corollulis anguſtiſſimis truncatis. *Syſt. veget.* 710. *Jacqu. ic. collect.* 1. *p.* 99. *porrifolium.*

Purple Goat's-beard.

Nat. of England.

Fl. May——July. H. ♂.

4. T. calycibus corollæ radio longioribus, foliis integris, radicalibus peduncuIiſque baſi villoſis. *Sp. pl.* 1110. *crocifolium.*

Crocus-leav'd Goat's-beard.

Nat. of Italy and Montpelier.

Cult. 1739, by Mr. Philip Miller. *Rand. chel. n.* 4.

Fl. June and July. H. ♂.

5. T. calycibus monophyllis corolla brevioribus inermibus, foliis runcinatis. *Sp. pl.* 1110. *Dalechampii.*

Great-flower'd Goat's-beard.

Nat. of the South of France and Spain.

Cult. 1739, by Mr. Philip Miller. *Rand. chel.* Hieracium 20.

Fl. June——October. H. ♃.

6. T.

Picroides. 6. T. calycibus monophyllis corolla brevioribus aculeatis, foliis runcinatis denticulatis. *Sp. pl.* 1111.
Prickly-cup'd Goat's-beard.
Nat. of the South of Europe.
Cult. 1683, by Mr. James Sutherland. *Sutherl. hort. edin.* 152. *n.* 2.
Fl. July and Auguſt. H. ☉.

aſperum. 7 T. calycibus corolla brevioribus hiſpidis, foliis integris: caulinis oblongis. *Sp. pl.* 1111.
Rough Goat's-beard.
Nat. of Montpelier.
Introd. 1774, by Monſ. Richard.
Fl. July and Auguſt. H. ☉.

SCORZONERA, *Gen. pl.* 906.

Recept. nudum. *Pappus* plumoſus. *Cal.* imbricatus ſquamis margine ſcarioſis.

humilis. 1. S. caule ſubnudo unifloro, foliis lato-lanceolabis nervoſis planis. *Sp. pl.* 1112. *Jacqu. auſtr.* 1. *p.* 24. *t.* 36.
Dwarf Viper's-graſs.
Nat. of Scotland.
Fl. Auguſt. H. ♃.

hiſpanica. 2. S. caule ramoſo, foliis amplexicaulibus integris ſerrulatis. *Sp. pl.* 1112.
Garden Viper's-graſs, or Scorzonera.
Nat. of Spain and Siberia.
Cult. 1596, by Mr. John Gerard. *Hort. Ger.*
Fl. June——September. H. ♃.

graminifolia. 3. S. foliis lineari-enſiformibus integris carinatis. *Sp. pl.* 1112.

Graſs-

Graſs-leav'd Viper's-graſs.
Nat. of Portugal.
Cult. 1759, by Mr. Ph. Miller. *Mill. dict. edit.* 7. *n.* 4.
Fl. June——Auguſt. H. ♃.

4. S. foliis obtuſe dentatis, caule divaricato, calycum apicibus tomentoſis. *Syſt. veget.* 711. *reſedifolia.*
Spreading Viper's-graſs.
Nat. of Spain.
Cult. 1729, by Mr. Philip Miller. *Mill. dict.* 7. *n:* 8.
Fl. June and July. H. ♂.

5. S. foliis linearibus dentatis acutis, caule erecto, ſquamis calycinis patulo-mucronatis. *Sp. pl.* 1114. *Jacqu. auſtr.* 4. *p.* 29. *t.* 356. *laciniata.*
Cut-leav'd Viper's-graſs.
Nat. of the South of Europe.
Cult. 1640, by Mr. John Parkinſon. *Park. theat.* 411. *f.* 3.
Fl. June and July. H. ♂.

6. S. foliis omnibus runcinatis amplexicaulibus. *Sp. pl.* 1114. *tingitana.*
Poppy-leav'd Viper's-graſs.
Nat. of Barbary.
Cult. 1713. *Philoſoph. tranſ. n.* 337. *p.* 183. *n.* 24.
Fl. June——September. H. ⊙.

7. S. foliis ſuperioribus amplexicaulibus integerrimis; inferioribus runcinatis, pedunculis ſquamatis. *Sp. pl.* 1114. *Picroides.*
Various-leav'd Viper's-graſs.
Nat. of Montpelier.
Introd. 1773, by Monſ. Richard.
Fl. June——Auguſt. H. ⊙.

PICRIS. *Gen. pl.* 907.

Recept. nudum. *Cal.* calyculatus. *Pappus* plumoſus. *Sem.* tranſverſim ſulcata.

Echioides. 1. P. perianthiis exterioribus pentaphyllis interiore ariſtato majoribus. *Sp. pl.* 1114. *Curtis lond.*
Rough Picris, or Ox-tongue.
Nat. of England.
Fl. July and Auguſt. H. ☉.

Hieracioides. 2. P. perianthiis laxis, foliis integris, pedunculis ſquamatis in calycem. *Syſt. veget.* 711.
Hawkweed Picris, or yellow Succory.
Nat. of England.
Fl. July. H. ☉.

SONCHUS. *Gen. pl.* 908.

Recept. nudum. *Cal.* imbricatus, ventricoſus. *Pappus* piloſus.

maritimus. 1. S. pedunculo nudo, foliis lanceolatis amplexicaulibus indiviſis retrorſum argute dentatis. *Sp. pl.* 1116.
Sea Sow-thiſtle.
Nat. of the South of Europe.
Introd. 1774, by Monſ. Richard.
Fl. July——September. H. ♃.

paluſtris. 2. S. pedunculis calycibuſque hiſpidis ſubumbellatis, foliis runcinatis baſi ariſtatis. *Syſt. veget.* 712. *Curtis lond.*
Marſh Sow-thiſtle.
Nat. of England.
Fl. Auguſt. H. ♃.

3. S.

3. S. pedunculis calycibusque hispidis subumbellatis, foliis runcinatis basi cordatis. *Sp. pl.* 1116. *Curtis lond.* *arvensis.*

Corn Sow-thistle.

Nat. of Britain.

Fl. August. H. ♃.

4. S. pedunculis tomentosis, calycibus glabris. *Sp. pl.* 1116. *Curtis lond.* *oleraceus.*

Common Sow-thistle.

Nat. of Britain.

Fl. June——August. H. ☉.

5. S. pedunculis tomentosis, calycibus pilosis. *Sp. pl.* 1117. *tenerrimus.*

Clammy Sow-thistle.

Nat. of Italy and the South of France.

Cult. 1713, in Chelsea Garden. *Philosoph. trans. n.* 337. *p.* 37. *n.* 15.

Fl. July and August. H. ☉.

6. S. pedunculis squamosis, floribus racemosis, foliis runcinatis. *Sp. pl.* 1117. *alpinus.*

Alpine Sow-thistle.

Nat. of England.

Fl. July and August. H. ☉.

7. S. pedunculis subsquamosis, foliis basi attenuatis lyratis: lobis rotundatis obtusis, calycibus florentibus squarrosis. *fruticosus.*

Sonchus fruticosus. *Linn suppl.* 346. *Jacqu. ic. collect.* 1. *p.* 83..

Shrubby Sow-thistle.

Nat. of Madeira. Mr. *Francis Masson.*

Introd. 1777.

Fl. April——July. G. H. ♄.

pinnatus. 8. S. pedunculis nudis, calycibus lævibus, foliis pinnatis: pinnis lineari-lanceolatis ſubdentatis.
Wing-leav'd Sow-thiſtle.
Nat. of Madeira. Mr. *Francis Maſſon.*
Introd. 1777.
Fl. G. H. ♄.

radica*tus.* 9. S. pedunculis nudis calycibuſque glabris, caule ſubnudo, foliis radicalibus lyratis utrinque lævibus: lobis triangulari-ovatis.
Long-rooted Sow-thiſtle.
Nat. of the Canary Iſlands. Mr. *Francis Maſſon.*
Introd. 1780.
Fl. July. G. H. ♄.

floridanus. 10. S. pedunculis ſquamoſis, foliis lyrato-haſtatis. *Syſt. veget.* 712.
Small-flower'd Sow-thiſtle.
Nat. of North America.
Cult. 1713. *Philoſoph. tranſ. n.* 337. *p.* 182. *n.* 20.
Fl. July. H. ♂.

ſibiricus. 11. S. pedunculis ſquamatis, foliis lanceolatis indiviſis ſeſſilibus. *Sp. pl.* 1118.
Willow-leav'd Sow-thiſtle.
Nat. of Sweden and Ruſſia.
Cult. 1759, by Mr. Philip Miller.
Fl. July and Auguſt. H. ♃.

tataricus. 12. S. pedunculis nudis, foliis lanceolatis dentatis runcinatis. *Syſt. veget.* 712.
Tartarian Sow-thiſtle.
Nat. of Siberia.
Introd. 1784, by William Pitcairn, M.D.
Fl. H. ♃.

13. S.

13. S. pedunculis hiſpidis, floribus racemoſis, foliis runcinatis. *Sp. pl.* 1115. *canadenſis.*
Canadian Sow-thiſtle.
Nat. of North America.
Introd. 1772, by John Hope, M.D.
Fl. July and Auguſt. H. ♃.

LACTUCA. *Gen. pl.* 909.

Recept. nudum. *Cal.* imbricatus, cylindricus, margine membranaceo. *Pappus* ſimplex, ſtipitatus. *Sem.* lævia.

1. L. foliis rotundatis: caulinis cordatis, caule corymboſo. *Syſt. veget.* 713. *ſativa.*
Garden Lettuce.
Nat.
Cult. 1562. *Turn. herb. part* 2. *fol.* 26.
Fl. June and July. H. ⊙.

2. L. foliis verticalibus carina aculeatis. *Sp. pl.* 1119. *Scariola.*
Prickly Lettuce.
Nat. of England.
Fl. July. H. ⊙.

3. L. foliis horizontalibus carina aculeatis dentatis. *Sp. pl.* 1119. *viroſa.*
Strong-ſcented Lettuce.
Nat. of Britain.
Fl. July and Auguſt. H. ♂.

4. L. foliis haſtato-linearibus ſeſſilibus carina aculeatis. *Sp. pl.* 1119. *Jacqu. auſtr.* 3 *p.* 28. *t.* 250. *ſaligna.*
Leaſt Lettuce.

Nat. of England.
Fl. July and Auguſt. H. ☉.

canadenſis. 5. L. foliis lanceolato-enſiformibus amplexicaulibus dentatis inermibus. *Syſt. veget.* 713.
Canadian Lettuce.
Nat. of Canada.
Cult. 1726, by Mr. Philip Miller. *R. S. n.* 222.
Fl. July and Auguſt. H. ♂

indica. 6. L. foliis lanceolato-enſiformibus ſeſſilibus inæqualiter dentatis. *Linn. mant.* 278.
Indian Lettuce.
Nat. of the Eaſt Indies.
Introd. 1784, by Sir Joſeph Banks, Bart.
Fl. July and Auguſt. S. ☉

perennis. 7. L. foliis linearibus dentato-pinnatis: laciniis ſurſum dentatis. *Sp. pl.* 1120.
Perennial Lettuce.
Nat. of Germany and France.
Cult. 1633. *Ger. emac.* 286. *f.* 1.
Fl. June—Auguſt. H. ♃.

CHONDRILLA. *Gen. pl.* 910.

Recept. nudum. *Cal.* calyculatus. *Pappus* ſimplex, ſtipitatus. *Floſculi* multiplici ſerie. *Sem.* muricata.

juncea. 1. C. foliis radicalibus runcinatis; caulinis linearibus integris. *Syſt. veget.* 713. *Jacqu. auſtr.* 5. *p.* 12. *t.* 427.
Common Gum-ſuccory.
Nat. of Switzerland, France, and Germany.
Cult. 1533. *Ger. emac.* 288. *f.* 5.
Fl. September and October. H. ♃.

PRENAN

PRENANTHES. *Gen. pl.* 911.

Recept. nudum. *Cal.* calyculatus. *Pappus* ſimplex, ſubſeſſilis. *Floſculi* ſimplici ſerie.

1. P. floſculis quinis, foliis lanceolatis denticulatis. *Sp. pl.* 1121. *Jacqu. auſtr.* 4. *p.* 9. *t.* 317. *purpurea.*
Purple Prenanthes.
Nat. of Germany, Switzerland, and Italy.
Cult. 1683, by Mr. James Sutherland. *Sutherl. hort. edin.* 180. *n.* 6.
Fl. July——September. H. ♃.

2. P. floſculis quinis, foliis runcinatis. *Sp. pl.* 1121. *Curtis lond.* *muralis.*
Wall Prenanthes.
Nat. of Britain.
Fl. July. H. ⊙.

3. P. floſculis quinis, foliis trilobis, caule erecto. *Sp. pl.* 1121. *altiſſima.*
Tall Prenanthes.
Nat. of Virginia and Canada.
Cult. 1696, in Chelſea Garden. *Pluk. alm.* 355. *t.* 317. *f.* 2.
Fl. July and Auguſt. H. ♃.

4. P. floſculis plurimis, floribus nutantibus ſubumbellatis, foliis haſtato-angulatis. *Sp. pl.* 1121. *alba.*
White Prenanthes.
Nat. of North America.
Introd. 1778, by Mr. William Young.
Fl. July and Auguſt. H. ♃.

LEONTODON. *Gen. pl.* 912.

Recept. nudum. *Cal.* imbricatus, ſquamis laxiuſculis. *Pappus* plumoſus.

Taraxacum. 1. L. calyce ſquamis inferne reflexis, foliis runcinatis denticulatis lævibus. *Syſt. veget.* 715. *Curtis lond.*
Common Dandelion.
Nat. of Britain.
Fl. April——June. H. ♃.

aureum. 2. L. foliis runcinatis, caule ſubunifolio, calyce hiſpido. *Syſt. veget.* 715. *Jacqu. auſtr.* 3. *p.* 53. *t.* 297.
Golden Dandelion.
Nat. of Italy, Switzerland, and Auſtria.
Introd. 1769, by Monſ. Richard.
Fl. May——July. H. ♃.

autumnale. 3. L. caule ramoſo, pedunculis ſquamoſis, foliis lanceolatis dentatis integerrimis glabris. *Sp. pl.* 1123.
Autumnal Dandelion.
Nat. of Britain.
Fl. Auguſt. H. ♃.

hiſpidum. 4. L. calyce toto erecto, foliis dentatis integerrimis hiſpidis: ſetis furcatis. *Syſt. veget.* 715. *Curtis lond.*
Rough Dandelion.
Nat. of Britain.
Fl. July——September. H. ♃.

HIERA-

HIERACIUM. *Gen. pl.* 913.

Recept. nudum. *Cal.* imbricatus, ovatus. *Pappus* simplex, sessilis.

* *Scapo nudo unifloro.*

1. H. foliis oblongis integris dentatis, scapo subnudo unifloro, calyce piloso. *Sp. pl.* 1124. *alpinum.*
Alpine Hawkweed.
Nat. of Scotland.
Fl. July and August. H. ♃.

2. H. foliis integerrimis ovatis subtus tomentosis, stolonibus repentibus, scapo unifloro. *Syst. veget.* 716. *Curtis lond.* *Pilosella.*
Mouse-ear Hawkweed.
Nat. of Britain.
Fl. May. H. ♄.

** *Scapo nudo multifloro.*

3. H. foliis integris ovato-oblóngis, stolonibus repentibus, scapo nudo multifloro. *Sp. pl.* 1125. *dubium.*
Creeping Hawkweed.
Nat. of England.
Fl. July and August. H. ♃.

4. H. foliis integerrimis lanceolatis, stolonibus repentibus, scapo nudo multifloro. *Syst. veget.* 716. *Auricula.*
Narrow-leav'd Hawkweed.
Nat. of England.
Fl. July. H. ♃.

5. H. foliis lanceolatis integris pilosis, scapo subnudo basi piloso, floribus subumbellatis. *Sp. pl.* 1126. *cymosum.*
Small-flower'd Hawkweed.

Nat.

Nat. of Europe.
Cult. 1739, by Mr. Ph. Miller. *Rand. chel. n.* 15.
Fl. May and June. H. ♃.

aurantiacum. 6. H. foliis integris, caule ſubnudo ſimpliciſſimo piloſo corymbifero. *Sp. pl.* 1126. *Jacqu. auſtr.* 5. *p.* 5. *t.* 410.
Orange-flower'd Hawkweed.
Nat. of Auſtria and Switzerland.
Cult. 1739, in Chelſea Garden. *Rand. chel. n.* 21.
Fl. June and July. H. ♃.

*** *Caule folioſa.*

porrifolium. 7. H. caule ramoſo folioſo, foliis lanceolato-linearibus integerrimis. *Sp. pl.* 1128. *Jacqu. auſtr.* 3. *p.* 47. *t.* 286.
Leek-leav'd Hawkweed.
Nat. of Italy and Auſtria.
Introd. 1775, by the Doctors Pitcairn and Fothergill.
Fl. July and Auguſt. H. ♃.

chondrilloides. 8. H. caule ramoſo, foliis caulinis elongato-dentatis glabris; radicalibus lanceolatis integris. *Sp. pl.* 1128. *Jacqu. auſtr.* 5. *p.* 13. *t.* 429.
Gum-ſuccory Hawkweed.
Nat. of Auſtria.
Cult. 1640. *Park. theat.* 796. *n.* 3.
Fl. June and July. H. ♃.

murorum. 9. H. caule ramoſo, foliis radicalibus ovatis dentatis; caulino minori. *Sp. pl.* 1128.
Wall Hawkweed.
Nat. of Britain.
Fl. July. H. ♃.

10. H.

10. H. caule paniculato, foliis amplexicaulibus dentatis glabris, calycibus hiſpidis. *Sp. pl.* 1129. *paludoſum.*
Marſh Hawkweed.
Nat. of Britain.
Fl. July and Auguſt. H. ♃.

11. H. caule multifloro, foliis lyratis glabris, calycibus pedunculiſque hiſpidis. *Sp. pl.* 1129. *lyratum.*
Siberian Hawkweed.
Nat. of Siberia.
Introd. 1777, by Monſ. Thouin.
Fl. July and Auguſt. H. ♃.

12. H. foliis radicalibus obovatis denticulatis; caulinis oblongis ſemiamplexicaulibus. *Sp. pl.* 1129. *cerinthoides.*
Honeywort Hawkweed.
Nat. of the Pyrenean Mountains.
Cult. 1739, by Mr. Ph. Miller. *Rand. chel. n.* 18.
Fl. July—September. H. ♃.

13. H. foliis amplexicaulibus cordatis ſubdentatis, pedunculis unifloris hirſutis, caule ramoſo. *Sp. pl.* 1129. *amplexicaule.*
Heart-leav'd Hawkweed.
Nat. of the Pyrenean Mountains.
Cult. 1739, by Mr. Ph. Miller. *Rand. chel. n.* 19.
Fl. July and Auguſt. H. ♃.

14. H. foliis amplexicaulibus obovato-lanceolatis retrorſum dentatis, caule ſimplici, calycibus laxis. *Syſt. veget.* 718. *pyrenaicum.*

α Hieracium foliis lanceolatis amplexicaulibus dentatis, floribus ſolitariis, calycibus laxis. *Sp. pl.* 1129. blattarioides.
Pyrenean Hawkweed.

pilosum. β Picris *pyrenaica*, perianthiis laxis, caule pilofo, foliis dentato-finuatis. *Sp. pl.* 1115.
Hairy Pyrenean Hawkweed.

austriacum. γ Crepis *auftriaca*, foliis oblongis denticulatis, involucro laxiffimo et calycibus hifpidis. *Jacqu. vind.* 270. *auftr.* 5. *p.* 20. *t.* 441.
Auftrian Hawkweed.
Nat. of the South of Europe.
Introd. about 1771, by Mr. James Gordon.
Fl. July and Auguft. H. ♃.

molle. 15. H. foliis lanceolatis fubintegerrimis mollibus: inferioribus petiolatis, floribus pedunculis fubcorymbofis. *Syft. veget.* 718. *Jacqu. auftr.* 2. *p.* 12. *t.* 119.
Soft-leav'd Hawkweed.
Nat. of Scotland. Mr. *James Dickfon.*
Fl. June. H. ♃.

villofum. 16. H. caule ramofo foliofo, foliis hirfutis: radicalibus lanceolato-ovatis dentatis; caulinis amplexicaulibus cordatis. *Syft. veget.* 718. *Jacqu. auftr.* 1. *p.* 55. *t.* 87.
Villous Hawkweed.
Nat. of Scotland. Mr. *James Dickfon.*
Fl. July. H. ♃.

undulatum. 17. H. caule ramofo foliofo, foliis ellipticis dentatis undulatis pilofis: pilis plumofis.
Wave-leav'd Hawkweed.
Nat. of Spain.
Introd. 1778, by Meffrs. Kennedy and Lee.
Fl. June and July. H. ♃.

18. H.

18. H. caule ramoſo folioſo, foliis ſemiamplexicaulibus oblongis repandis hiſpidis. *Sp. pl.* 1130. *ſprengerianum.*
Branch'd Hawkweed.
Nat. of Portugal.
Introd. 1783, by Mr. John Græfer.
Fl. H. ♃.

19. H. caule multifloro, foliis amplexicaulibus piloſis rariter dentatis. *Hall. hiſt.* 43. *Allion. pedem.* 1. *p.* 218. *tab.* 27. *fig.* 1. & 3. *ſpicatum.*
Hairy Hawkweed.
Nat. of Scotland. Mr. *James Dickſon.*
Fl. June. H. ♃.

20. H. caule erecto multifloro, foliis ovato-lanceolatis dentatis ſemiamplexicaulibus. *Sp. pl.* 1131. *ſabaudum.*
Shrubby Hawkweed.
Nat. of Britain.
Fl. July and Auguſt. H. ♃.

21. H. foliis linearibus ſubdentatis ſparſis, floribus ſubumbellatis. *Sp. pl.* 1131. *umbellatum.*
Umbel'd Hawkweed.
Nat. of Britain.
Fl. July and Auguſt. H. ♃.

CREPIS. *Gen. pl.* 914.

Recept. nudum. *Cal.* calyculatus ſquamis deciduis. *Pappus* plumoſus, ſtipitatus.

1. C. involucris calyce longioribus: ſquamis ſetaceis ſparſis. *Sp. pl.* 1191. *Curtis magaz.* 35. *barbata.*
Spaniſh Crepis, or Hawkweed.
Nat. of the South of France and Sicily.

Introd.

Introd. about 1620, by Mr. William Boel. *Philosoph. transf n.* 332. *p.* 381. *n.* 16.
Fl. June and July. H. ☉.

alpina. 2. C. involucris fcariofis longitudine calycis, floribus folitariis. *Syft. veget.* 719.
Alpine Crepis.
Nat. of the Alps of Italy.
Cult. 1739, by Mr. Philip Miller. *Rand. chel.* Hieracium 27.
Fl. July. H. ☉.

rubra. 3. C. foliis amplexicaulibus lyrato-runcinatis. *Sp. pl.* 1132.
Purple Crepis, or Hawkweed.
Nat. of Italy.
Cult. 1632, by Mr. Nicholas Swayton. *Johnf. it. cant.*
Fl. June and July. H. ☉.

fœtida. 4. C. foliis runcinato-pinnatis hirtis, petiolis dentatis. *Sp. pl.* 1133.
Stinking Crepis.
Nat. of England.
Fl. June and July. H. ☉.

fibirica. 5. C. foliis amplexicaulibus oblongis rugofis inferne dentatis, caule hirto, calycibus carina ciliatis. *Syft. veget.* 719.
Siberian Crepis.
Nat. of Siberia.
Introd. 1773, by John Earl of Bute.
Fl. July and Auguft. H. ♃.

tectorum. 6. C. foliis lanceolato-runcinatis feffilibus lævibus: inferioribus dentatis. *Syft. veget.* 719. *Curtis lond.*

Smooth

Smooth Crepis.
Nat. of Britain.
Fl. June——September. H. ☉.

7. C. foliis runcinato-pinnatifidis ſcabris: baſi ſuperne dentatis, calycibus muricatis. *Sp. pl.* 1136. *biennis.*
Biennial Crepis.
Nat. of England.
Fl. July and Auguſt. H. ♂.

8. C. foliis radicalibus runcinatis; caulinis haſtatis, calycibus ſubtomentoſis. *Sp. pl.* 1133. *Dioſcoridis.*
Dioſcorides's Crepis.
Nat. of France.
Introd. 1772, by Monſ. Richard.
Fl. June and July. H. ☉.

9. C. foliis runcinato-dentatis ſubincanis, pedunculis nudis unifloris, ſquamis calycinis margine albicantibus. *De Lamarck encycl.* 2. *p.* 179. *albida.*
Crepis albida. *Jacqu. ic. collect.* 1. *p.* 81. *Allion. pedem.* 1. *p.* 219. *t.* 32. *f.* 3.
Pale-flower'd Crepis.
Nat. of France and Italy.
Fl. July——October. H. ♃.

10. C. foliis oblongis duplicato-ſerratis ſetoſis, caule nudo ramoſo, floribus paniculatis, calycibus cylindraceis glabris, pappo ſeſſili. *rigens.*
Briſtly-leav'd Crepis.
Nat. of the Azores. Mr. *Francis Maſſon.*
Introd. 1778.
Fl. July and Auguſt. G. H. ♃.

11. C.

pulchra. 11. C. foliis sagittatis dentatis, caule paniculato, calycibus pyramidatis glabris. *Sp. pl.* 1134.
Small-flower'd Crepis.
Nat. of France and Italy.
Cult. 1739, in Chelsea Garden. *Rand. chel.* Chondrilla 2.
Fl. July and August. H. ⊙.

filiformis. 12. C. foliis lineari-filiformibus integerrimis glabris, pappo sessili.
Fine-leav'd Crepis.
Nat. of Madeira. Mr. *Francis Masson.*
Introd. 1777.
Fl. June. G. H. ♂.

succulenta. 13 C. foliis pinnatifidis dentatisve subcarnosis lævibus, calycibus subtomentosis, pappo sessili.
Fleshy-leav'd Crepis.
Nat. of Madeira. Mr. *Francis Masson.*
Introd. 1777.
Fl. August and September. G. H. ⊙.

ANDRYALA. *Gen. pl.* 915.

Recept. villosum. *Cal.* multipartitus, subæqualis, rotundatus. *Pappus* simplex, sessilis.

integrifolia. 1. A. foliis inferioribus runcinatis; superioribus ovato-oblongis tomentosis. *Syst. veget.* 720.
Hoary Andryala.
Nat. of the South of Europe.
Cult. 1711, in Chelsea Garden. *Philosoph. transact.* n. 332. p. 381. n. 15.
Fl. July and August. H. ♂.

2. A.

2. A. foliis runcinatis; summis lanceolatis integris, villo glandulifero. *L'Herit. stirp. nov. p.* 35. *tab.* 18. *cheiranthifolia.*
Andryala glandulosa. *De Lamarck encycl.* 1. *p.* 154.
Andryala tomentosa. *Scop. insubr.* 2. *p.* 12. *tab.* 6.
Various-leav'd Andryala.
Nat. of Madeira. Mr. *Francis Masson.*
Introd. 1777.
Fl. May——October. G. H. ♃.

3. A. foliis tomentosis pinnatifidis, calycibus tomentosis pilosis: pilis rigidiusculis. *pinnatifida.*
α foliis pinnatifidis: pinnis distantibus dentatis.
Dented-leav'd Andryala.
β foliis profunde pinnatifidis: pinnis brevibus integris.
Wing-leav'd Andryala.
Nat. α. of Madeira; β. of the Canary Islands. Mr. *Francis Masson.*
Introd. 1778.
Fl. July and August. G. H. ♂.

4. A. foliis pinnatis linearibus tomentosis. *crithmifolia.*
Sampire-leav'd Andryala.
Nat. of Madeira. Mr. *Francis Masson.*
Introd. 1778.
Fl. June——August. G. H. ♂.

5. A. foliis lanceolatis indivisis denticulatis acutis tomentosis, floribus solitariis. *Sp. pl.* 1136. *ragusina.*
Downy Andryala.
Nat. of the Archipelago.
Cult. 1757, by Mr. Ph. Miller. *Mill. ic.* 97. *t.* 146. *f.* 2.
Fl. June——August. G. H. ♃.

6. A. foliis oblongo-ovatis subdentatis lanatis, pedunculis ramosis. *Sp. pl.* 1137. *lanata.*
Woolly Andryala.

Nat. of the South of Europe.
Cult. 1732, by James Sherard, M. D. *Dill. elth.* 181. *t.* 150. *f.* 180.
Fl. May and June. H. ♃.

HYOSERIS. *Gen. pl.* 916.

Recept. nudum. *Cal.* subæqualis. *Pappus* pilosus calyculatusque.

fœtida. 1. H. scapis unifloris, foliis pinnatifidis, seminibus nudis. *Syst. veget.* 720.
Stinking Hyoseris.
Nat. of the Alps of Italy and Austria.
Introd. 1775, by the Doctors Pitcairn and Fothergill.
Fl. July. H. ♃.

radiata. 2. H. scapis unifloris, foliis glabris runcinatis angulis dentatis: apice laciniatis. *Syst. veget.* 720.
Starry Hyoseris.
Nat. of the South of France and Spain.
Cult. 1640. *Park. theat.* 780. *n.* 4.
Fl. June and July. H. ♃.

lucida. 3. H. scapis unifloris, foliis subcarnosis runcinatis angulatis dentatis. *Syst. veget.* 720. *Jacqu. hort.* 2. *p.* 70. *t.* 150.
Shining Hyoseris.
Nat. of the Levant.
Introd. 1770, by Monf. Richard.
Fl. June——August. H. ♃.

pygmæa. 4. H. scapis unifloris, foliis spathulatis dentatis ciliatis, calycibus pilosis: pilis ciliisque furcatis, pappo stipitato plumoso.
Dwarf Hyoseris.
Nat. of Madeira. Mr. *Francis Masson.*

Introd.

Introd. 1778.
Fl. June and July. G. H. ⊙.

5. H. caule diviſo nudo, pedunculis incraſſatis. *Sp. pl.* 1138. *minima.*
Leaſt Hyoſeris.
Nat. of Britain.
Fl. May and June. H. ⊙

6. H. fructibus ovatis glabris, caule ramoſo. *Sp. pl.* 1138. *Hedypnois.*
Branching Hyoſeris.
Nat. of the South of Europe.
Cult. 1683, by Mr. James Sutherland. *Sutherl. hort. edin.* 153. *n.* 1.
Fl. June. H. ⊙.

7. H. fructibus ovatis piloſis, caule ramoſo. *Sp. pl.* 1139. *Rhagadioloides.*
Nipple-wort Hyoſeris.
Nat. of the South of Europe.
Introd. 1773, by Monſ. Richard.
Fl. July and Auguſt. H. ⊙.

8. H. fructibus ovatis ſcabris, caule ramoſo. *Sp. pl.* 1139. *cretica.*
Cretan Hyoſeris.
Nat. of Candia.
Cult. 1739, by Mr. Ph. Miller. *Rand. chel.* Hedypnois 2.
Fl. June and July. H. ⊙.

SERIOLA. *Gen. pl.* 917.

Recept. paleaceum. *Cal.* ſimplex. *Pappus* ſubplumoſus.

1. S. læviuſcula, foliis obovatis dentatis. *Sp. pl.* 1139. *lævigata.*
Smooth Seriola.
Nat. of the Iſland of Candia.
Introd. 1772, by Monſ. Richard.
Fl. July and Auguſt. H. ⊙.

æthnensis. 2. S. hiſpida, foliis obovatis ſubdentatis. *Sp. pl.* 1139.
Rough Seriola.
Nat. of Italy.
Introd. 1771, by John Earl of Bute.
Fl. July and Auguſt. H. ⊙.

urens. 3. S. urens, foliis dentatis, caule ramoſo. *Sp. pl.* 1139.
Stinging Seriola.
Nat. of the South of Europe.
Introd. 1773, by John Earl of Bute.
Fl. July and Auguſt. H. ⊙.

HYPOCHÆRIS. *Gen. pl.* 918.

Recept. paleaceum. *Cal.* ſubimbricatus. *Pappus* plumoſus.

pontana. 1. H. caule ſimplici folioſo unifloro, foliis lanceolatis dentatis. *Sp. pl.* 1140.
Endive-leav'd Hypochæris.
Nat. of the South of Europe.
Introd. 1775, by the Doctors Pitcairn and Fothergill.
Fl. June and July. H. ♂.

maculata. 2. H. caule ſubnudo: ramo ſolitario, foliis ovato-oblongis integris dentatis. *Sp. pl.* 1140.
Spotted Hypochæris.
Nat. of England.
Fl. June and July. H. ♃.

glabra. 3. H. glabra, calycibus oblongis imbricatis, caule ramoſo nudo, foliis dentato-ſinuatis. *Sp. pl.* 1140. *Curtis lond.*
Smooth Hypochæris.
Nat. of Britain.
Fl. July. H. ⊙.

4. H.

4. H. foliis runcinatis obtusis scabris, caule ramoso nudo lævi, pedunculis squamosis. *Sp. pl.* 1140. *Curtis lond.* *radicata.*
Long-rooted Hypochæris.
Nat. of Britain.
Fl. July——September. H. ♃.

LAPSANA. *Gen. pl.* 919.

Recept. nudum. *Cal.* calyculatus, squamis singulis interioribus canaliculatis.

1. L. calycibus fructus angulatis, pedunculis tenuibus ramosissimis. *Sp. pl.* 1141. *Curtis lond.* *communis.*
Common Nipple-wort.
Nat. of Britain.
Fl. June and July. H. ☉.

2. L. calycibus fructus torulosis depressis obtusis sessilibus. *Sp. pl.* 1141. *Zacintha.*
Warted Nipple-wort.
Nat. of the South of Europe.
Cult. 1640. *Park. theat.* 778. *f.* 8.
Fl. June and July. H. ☉.

3. L. calycibus fructus undique patentibus: radiis subulatis, foliis caulinis lanceolatis indivisis. *Sp. pl.* 1141. *stellata.*
Starry Nipple-wort.
Nat. of the South of France and Italy.
Cult. before 1633. *Ger. emac.* 1625. Hieracium stellatum.
Fl. June and July. H. ☉.

Kölpinia. 4. L. calycibus fructus undique patentibus, radiis ſubulatis incurvis echinatis, foliis caulinis lanceolatis indiviſis. *Linn. ſuppl.* 348.
Small Nipple-wort.
Nat. of Siberia and the Levant.
Introd. 1788, by John Sibthorp, M. D.
Fl. July. H. ⊙.

Rhagadiolus. 5. L. calycibus fructus undique patentibus: radiis ſubulatis, foliis lyratis. *Sp. pl.* 1141.
Heart-leav'd Nipple-wort.
Nat. of the Levant.
Cult. 1739, by Mr. Philip Miller. *Rand. chel.* Rhagadiolus 2.
Fl. June and July. H. ⊙.

CATANANCHE. *Gen. pl.* 920.

Recept. paleaceum. *Cal.* imbricatus. *Pappus* ariſtatus calyculo 5-ſeto.

cærulea. 1. C. ſquamis calycinis inferioribus ovatis. *Sp. pl.* 1142.
Blue Catananche.
Nat. of the South of Europe.
Cult. 1640. *Park. theat.* 786. *f.* 5.
Fl. July——October. H. ♃.

lutea 2. C. ſquamis calycinis inferioribus lanceolatis. *Sp. pl.* 1142.
Yellow Catananche.
Nat. of the Iſland of Candia.
Cult. 1714, in Chelſea Garden. *Philoſoph. tranſ. n.* 343. *p.* 232. *n.* 8.
Fl. June and July. H. ⊙.

CICHO-

CICHORIUM. *Gen. pl.* 921.

Recept. ſubpaleaceum. *Cal.* calyculatus. *Pappus* ſub-5-dentatus, obſolete piloſus.

1. C. floribus geminis ſeſſilibus, foliis runcinatis. *Sp. pl.* 1142. *Curtis lond.* *Intybus.*
Wild Endive, or Succory,
Nat. of Britain.
Fl. July and Auguſt. H. ♃.

2. C. floribus ſolitariis pedunculatis, foliis integris crenatis. *Sp. pl.* 1142. *Endivia.*
α Intybus ſativa latifolia ſive Endivia vulgaris. *Bauh. pin.* 125.
Common broad-leav'd Endive.
β Intybus criſpa. *Bauh. pin.* 125.
Curl'd Endive.
Nat.
Cult. 1562. *Turn. herb. part* 2. *fol.* 21.
Fl. July and Auguſt. H. ⊙.

3. C. caule dichotomo ſpinoſo, floribus axillaribus ſeſſilibus. *Sp. pl.* 1143. *ſpinoſum.*
Prickly Endive.
Nat. of the Iſlands of Candia and Sicily.
Cult. 1633. *Ger. emac.* 283. *f.* 5.
Fl. July and Auguſt. H. ♂.

SCOLYMUS. *Gen. pl.* 922.

Recept. paleaceum. *Cal.* imbricatus, ſpinoſus. *Pappus* nullus.

1. S. floribus ſolitariis. *Syſt. veget.* 722. *maculatus.*
Annual Golden-thiſtle.
Nat. of the South of Europe.

 Cult.

Cult. 1633, by Mr. John Tradescant, Sen. *Ger. emac.* 1155. *f.* 2.
Fl. July and August. H. ☉.

hispanicus. 2. S. floribus congestis. *Syst. veget.* 722.
Perennial Golden-thistle.
Nat. of the South of Europe.
Cult. 1658, in Oxford Garden. *Hort. oxon. edit.* 2. *p.* 39.
Fl. July——September. H. ♃.

ARCTIUM. *Gen. pl.* 923.

Cal. globosus: squamis apice hamis inflexis.

Lappa. 1. A. foliis cordatis inermibus petiolatis. *Sp. pl.* 1143. *Curtis lond.*
α Lappa major capitulo glabro maximo. *Raj. syn.* 196.
Smooth-headed Common Burdock.
β Lappa major montana, capitulis tomentosis. *Bauh. pin.* 198.
Woolly-headed Burdock.
Nat. of Britain.
Fl. July and August. H. ♂.

Personata. 2. A. foliis decurrentibus ciliato-spinosis: radicalibus pinnatis; caulinis oblongo-ovatis. *Syst. veget.* 723.
Carduus Personata. *Jacqu. austr.* 4. *p.* 25. *t.* 348.
Cut-leav'd Burdock.
Nat. of the Alps of Austria and Switzerland.
Introd. 1776, by Joseph Nicholas de Jacquin, M.D.
Fl. July and August. H. ♂.

SERRATULA. *Gen. pl.* 924.

Cal. subcylindricus, imbricatus, muticus.

tinctoria. 1. S. foliis lyrato-pinnatifidis: pinna terminali maxima, flosculis conformibus. *Sp. pl.* 1144.

Common

Common Saw-wort.
Nat. of Britain.
Fl. August——October. H. ♃.

2. S. foliis lyrato-pinnatifidis: pinna terminali maxima, flosculis radii femineis longioribus. *Sp. pl.* 1144. *coronata.*
Siberian Saw-wort.
Nat. of Italy and Siberia.
Cult. 1748, by Mr. Ph. Miller. *Mill. dict. edit.* 5. *n.* 5.
Fl. July and August. H. ♃.

3. S. calycibus subhirsutis ovatis, foliis indivisis. *Sp. pl.* 1145. *alpina.*
Mountain Saw-wort.
Nat. of Britain.
Fl. July. H. ♃.

4. S. foliis lanceolato-oblongis serratis pendulis. *Sp. pl.* 1146. *noveboracensis.*
Long-leav'd Saw-wort.
Nat. of North America.
Cult. 1732, by James Sherard, M.D. *Dill. elth.* 355. *t.* 263. *f.* 342.
Fl. September——November. H. ♃.

5. S. foliis lanceolato-oblongis serratis patentibus subtus hirsutis. *Sp. pl.* 1146. *præalta.*
Tall Saw-wort.
Nat. of North America.
Cult. 1732, by James Sherard, M.D. *Dill. elth.* 356. *t.* 264. *f.* 343.
Fl. September——November. H. ♃.

6. S. foliis linearibus, calycibus squarrosis subsessilibus acuminatis lateralibus. *Sp. pl.* 1146. *squarrosa.*
Rough-headed Saw-wort.

Nat.

Nat. of Virginia.
Cult. 1732, by James Sherard, M.D. *Dill. elth.* 83. *t.* 71. *f.* 82.
Fl. July and Auguſt. H. ♃.

ſcarioſa. 7. S. foliis lanceolatis integerrimis, calycibus ſquarroſis pedunculatis obtuſis lateralibus. *Sp. pl.* 1147.
Ragged-cup'd Saw-wort.
Nat. of Virginia.
Cult. 1759, by Mr. Ph. Miller. *Mill. dict. edit.* 7. *n.* 5.
Fl. September and October. H. ♃.

piloſa. 8. S. foliis linearibus piloſis, floribus axillaribus longe pedunculatis.
Hairy-leav'd Saw-wort.
Nat. of North America.
Introd. 1783, by Mr. William Young.
Fl. September and October. H. ♃.

ſpecioſa. 9. S. foliis lineari-falcatis, floribus ſeſſilibus ſpicatis, foliolis calycinis hirtis acutis: interioribus elongatis apice coloratis.
Stæhelina elegans. *Walt. carol.* 202.
Hairy-cup'd Saw-wort.
Nat. of Carolina and Georgia.
Introd. 1787, by Meſſrs. Watſons.
Fl. October. G. H. ♄.

ſpicata. 10. S. foliis linearibus baſi ciliatis, floribus ſpicatis ſeſſilibus lateralibus, caule ſimplici. *Sp. pl.* 1147.
Spik'd Saw-wort.
Nat. of North America.
Cult. 1732, by James Sherard, M.D. *Dillen. elth.* 85. *t.* 72. *f.* 83.
Fl. Auguſt—October. H. ♃.

§ 11. S.

11. S. foliis dentatis ſpinoſis. *Sp. pl.* 1149. *arvenſis.*
Corn Saw-wort, or Way Thiſtle.
Nat. of Britain.
Fl. July. H. ♃.

CARDUUS. *Gen. pl.* 925.

Cal. ovatus, imbricatus ſquamis ſpinoſis. *Recept.* piloſum.

* *Foliis decurrentibus.*

1. C. foliis decurrentibus pinnatifidis hiſpidis: laciniis divaricatis, calycibus ovatis ſpinoſis villoſis, caule piloſo. *Sp. pl.* 1149. *lanceolatus.*
Spear Thiſtle.
Nat. of Britain.
Fl. July. H. ♂.

2. C. foliis ſemidecurrentibus ſpinoſis, floribus cernuis: ſquamis calycinis ſuperne patentibus. *Sp. pl.* 1150. *nutans.*
Muſk Thiſtle.
Nat. of Britain.
Fl. July. H. ♂.

3. C. foliis decurrentibus ſinuatis margine ſpinoſis, calycibus pedunculatis ſolitariis erectis villoſis. *Sp. pl.* 1150. *Jacqu. auſtr.* 3. *p.* 28. *t.* 249. *acanthoides.*
Welted Thiſtle.
Nat. of Britain.
Fl. July and Auguſt. H. ☉.

4. C. foliis decurrentibus decurſive pinnatis: pinnulis palmato-quadrifidis aculeatis lanatis, caule corymboſo multifloro, floribus glomeratis. *Gouan illuſtr.* 62. *tab.* 23. *carlinoides.*

Pyrenèan

Pyrenean Thiſtle.
Nat. of the Pyrenees.
Introd. 1784, by Caſimir Gomez Ortega, M.D.
Fl. July and Auguſt. H. ♂.

criſpus. 5. C. foliis decurrentibus ſinuatis margine ſpinoſis, floribus aggregatis terminalibus: ſquamis inermibus ſubariſtatis patulis. *Syſt. veget.* 724.
Curl'd Thiſtle.
Nat. of Britain.
Fl. June. H. ☉.

paluſtris. 6. C. foliis decurrentibus dentatis margine ſpinoſis, floribus racemoſis erectis, pedunculis inermibus. *Sp. pl.* 1151.
Marſh Thiſtle.
Nat. of Britain.
Fl. July. H. ♃.

pycnocephalus. 7. C. foliis decurrentibus pinnatifido-ſinuatis pubeſcentibus ſpinoſis, pedunculis nudis tomentoſis, calycibus deciduis. *Sp. pl.* 1151. *Jacqu. hort.* 1. *p.* 17. *t.* 44.
Italian Thiſtle.
Nat. of the South of Europe.
Cult. 1739, by Mr. Ph. Miller. *Rand. chel. n.* 14.
Fl. July——September. H. ♃.

diſſectus. 8. C. foliis decurrentibus lanceolatis: denticulis inermibus, calyce ſpinoſo. *Sp. pl.* 1151.
Meadow Thiſtle.
Nat. of England.
Fl. June. H. ♃.

cyanoides. 9. C. foliis decurrentibus pinnatifidis linearibus integerrimis inermibus petiolatis ſubtus tomentoſis. *Sp. pl.* 1152.

Blue-

Blue-bottle-leav'd Thiſtle.
Nat. of Siberia.
Introd. 1778, by Mr. William Malcolm.
Fl. July and Auguſt. H. ♃.

10. C. foliis decurrentibus lanceolatis eroſo-dentatis ciliato-aculeatis utrinque arachnoideo-ſubvilloſis. *Syſt. veget.* 725. *Jacqu. auſtr.* 1. *p.* 27. *t.* 42, 43. *canus.*
Hoary Thiſtle.
Nat. of Auſtria.
Introd. 1775, by Joſeph Nicholas de Jacquin, M.D.
Fl. July and Auguſt. H. ♃.

11. C. foliis decurrentibus lanceolatis ſerratis ſubſpinoſo-ciliatis nudis, pedunculis longiſſimis lanuginoſis unifloris. *Sp. pl.* 1152. *Jacqu. auſtr.* 1. *p.* 56. *t.* 89. *defloratus.*
Various-leav'd Thiſtle.
Nat. of France, Switzerland, and Auſtria.
Introd. 1775, by Joſeph Nicholas de Jacquin, M.D.
Fl. July——September. H. ♃.

12. C. foliis decurrentibus lanceolatis ſubrepandis glabris inæqualiter ciliatis, pedunculis alternis, calycibus inermibus. *Sp. pl.* 1152. *monſpeſſulanus.*
Montpelier Thiſtle.
Nat. of Montpelier.
Cult. 1570, by Mr. Hugh Morgan. *Lobel. adv.* 251.
Fl. June and July. H. ⊙.

13. C. foliis ſubdecurrentibus petiolatis ſubpinnatifidis ſpinoſis, caule inermi, floribus ſolitariis. *Sp. pl.* 1154. *tuberoſus.*
Tuberous Thiſtle.
Nat. of Germany, France, and Switzerland.

Cult.

Cult. 1683, by Mr. James Sutherland. *Sutherl. hort. edin.* 67. *n.* 7.

Fl. August——October. H. ♃.

parviflorus. 14. C. foliis basi adnatis lanceolatis nudis erosis ciliato-spinulosis inermibus. *Linn. mant.* 279.

Small-flower'd Thistle.

Nat. of the South of Europe.

Introd. 1781, by Monf. Thouin.

Fl. June and July. H. ♃.

** *Foliis sessilibus.*

casabonæ. 15. C. foliis sessilibus lanceolatis integerrimis subtus tomentosis: margine spinis ternatis. *Sp. pl.* 1153.

Fish Thistle.

Nat. of the South of Europe.

Cult. 1714. *Philosoph. trans. n.* 343. *p.* 235. *n.* 18.

Fl. June——August. G. H. ♂.

stellatus. 16. C. foliis sessilibus integris lanceolatis inermibus subtus tomensis, spinis ramosis axillaribus, floribus sessilibus lateralibus. *Syst. veget.* 726.

Starry Thistle.

Nat.

Introd. 1771, by John Earl of Bute.

Fl. June and July. H. ☉.

marianus. 17. C. foliis amplexicaulibus hastato-pinnatifidis spinosis, calycibus aphyllis: spinis canaliculatis duplicato-spinosis. *Sp. pl.* 1153. *Curtis lond.*

Milk Thistle.

Nat. of Britain.

Fl. July. H. ☉.

18. C.

18. C. foliis amplexicaulibus angulato-spinosis, floribus solitariis subsessilibus obvallatis foliolis subquinis. *Sp. pl.* 1153. *syriacus.*
Syrian Thistle.
Nat. of Spain and the Levant.
Cult. 1640. *Park. theat.* 974.
Fl. July and August. H. ☉.

19. C. foliis sessilibus bifariam pinnatifidis: laciniis alternis erectis, calycibus globosis villosis. *Sp. pl.* 1153. *Jacqu. austr.* 2. *p.* 45. *t.* 171. *eriophorus.*
Woolly-headed Thistle.
Nat. of Britain.
Fl. July. H. ♂.

20. C. foliis amplexicaulibus lanceolatis dentatis: spinulis inæqualibus ciliatis, caule inermi. *Sp. pl.* 1155. *helenioides.*
Melancholy Thistle.
Nat. of Britain.
Fl. June and July. H. ♃.

21. C. foliis subamplexicaulibus lanceolatis integris: serraturis spinoso-setaceis, pedunculis unifloris. *Sp. pl.* 1155. *Jacqu. austr.* 2. *p.* 16. *t.* 127. *serratuloides.*
Saw-wort Thistle.
Nat. of Siberia, Austria, and Switzerland.
Introd. 1765, by John Earl of Bute.
Fl. June——October. H. ♃.

22. C. foliis semidecurrentibus oblongo-lanceolatis inæqualiter ciliatis glabris: inferioribus lyratis undulatis, floribus paniculatis. *paniculatus.*
Panicl'd Thistle.
Nat. of the South of Europe.
Introd. 1781, by Monf. Thouin.
Fl. June and July. H. ♃.

rigens. 23. C. foliis oblongo-lanceolatis glabris margine ſpinoſis pinnatifidis: laciniis obliquis lobatis, calycibus oblongis bracteatis.

Carduus foliis lanceolatis amplexicaulibus, ſerraturis ſpinoſo-ſetaceis, capitulis triphyllis. *La Chenal in act. helv.* 4. *p.* 294. *tab.* 16.

Cirſium foliis ciliatis ſemipinnatis, pinnis anguloſis ſpinoſis, ſuperioribus amplexicaulibus. *Hall. hiſt.* 176.

Upright Alpine Thiſtle.

Nat. of Switzerland.

Introd. 1775, by the Doctors Pitcairn and Fothergill.

Fl. July and Auguſt. H. ♃.

tataricus. 24. C. foliis amplexicaulibus lanceolatis: ſerraturis ſpinoſo-ſetaceis, floribus triphyllis. *Sp. pl.* 1155. *Jacqu. auſtr.* 1. *p.* 56. *t.* 90.

Tartarian Thiſtle.

Nat. of Siberia.

Introd. 1775, by Joſeph Nicholas de Jacquin, M.D.

Fl. July and Auguſt. H. ♃.

ciliatus. 25. C. foliis ſemiamplexicaulibus pinnatifidis laciniatis ſpinoſis ſubtus tomentoſis, calycis ſquamis ciliatis baſi reflexis. *Murray in commentat. gotting.* 6. (1784) *p.* 35. *tab.* 5.

Fringed-cup'd Thiſtle.

Nat. of Siberia.

Introd. 1787, by Chevalier Murray.

Fl. Auguſt. H. ♃.

acaulis. 26. C. acaulis, calyce glabro. *Sp. pl.* 1156. *Jacqu. ic. vol.* 2.

Dwarf Thiſtle.

Nat. of Britain.

Fl. July. H. ♃.

CNICUS.

C N I C U S. *Gen. pl.* 926.

Cal. ovatus, imbricatus ſquamis ramoſo-ſpinoſis, obvallatus bracteis. *Corollulæ* æquales.

1. C. foliis pinnatifidis carinatis nudis, bracteis ſubcoloratis integris concavis. *Sp. pl.* 1156. *oleraceus.*
Pale-flower'd Cnicus.
Nat. of Europe.
Cult. 1570, by Mr. Hugh Morgan. *Lobel. adv.* 371.
Fl. July. H. ♃.

2. C. foliis amplexicaulibus pinnatifidis ariſtato-ſerratis, pedunculis cernuis, calycibus glutinoſis. *Syſt. veget.* 727. *Eriſithales.*
Carduus Eriſithales. *Jacqu. auſtr.* 4. *p.* 5. *t.* 310.
Clammy Cnicus.
Nat. of France and Auſtria.
Introd. 1787, by Monſ. Vare.
Fl. June——Auguſt. H. ♃.

3. C. foliis decurrentibus ligulatis dentato-ſpinoſis, caule ramoſo erecto. *Linn. mant.* 109. *ferox.*
Prickly Cnicus.
Nat. of the South of Europe.
Introd. 1775, by Monſ. Thouin.
Fl. July and Auguſt. H. ♂.

4. C. foliis decurrentibus lanceolatis indiviſis, calycibus pinnato-ſpinoſis. *Sp. pl.* 1158. *Acarna.*
Yellow Cnicus.
Nat. of Spain.
Cult. 1683, by Mr. James Sutherland. *Sutherl. hort. edin.* 5. *n.* 2.
Fl. July——September. H. ⊙.

Centauroides. 5. C. foliis pinnatifidis, calycibus ſcarioſis: ſquamis acuminatis. *Syſt. veget.* 727.
Artichoke-leav'd Cnicus.
Nat. of the Pyrenean Mountains.
Cult. 1640. *Park. theat.* 466. *f.* 2.
Fl. July and Auguſt. H. ♃.

cernuus. 6. C. foliis cordatis, petiolis criſpis ſpinoſis amplexicaulibus, floribus cernuis, calycibus ſcarioſis. *Syſt. veget.* 727.
Siberian Cnicus.
Nat. of Siberia.
Cult. 1758, by Mr. Ph. Miller. *Mill. ic.* 165. *t.* 248.
Fl. June and July. H. ♃.

ONOPORDUM. *Gen. pl.* 927.

Recept. favoſum. *Cal.* ſquamæ mucronatæ.

Acanthium. 1. O. calycibus ſquarroſis: ſquamis patentibus, foliis ovato oblongis ſinuatis. *Syſt. veget.* 728. *Curtis lond.*
Woolly Onopordum, or Cotton-thiſtle.
Nat. of Britain.
Fl. July and Auguſt. H. ♂.

illyricum. 2. O. calycibus ſquarroſis: ſquamis inferioribus uncinatis, foliis lanceolatis pinnatifidis. *Syſt. veget.* 728. *Jacqu. hort.* 2. *p.* 69. *t.* 148.
Illyrian Onopordum.
Nat. of the South of Europe.
Cult. 1640. *Park. theat.* 979. *f.* 2.
Fl. July and Auguſt. H. ♂

deltoides. 3. O. calycibus ſquarroſis arachnoideo-tomentoſis, foliis petiolatis ovatis angulatis ſubtus tomentoſis.

Siberian

Siberian Onopordum.
Nat. of Siberia.
Introd. 1784, by Mr. John Bell.
Fl. Auguſt. H. ♃.

4. O. calycibus imbricatis. *Sp. pl.* 1159. *Jacqu. hort.* 2. *p.* 70. *t.* 149. *arabicum.*
Arabian Onopordum.
Nat. of the South of Europe.
Cult. 1692, in Oxford Garden. *Pluk. phyt. t.* 154. *f.* 5.
Fl. July. H. ♂.

5. O. acaule. *Syſt. veget.* 728. *Jacqu. ic. miſcell.* 2. *p.* 310. *acaulon.*
Dwarf Onopordum.
Nat.
Cult. 1739, by Mr. Ph. Miller. *Rand. chel. n.* 7.
Fl. July and Auguſt. H. ♂.

CYNARA. *Gen. pl.* 928.

Cal. dilatatus, imbricatus ſquamis carnoſis, emarginatis cum acumine.

1. C. foliis ſubſpinoſis pinnatis indiviſiſque, calycinis ſquamis ovatis. *Sp. pl.* 1159. *Scolymus.*
α Cinara hortenſis aculeata. *Bauh. pin.* 383.
French Artichoke.
β Cinara hortenſis foliis non aculeatis. *Bauh. pin.* 383.
Globe Artichoke.
Nat. of the South of Europe.
Cult. 1551. *Turn. herb. part* 1. *ſign. Hiiij.*
Fl. Auguſt and September. H. ♃.

horrida. 2. C. foliis pinnatifidis ſubtus tomentoſis ſpinoſis: ſpinis baſeos foliorum pinnarumque baſi connatis.
Madeira Artichoke.
Nat. of the Iſland of Porto Santo, near Madeira. Mr. *Francis Maſſon.*
Introd. 1778.
Fl. H. ♃.

Cardunculus. 3. C. foliis ſpinoſis: omnibus pinnatifidis, calycinis ſquamis ovatis. *Sp. pl.* 1159.
Cardoon Artichoke.
Nat. of Candia.
Cult. 1683, by Mr. James Sutherland. *Sutherl. hort. edin.* 85. *n.* 5.
Fl. Auguſt. H. ♂.

humilis. 4. C. foliis ſpinoſis pinnatifidis ſubtus tomentoſis, calycinis ſquamis ſubulatis. *Sp. pl.* 1159.
Dwarf Artichoke.
Nat. of Spain and Barbary.
Cult. before 1683, in Oxford Garden. *Moriſ. hiſt.* 3. *p.* 158. *n.* 9.
Fl. July and Auguſt. H. ♃.

CARLINA. *Gen. pl.* 929.

Cal. radiatus ſquamis marginalibus longis, coloratis.

acaulis. 1. C. caule unifloro flore breviore. *Sp. pl.* 1160.
Dwarf Carlina.
Nat. of Italy and Germany.
Cult. 1640. *Park. theat.* 968. *f.* 2.
Fl. June. H. ♃.

lanata. 2. C. caule bifido, calycibus ſanguineis terminalibus: primo axillari ſeſſili. *Syſt. veget.* 728.
Woolly Carlina.

Nat.

Nat. of the South of Europe.
Cult. 1683, by Mr. James Sutherland. *Sutherl. hort. edin.* 5. *n.* 1.
Fl. June and July. H. ⊙.

3. C. caule multifloro corymboſo, floribus terminalibus, calycibus radio albo. *Sp. pl.* 1161. *vulgaris.*
Common Carlina.
Nat. of Britain.
Fl. July——September. H. ♂.

4. C. caule multifloro, foliis decurrentibus runcinatis. *Syſt. veget.* 729. *pyronaica.*
Pyrenean Carlina.
Nat. of the Pyrenees.
Introd. 1788, by John Sibthorp, M.D.
Fl. H. ♃.

ATRACTYLIS. *Gen. pl.* 930.

Cor. radiatæ: corollulis radii 5-dentatis. *Pappus* plumoſus.

1. A. involucris cancellatis ventricoſis linearibus dentatis, calycibus ovatis, floribus floſculoſis. *Sp. pl.* 1162. *cancellata.*
Netted Atractylis.
Nat. of the South of Europe.
Cult. 1640. *Park. theat.* 965. *n.* 4.
Fl. June and July. H. ⊙.

STOKESIA. *L'Herit. ſert. angl.*

Cor. radii infundibuliformes, longiores, irregulares. *Pappus* 4-ſetus.

1. STOKESIA. *L'Herit. ſert. angl. tab.* 38. *cyanea.*
Blue-flower'd Stokeſia.

Nat. of South Carolina.
Introd. about 1766, by Mr. James Gordon.
Fl. Auguſt. G. H. ♃.

CARTHAMUS. *Gen. pl.* 931.

Cal. ovatus, imbricatus ſquamis apice ſubovato-foliaceis.

tinctorius. 1. C. foliis ovatis integris ſerrato-aculeatis. *Sp. pl.* 1162.
Baſtard Saffron.
Nat. of Egypt.
Cult. 1551. *Turn. herb. part* 1. *ſign. Lij. verſo.*
Fl. June and July. H. ☉.

lanatus. 2. C. caule piloſo: ſuperne lanato, foliis inferioribus pinnatifidis; ſummis amplexicaulibus dentatis. *Sp. pl.* 1163.
Woolly Carthamus.
Nat. of the South of Europe.
Cult. 1596, by Mr. John Gerard. *Hort. Ger.*
Fl. July and Auguſt. H. ☉.

creticus. 3. C. caule læviuſculo, calycibus ſublanatis, floſculis ſubnovenis, foliis inferiobus lyratis. *Syſt. veget.* 730.
Cretan Carthamus.
Nat. of the Iſland of Candia.
Cult. 1739, by Mr. Ph. Miller. *Rand. chel.* Cnicus 6.
Fl. June and July. H. ☉.

tingitanus. 4. C. foliis radicalibus pinnatis; caulinis pinnatifidis, caule unifloro. *Sp. pl.* 1163.
Tangier Carthamus.
Nat. of Barbary.
Cult. 1768, by Mr. Ph. Miller. *Mill. dict. edit.* 8.
Fl. June and July. H. ♃.

5. C. foliis lanceolatis ſpinoſo-dentatis, caule ſubunifloro. *Sp. pl.* 1163. *cæruleus.*
Blue-flower'd Carthamus.
Nat. of Spain.
Cult. 1640. *Park. theat.* 260. *f.* 3.
Fl. June and July. H. ♃.

6. C. foliis inermibus: radicalibus dentatis; caulinis pinnatis. *Sp. pl.* 1164. *mitiſſimus.*
Small Carthamus.
Nat. of France.
Introd. 1776, by Monſ. Thouin.
Fl. June and July. H. ♃.

7. C. foliis caulinis linearibus pinnatis longitudine plantæ. *Sp. pl.* 1164. *Carduncellus.*
Mountain Carthamus.
Nat. of the South of France.
Cult. 1739, by Mr. Ph. Miller. *Rand. chel.* Cnicus 2.
Fl. May. H. ♃.

8. C. fruticoſus, petiolis ſpinoſis, foliis lanceolatis integris ſubtus tomentoſis apice pungentibus, ramis unifloris. *Linn. ſuppl.* 350. *ſalicifolius.*
Willow-leav'd Carthamus.
Nat. of Madeira.
Introd. 1784, by Meſſrs. Lee and Kennedy.
Fl. Auguſt. G. H. ♄.

9. C. floribus corymboſis numeroſis. *Sp. pl.* 1164. *corymboſus.*
Umbel'd Carthamus.
Nat. of the South of Europe.
Cult. 1714, by Charles Dubois, Eſq. *Philoſoph. tranſact. n.* 343. *p.* 234. *n.* 16.
Fl. June and July. H. ♃.

SPILANTHUS. *Syst. veget.* 731.

Recept. paleaceum, conicum. *Pappus* 2-dentatus. *Cal.* ſubæqualis.

Pſeudo-Acmella. 1. S. foliis lanceolatis ſerratis, caule erecto. *Syſt. veget.* 731.
Verbeſina Pſeudo-Acmella. *Sp. pl.* 1270.
Spear-leav'd Spilanthus.
Nat. of Ceylon.
Cult. 1768, by Mr. Ph. Miller. *Mill. dict. edit.* 8.
Fl. July. S. ☉.

albus. 2. S. foliis ovatis ſubintegris: imis alternis; ſummis oppoſitis, caule paniculato. *L'Herit. ſtirp. nov. p.* 7, *tab.* 4.
Spilanthus ſalivaria. *Murray in commentat. gotting.* 6. (1783) *p.* 3. *tab.* 1.
White-flower'd Spilanthus.
Nat. of Peru.
Introd. 1783, by Monſ. Thouin.
Fl. June and July. S. ☉.

Acmella. 3. S. foliis ovatis ſerratis, caule erecto, floribus radiatis. *Syſt. veget.* 731.
Verbeſina Acmella. *Sp. pl.* 1271.
Balm-leav'd Spilanthus.
Nat. of Ceylon.
Cult. 1768, by Mr. Philip Miller. *Mill. dict. edit.* 8.
Fl. July and Auguſt. S. ☉.

oleracea. 4. S. foliis ſubcordatis ſerrulatis petiolatis. *Syſt. veget.* 731. *Jacqu. hort.* 2. *p.* 63. *t.* 135.
Eatable Spilanthus.
Nat. of the Eaſt Indies.

Introd.

Introd. 1770, by Monſ. Richard.
Fl. July——November. S. ☉.

B I D E N S. *Gen. pl.* 932.

Recept. paleaceum. *Pappus* ariſtis erectis, ſcabris. *Cal.* imbricatus. *Cor.* rarius floſculo uno alterove radiante inſtruitur.

1. B. foliis trifidis, calycibus ſubfolioſis, ſeminibus erectis. *Sp. pl.* 1165. *Curtis lond.* *tripartita.*
Trifid Bidens, or Water Hemp Agrimony.
Nat. of Britain.
Fl. July and Auguſt. H. ☉.

2. B. foliis lanceolatis ſeſſilibus, floribus ſeminibuſque erectis. *Sp. pl.* 1165. *minima.*
Leaſt Bidens.
Nat. of Britain.
Fl. July and Auguſt. H. ☉.

3. B. foliis oblongis integerrimis unidentatis, caule dichotomo, floribus ſolitariis ſeſſilibus. *Sp. pl.* 1165. *nodiflora.*
Seſſile-flower'd Bidens.
Nat. of the Eaſt Indies.
Cult. 1732, by James Sherard, M.D. *Dill. elth.* 52. *t.* 44. *f.* 52.
Fl. July. S. ☉.

4. B. foliis lanceolatis amplexicaulibus, floribus cernuis, ſeminibus erectis. *Sp. pl.* 1165. *Curtis lond.* *cernua.*
Nodding Bidens.
Nat. of Britain.
Fl. July and Auguſt. H. ☉.

5. B.

frondosa. 5. B. foliis pinnatis ſerratis lineatis glabris, ſeminibus erectis, calycibus frondoſis, caule lævi. *Syſt. veget.* 732.
Smooth-ſtalk'd Bidens.
Nat. of North America.
Cult. 1752, by Mr. Ph. Miller. *Mill. dict. edit.* 6. *n.* 3.
Fl. July and Auguſt. H. ☉.

piloſa. 6. B. foliis pinnatis ſubpiloſis, caulis geniculis barbatis, calycibus involucro ſimplici, ſeminibus divergentibus. *Sp. pl.* 1166.
Hairy Bidens.
Nat. of America.
Cult. 1732, by James Sherard, M.D. *Dill. elth.* 51. *t.* 43. *f.* 51.
Fl. July. H. ☉.

bipinnata. 7. B. foliis bipinnatis inciſis, calycibus involucratis, corollis ſemiradiatis, ſeminibus divergentibus. *Sp. pl.* 1166.
Hemlock-leav'd Bidens.
Nat. of North America.
Cult. 1699. *Moriſ. hiſt.* 3. *p.* 17. *n.* 24. *ſ.* 6. *t.* 7. *f.* 23.
Fl. July and Auguſt. H. ☉.

nivea. 8. B. foliis ſimplicibus ſubhaſtatis ſerratis petiolatis, floribus globoſis, pedunculis elongatis, ſeminibus lævibus. *Sp. pl.* 1167.
Snowy Bidens.
Nat. of North America.
Cult. 1732, by James Sherard, M. D. *Dill. elth.* 55. *t.* 47. *f.* 55.
Fl. June and July. G. H. ♃.

9. B.

9. B. foliis ovatis ſerratis: inferioribus oppoſitis; ſuperioribus ternatis: intermedio majore. *Sp. pl.* 1167. *bullata.*
Various-leav'd Bidens.
Nat. of America.
Cult. 1759, by Mr. Ph. Miller. *Mill. dict. edit.* 7. *n.* 7.
Fl. July——September. H. ⊙.

C A C A L I A. *Gen. pl.* 933.

Recept. nudum. *Pappus* piloſus. *Cal.* cylindricus, oblongus, baſi tantum ſubcalyculatus.

* *Fruteſcentes.*

1. C. caule fruticoſo obvallato ſpinis petiolaribus truncatis. *Sp. pl.* 1168. *papillaris.*
Rough-ſtalk'd Cacalia.
Nat. of the Cape of Good Hope.
Cult. 1727, by James Sherard, M.D. *Dill. elth.* 63. *t.* 55. *f.* 63.
Fl. D. S. ♄.

2. C. carnoſa, caule decumbente articulato, foliis inferioribus haſtatis; ſuperioribus lyratis. *Linn. ſuppl.* 354. *Syſt. veget.* 734. *articulata.*
Cacalia laciniata. *Jacqu. ic. collect.* 1. *p.* 77. *Syſt. veget.* 733.
Jointed-ſtalk'd Cacalia.
Nat. of the Cape of Good Hope. Mr. *Fr. Maſſon.*
Introd. 1775.
Fl. November. G. H. ♄.

3. C. caule fruticoſo, foliis ovato-oblongis planis, petiolis baſi linea triplici deductis. *Syſt. veget.* 733. *Anteuphorbium.*
Oval-leav'd Cacalia.
Nat. of the Cape of Good Hope.

Cult.

Cult. 1596, by Mr. John Gerard. *Hort. Ger.*
Fl. D. S. ♄.

Kleinia. 4. C. caule fruticoso composito, foliis lanceolatis planis, petiolorum cicatricibus obsoletis. *Sp. pl.* 1168.
Oleander-leav'd Cacalia, or Cabbage-tree.
Nat. of the Canary Islands.
Cult. 1732, by James Sherard, M. D. *Dill. elth.* 61. *t.* 54. *f.* 62.
Fl. September and October. D. S. ♄.

Ficoides. 5. C. caule fruticoso, foliis compressis carnosis. *Sp. pl.* 1168.
Flat-leav'd Cacalia.
Nat. of the Cape of Good Hope.
Cult. 1727, by Professor Richard Bradley. *Bradl. succ.* 5. *p.* 11. *t.* 49.
Fl. June——November. D. S. ♄.

carnosa. 6. C. caule fruticoso, foliis teretiusculis carnosis incurvis, pedunculis terminalibus unifloris nudis.
Narrow-leav'd Cacalia.
Nat. of the Cape of Good Hope.
Cult. 1757, by Mr. Ph. Miller.
Fl. June. G. H. ♄.

repens. 7. C. caule fruticoso, foliis depressis carnosis. *Linn. mant.* 110.
Glaucous-leav'd Cacalia.
Nat. of the Cape of Good Hope.
Cult. 1759, by Mr. Philip Miller.
Fl. June. D. S. ♄.

** *Herbaceæ.*

Porophyllum. 8. C. caule herbaceo indiviso, foliis ellipticis subcrenatis. *Sp. pl.* 1169.

Perforated

Perforated Cacalia.
Nat. of America.
Introd. 1780, by Benjamin Bewick, Esq.
Fl. June——October. S. ⊙.

9. C. caule herbaceo, foliis lyratis amplexicaulibus dentatis. *Sp. pl.* 1169. *sonchifolia.*
Sow-thistle-leav'd Cacalia.
Nat. of the East Indies.
Cult. 1768, by Mr. Philip Miller. *Mill. dict. edit.* 8.
Fl. July. S. ⊙.

10. C. caule herbaceo, foliis lanceolatis serratis decurrentibus. *Sp. pl.* 1169. *saracenica.*
Creeping-rooted Cacalia.
Nat. of the South of France.
Cult. 1772, by Mr. James Gordon, Sen.
Fl. August——October. H. ♃.

11. C. caule herbaceo, foliis trilobis acuminatis serratis, floribus nutantibus. *Syst. veget.* 734. *hastata.*
Spear-leav'd Cacalia.
Nat. of Siberia.
Introd. 1780, by Peter Simon Pallas, M. D.
Fl. H. ♃.

12. C. caule herbaceo scandente, foliis sagittatis dentatis, petiolis simplicibus. *scandens.*
Climbing Cacalia.
Nat. of the Cape of Good Hope. Mr. *Fr. Masson.*
Introd. 1774.
Fl. April. G. H. ♃.

13. C. caule herbaceo erecto, foliis hastato-sagittatis denticulatis, petiolis superne dilatatis. *suaveolens.*
Cacalia suaveolens. *Sp. pl.* 1170.

Sweet-

Sweet-ſcented Cacalia.
Nat. of Virginia and Canada.
Cult. 1752, by Mr. Ph. Miller. *Mill. dict. edit.* 6. *n.* 8.
Fl. July——September. H. ♃.

atriplicifolia. 14. C. caule herbaceo, foliis ſubcordatis dentato-ſinuatis, calycibus quinquefloris. *Sp. pl.* 1170.
Orache-leav'd Cacalia.
Nat. of Virginia and Canada.
Cult. 1739, by Mr. Ph. Miller. *Mill. dict. vol.* 2. *n.* 6.
Fl. Auguſt. H. ♃.

alpina. 15. C. foliis reniformi-cordatis acutis denticulatis. *Syſt. veget.* 734. *Jacqu. auſtr.* 3. *p.* 20. *t.* 234.
Alpine Cacalia.
Nat. of Auſtria and Switzerland.
Cult. 1739, by Mr. Philip Miller. *Mill. dict. vol.* 2. *n.* 3.
Fl. July and Auguſt. H. ♃.

ETHULIA. *Gen. pl.* 934.

Recept. nudum. *Pappus* nullus.

conyzoides. 1. E. floribus paniculatis. *Sp. pl.* 1171.
Panicl'd Ethulia.
Nat. of India.
Introd. 1776, by Monſ. Thouin.
Fl. July and Auguſt. S. ☉.

EUPATORIUM. *Gen. pl.* 935.

Recept. nudum. *Pappus* plumoſus. *Cal.* imbricatus, oblongus. *Stylus* ſemibifidus, longus.

* *Calycibus quadrifloris.*

Dalea. 1. E. foliis lanceolatis venoſis obſolete ſerratis glabris, calycibus quadrifloris, caule fruticoſo. *Sp. pl.* 1171.
Shrubby

Shrubby Eupatorium.
Nat. of Jamaica.
Introd. 1773, by Monf. Richard.
Fl. Auguft. S. ♄.

2. E. caule volubili, foliis cordatis dentatis acutis. *Sp. pl.* 1171. *fcandens.*
Climbing Eupatorium.
Nat. of Virginia.
Cult. 1714. *Philofoph. tranfact. n.* 346. *p.* 353. *n.* 95.
Fl. Auguft and September. H. ♃.

** *Calycibus quinquefloris.*

3. E. foliis feffilibus amplexicaulibus diftinctis lanceolatis. *Sp. pl.* 1172. *feffilifolium.*
Seffile-leav'd Eupatorium.
Nat. of Virginia.
Introd. 1777, by Monf. Thouin.
Fl. September and October. H. ♃.

4. E. foliis feffilibus diftinctis fubrotundo-cordatis. *Sp. pl.* 1173. *rotundifolium.*
Round-leav'd Eupatorium.
Nat. of Virginia and Canada.
Cult. 1731, by Mr. Ph. Miller. *Mill. dict. edit.* 1. *n.* 7.
Fl. July and Auguft. H. ♃.

5. E. foliis lanceolatis nervofis: inferioribus extimo fubferratis, caule fuffruticofo. *Sp. pl.* 1173. *Jacqu. hort.* 2. *p.* 77. *t.* 164. *altiffimum.*
Tall Eupatorium.
Nat. of Penfylvania.
Cult. 1768, by Mr. Philip Miller. *Mill. dict. edit.* 8.
Fl. September and October. H. ♃.

6. E.

cannabinum. 6. E. foliis digitatis. *Sp. pl.* 1173.
Common Eupatorium, or Hemp Agrimony.
Nat. of Britain.
Fl. August——October. H. ♃.

*** *Calycibus octofloris.*

purpureum. 7. E. foliis quaternis scabris lanceolato-ovatis inæqualiter serratis rugosis petiolatis. *Sp. pl.* 1173.
Purple Eupatorium.
Nat. of North America.
Cult. 1731, by Mr. Ph. Miller. *Mill. dict. edit.* 1. *n.* 4.
Fl. September and October. H. ♃.

maculatum. 8. E. foliis quinis subtomentosis lanceolatis æqualiter serratis venosis petiolatis. *Sp. pl.* 1174.
Spotted Eupatorium.
Nat. of North America.
Cult. 1656, by Mr. John Tradescant, Jun. *Mus. Trad.* 112.
Fl. August and September. H. ♃.

**** *Calycibus quindecim pluribusve flosculis.*

perfoliatum. 9. E. foliis connato-perfoliatis tomentosis. *Sp. pl.* 1174.
Perfoliate Eupatorium.
Nat. of North America.
Cult. 1699, in Chelsea Garden. *Moris. hist.* 3. *p.* 97. *n.* 6.
Fl. August and September. H. ♃.

cœlestinum. 10. E. foliis cordato-ovatis obtuse serratis petiolatis, calycibus multifloris. *Sp. pl.* 1174.
Blue-flower'd Eupatorium.
Nat. of North America.
Cult. 1732, by James Sherard, M. D. *Dill. elth.* 140. *t.* 114. *f.* 139.
Fl. July——October. H. ♃.

11. E.

11. E. foliis ovatis obtuſe ſerratis petiolatis trinerviis, calycibus ſimplicibus. *Sp. pl.* 1175. *aromaticum.*
Aromatic Eupatorium.
Nat. of Virginia.
Cult. 1758, by Mr. Philip Miller.
Fl. July and Auguſt. H. ♃.

12. E. foliis deltoidibus inferne dentatis ſubtus tomentoſis, calycibus multifloris. *Sp. pl.* 1174. *odoratum.*
Sweet-ſcented Eupatorium.
Nat. of Jamaica.
Introd. 1780, by William Wright, M.D.
Fl. Auguſt and September. S. ♄.

13. E. foliis ovatis ſerratis petiolatis, caule glabro. *Syſt. veget.* 736. *Ageratoides.*
Ageratum altiſſimum. *Sp. pl.* 1176.
Nettle-leav'd Eupatorium.
Nat. of Virginia and Canada.
Introd. before 1640, by Mr. John Tradeſcant, Sen. *Park. theat.* 1221. *f.* 3.
Fl. Auguſt—October. H. ♃.

AGERATUM. *Gen. pl.* 936.

Recept. nudum. *Pappus* 5-ariſtatus. *Cal.* oblongus, ſubæqualis. *Corollulæ* quadrifidæ.

1. A. foliis ovatis, caule piloſo. *Sp. pl.* 1175. *Conyzoides.*
Hairy Ageratum.
Nat. of America.
Cult. 1714, by the Dutcheſs of Beaufort. *Br. Muſ. H. S.* 137. *fol.* 63.
Fl. July and Auguſt. H. ☉.

PTERONIA. *Gen. pl.* 937.

Recept. ſetis multipartitis. *Pappus* ſubplumoſus. *Cal.* imbricatus.

camphorata. 1. P. foliis ſparſis faſciculatiſque filiformibus ciliatis, foliolis calycinis ſerrulatis, receptaculi pilis ſubfaſciculatis.
Pteronia camphorata. *Sp. pl.* 1176.
Aromatic Pteronia.
Nat. of the Cape of Good Hope.
Introd. 1774, by Mr. Francis Maſſon.
Fl. June——September. G. H. ♄.

ſtricta. 2. P. foliis ſparſis faſciculatiſque filiformibus baſi ſubciliatis, foliolis calycinis integris, receptaculi foveis multipartito-ſetaceis.
Cluſter-flower'd Pteronia.
Nat. of the Cape of Good Hope. Mr. *Fr. Maſſon.*
Introd. 1774.
Fl. April——June. G. H. ♄.

oppoſitifolia. 3. P. foliis oppoſitis, ramis dichotomis divaricatis. *Syſt. veget.* 737.
Forked Pteronia.
Nat. of the Cape of Good Hope.
Introd. 1774, by Mr. Francis Maſſon.
Fl. G. H. ♄.

CHRYSOCOMA. *Gen. pl.* 939.

Recept. nudum. *Pappus* ſimplex. *Cal.* hemiſphæricus, imbricatus. *Stylus* vix floſculis longior.

Coma aurea. 1. C. fruticoſa, foliis linearibus rectis glabris dorſo decurrentibus. *Syſt. veget.* 739.
Great ſhrubby Goldy-locks.

Nat.

Nat. of the Cape of Good Hope.
Cult. 1748, by Mr. Philip Miller. *Mill. dict. edit.* 5. Coma aurea 2.
Fl. June——Auguſt. G. H. ♄.

2. C. fruticoſa, foliis linearibus recurvis ſubſcabris, floribus nubilibus cernuis. *Syſt. veget.* 739. *cernua.*
Small ſhrubby Goldy-locks.
Nat. of the Cape of Good Hope.
Cult. 1739, by Mr. Philip Miller. *Rand. chel.* Coma aurea 3.
Fl. Moſt part of the Summer. G. H. ♄.

3. C. ſuffruticoſa, foliis linearibus rectis ciliatis, ramis pubeſcentibus. *Sp. pl.* 1177. *ciliata.*
Heath-leav'd Goldy-locks.
Nat. of the Cape of Good Hope.
Introd. 1774, by Mr. Francis Maſſon.
Fl. July——October. G. H. ♄.

4. C. herbacea, foliis linearibus glabris, calycibus laxis. *Sp. pl.* 1178. *Linoſyris.*
German Goldy-locks.
Nat. of Europe.
Cult. 1683, by Mr. James Sutherland. *Sutherl. hort. edin.* 360. *n.* 5.
Fl. September and October. H. ♃.

5. C. herbacea paniculata, foliis lanceolatis trinerviis punctatis nudis. *Sp. pl.* 1178. *biflora.*
Two-flower'd Goldy-locks.
Nat. of Siberia.
Cult. 1759, by Philip Miller. *Mill. dict. edit.* 7. *n.* 2.
Fl. Auguſt and September. H. ♃.

TARCHONANTHUS. *Gen. pl.* 940.

Recept. villoſum. *Pappus* plumoſus. *Cal.* 1-phyllus, ſemi 7-fidus, turbinatus.

camphoratus. 1. T. foliis oblongis planis, calyce monophyllo quinquefido. *Syſt. veget.* 740.
Shrubby African Flea-bane.
Nat. of the Cape of Good Hope.
Cult. 1690, in the Royal Garden at Hampton-court. *Catal. mſſ.*
Fl. June——October. G. H. ♄.

C A L E A. *Gen. pl.* 941.

Recept. paleaceum. *Pappus* piloſus vel 0. *Cal.* imbricatus.

lobata. 1. C. corymbis congeſtis, foliis alternis: ſuperioribus ovato-lanceolatis; inferioribus dentato-haſtatis ſinuato-ſerratis. *Swartz prodr.* 113.
Conyza lobata. *Sp. pl.* 1207.
Halbert-weed.
Nat. of the Weſt Indies.
Introd. 1778, by William Pitcairn, M.D.
Fl. June and July. S. ♄.

SANTOLINA. *Gen. pl.* 942.

Recept. paleaceum. *Pappus* nullus. *Cal.* imbricatus, hemiſphæricus.

Chamæ-Cypariſſus. 1. S. pedunculis unifloris, foliis quadrifariam dentatis. *Sp. pl.* 1179.
α foliis incanis: denticulis imbricatis obtuſis, floribus minoribus.
Common Lavender Cotton.

β foliis

β foliis incanis: denticulis ſubimbricatis obtuſis, floribus majoribus.
Santolina villoſa. *Mill. dict.*
Hoary Lavender Cotton.

γ foliis glabris: dentibus ſubulatis ſubbifariis.
Santolina virens. *Mill. dict.*
Dark-green Lavender Cotton.
Nat. of the South of Europe.
Cult. 1596, by Mr. John Gerard. *Hort. Ger.*
Fl. July. H. ♄.

2. S. pedunculis unifloris, foliis linearibus margine tuberculatis. *Sp. pl.* 1180. *roſmarinifolia.*
Roſemary-leav'd Lavender Cotton.
Nat. of Spain.
Cult. 1683, by Mr. James Sutherland. *Sutherl. hort. edin.* 1. *n.* 6.
Fl. July—September. H. ♄.

ATHANASIA. *Gen. pl.* 943.

Recept. paleaceum. *Pappus* paleaceus, breviſſimus. *Cal.* imbricatus.

1. A. floribus terminalibus ſubſeſſilibus, foliis lanceolatis hirſutis. *Sp. pl.* 1181. *capitata.*
Hairy Athanaſia.
Nat. of the Cape of Good Hope.
Introd. 1774, by Mr. Francis Maſſon.
Fl. January—March. G. H. ♄.

2. A. pedunculis bifloris, foliis lanceolatis crenatis obtuſis tomentoſis. *Syſt. veget.* 741. *maritima.*
Sea Athanaſia.
Nat. of England.
Fl. July. H. ♃.

 3. A.

annua. 3. A. corymbis ſimplicibus coarctatis, foliis pinnatifidis dentatis. *Sp. pl.* 1182.
Annual Athanaſia.
Nat. of Africa.
Cult. 1768, by Mr. Philip Miller. *Mill. dict. edit.* 8.
Fl. July and Auguſt. G. H. ⊙.

trifurcata. 4. A. corymbis ſimplicibus, foliis trilobis cuneiformibus. *Sp. pl.* 1181.
Trifid-leav'd Athanaſia.
Nat. of the Cape of Good Hope.
Cult. 1714. *Philoſoph. tranſact. n.* 346. *p.* 354. *n.* 97.
Fl. July and Auguſt. G. H. ♄.

crithmifolia. 5. A. corymbis ſimplicibus, foliis ſemitrifidis linearibus. *Sp. pl.* 1181.
Samphire-leav'd Athanaſia.
Nat. of the Cape of Good Hope.
Cult. 1726, by Mr. Philip Miller. *R. S. n.* 219.
Fl. July and Auguſt. G. H. ♄.

linifolia. 6. A. corymbis ſubfaſtigiatis, foliis linearibus integerrimis acutis, receptaculo ſeminibuſque nudis.
Athanaſia linifolia. *Linn. ſuppl.* 361.
Flax-leav'd Athanaſia.
Nat. of the Cape of Good Hope. Mr. *Fr. Maſſon.*
Introd. 1774.
Fl. Auguſt. G. H. ♄.

dentata. 7. A. corymbis compoſitis, foliis recurvis: inferioribus linearibus dentatis; ſuperioribus ovatis ſerratis. *Syſt. veget.* 742.
Notch'd-leav'd Athanaſia.
Nat. of the Cape of Good Hope.

Introd.

Introd. 1780, by the Counteſs of Strathmore.
Fl. July and Auguſt. G. H. ♄.

8. A. corymbis compoſitis, foliis pinnatis linearibus. *Syſt. veget.* 742. *parviflora.*
Tanacetum crithmifolium. *Sp. pl.* 1182.
Small-flower'd Athanaſia.
Nat. of the Cape of Good Hope.
Introd. 1774, by Mr. Francis Maſſon.
Fl. April. G. H. ♄.

9. A. corymbo compoſito, foliis linearibus glabris patulis. *Linn. ſuppl.* 361. *filiformis.*
Fine-leav'd Athanaſia.
Nat. of the Cape of Good Hope.
Introd. 1787, by Mr. Francis Maſſon.
Fl. Auguſt. G. H. ♄.

10. A. corymbis compoſitis, foliis linearibus integerrimis tomentoſis. *cinerea.*
Athanaſia cinerea. *Linn. ſuppl.* 361.
Lavender-leav'd Athanaſia.
Nat. of the Cape of Good Hope. Mr. *Fr. Maſſon.*
Introd. 1774.
Fl. May and June. G. H. ♄.

POLYGAMIA SUPERFLUA.

TANACETUM. *Gen. pl.* 944.

Recept. nudum. *Pappus* submarginatus. *Cal.* imbricatus, hemisphæricus. *Cor.* radii obsoletæ, 3-fidæ (*interdum nullæ, omnesque flosculi hermaphroditi.*)

suffruticosum. 1. T. foliis pinnato-multifidis: laciniis linearibus subdivisis acutis, caule suffruticoso. *Sp. pl.* 1183.
Shrubby Tansy.
Nat. of the Cape of Good Hope.
Cult. 1759, by Mr. Ph. Miller. *Mill. dict. edit.* 7. *n.* 5.
Fl. Most part of the Summer. G. H. ♄.

annuum. 2. T. foliis bipinnatifidis linearibus acutis, corymbis tomentosis. *Sp. pl.* 1184.
Annual Tansy.
Nat. of Spain and Italy.
Cult. 1758, by Mr. Philip Miller. *Mill. ic.* 152. *t.* 227. *f.* 1.
Fl. July. H. ⊙.

vulgare. 3. T. foliis bipinnatis incisis serratis. *Sp. pl.* 1184.
α Tanacetum vulgare luteum. *Bauh. pin.* 132.
Common Tansy.
β Tanacetum foliis crispis. *Bauh. pin.* 132.
Curl'd Tansy.
Nat. of Britain.
Fl. August. H. ♃.

Balsamita. 4. T. foliis ovatis integris serratis. *Sp. pl.* 1184.
Cost-mary.
Nat. of the South of France and Italy.

Cult.

Cult. 1568. *Turn. herb. part* 3. *p.* 42.
Fl. August and September. H. ♃.

5. T. corymbis simplicibus, foliis deltoidibus apice serratis. *L'Herit. sert. angl. tab.* 27. *flabelliforme.*
Fan-leav'd Tansy.
Nat. of the Cape of Good Hope. Mr. *Fr. Masson.*
Introd. 1774.
Fl. May——August. G. H. ♄.

ARTEMISIA. *Gen. pl.* 945.

Recept. subvillosum vel nudiusculum. *Pappus* nullus. *Cal.* imbricatus squamis rotundatis, conniventibus. *Cor.* radii nullæ.

* *Fruticosæ, erectæ.*

1. A. fruticosa, foliis setaceis ramosissimis. *Syst. veget.* 743. *Abrotanum.*
α caule erecto,
Common Southernwood.
β Artemisia *humilis*, foliis setaceis pinnatifidis, caule decumbente suffruticoso. *Mill. dict.*
Dwarf Southernwood.
Nat. of France, Spain, and Italy.
Cult. 1596, by Mr. John Gerard. *Hort. Ger.*
Fl. August——October. H. ♄.

2. A. fruticosa, foliis tripinnatifidis sericeis cinereis: foliolis linearibus, floribus globosis, ramulis floriferis simplicibus. *arborescens.*
Artemisia arborescens. *Sp. pl.* 1188.
Common-narrow-leav'd Tree Wormwood.
Nat. of the Levant.
Cult. 1640. *Park. theat.* 93. *f.* 3.
Fl. June——August. H. ♄.

3. A.

argentea. 3. A. fruticosa, foliis bipinnatifidis sericeis candidis: foliolis lanceolato-linearibus, floribus globosis, ramulis floriferis virgatis.
Artemisia argentea. *L'Herit. sert. angl. tab.* 28.
Broad-leav'd Tree Wormwood.
Nat. of Madeira. Mr. *Francis Masson.*
Introd. 1777.
Fl. June and July. G. H. ♄.

** *Procumbentes ante florescentiom.*

santonica. 4. A. foliis caulinis linearibus pinnato-multifidis, ramis indivisis, spicis secundis reflexis, floribus quinquefloris. *Syst. veget.* 743.
Tartarian Southernwood.
Nat. of Siberia.
Cult. 1768, by Mr. Ph. Miller. *Mill. dict. edit.* 8.
Fl. September——November, H. ♄.

campestris. 5. A. foliis multifidis linearibus, caulibus procumbentibus virgatis. *Sp. pl.* 1185.
Field Southernwood.
Nat. of England.
Fl. August. H. ♃.

crithmifolia. 6. A. foliis compositis divaricatis linearibus carnosis glabris, caule adscendente paniculato. *Sp. pl.* 1186.
Samphire-leav'd Southernwood.
Nat. of Portugal.
Cult. 1768, by Mr. Philip Miller. *Mill. dict. edit.* 8.
Fl. August——October. H. ♄.

vallesiaca. 7. A. foliis pinnatis multipartitis filiformibus tomentosis, floribus sessilibus erectis subcylindraceis paucifloris,
Artemisia vallesiaca. *Allion. pedem.* 1. *p.* 169.

Artemisia

Artemisia foliis tomentosis, multifidis, floribus erectis, longe spicatis, pene sessilibus. *Hall. hist.* 128.

Absinthium seriphium montanum candidum. *Bauh. pin.* 139.

Downy Southernwood.

Nat. of Switzerland and Italy.

Introd. 1775, by the Doctors Pitcairn and Fothergill.

Fl. July and August. H. ♃.

8. A. foliis multipartitis tomentosis, racemis cernuis, flosculis femineis ternis. *Sp. pl.* 1186. *maritima.*

Sea Wormwood.

Nat. of Britain.

Fl. August. H. ♃.

9. A. foliis palmatis multifidis sericeis, caulibus adscendentibus, floribus glomeratis fastigiatis. *Sp. pl.* 1187. *glacialis.*

Silky Wormwood.

Nat. of Switzerland.

Cult. 1748, by Mr. Philip Miller. *Mill. dict. edit.* 5. Absinthium 19.

Fl. July and August. H. ♃.

10. A. foliis pinnatis, caulibus adscendentibus, floribus globosis cernuis: receptaculo papposo. *Sp. pl.* 1186. *rupestris.*

Creeping Wormwood.

Nat. of Siberia and Oeland.

Cult. 1748, by Mr. Philip Miller. *Mill. dict. edit.* 5. Absinthium 18.

Fl. August. H. ♃.

*** *Erectæ herbaceæ, foliis compositis.*

11. A. foliis multipartitis subtus tomentosis, floribus subrotundis nutantibus: receptaculo nudo. *Sp. pl.* 1187. *Jacqu. austr.* 1. *p.* 61. *t.* 99. *pontica.*

Roman

Roman Wormwood.
Nat. of Austria and Hungary.
Cult. 1683, by Mr. James Sutherland. *Sutherl. hort. edin.* 2. *n.* 4.
Fl. September. H. ♃.

annua. 12. A. foliis triplicato-pinnatis utrinque glabris, floribus subglobosis nutantibus: receptaculo glabro conico. *Sp. pl.* 1187.
Annual Wormwood.
Nat. of Siberia.
Cult. 1759, by Mr. Philip Miller. *Mill. dict. edit.* 7. Absinthium 33.
Fl. July and August. H. ⊙.

Absinthium. 13. A. foliis compositis multifidis, floribus subglobosis pendulis: receptaculo villoso. *Sp. pl.* 1188.
Common Wormwood.
Nat. of Britain.
Fl. July——October. H. ♃.

vulgaris. 14. A. foliis pinnatifidis planis incisis subtus tomentosis, racemis simplicibus recurvatis, floribus radio quinquefloro. *Syst. veget.* 744.
Mug-wort.
Nat. of Britain.
Fl. August and September. H. ♃.

**** *Foliis simplicibus.*

cærulescens. 15. A. foliis caulinis lanceolatis integris; radicalibus multifidis, flosculis femineis ternis. *Sp. pl.* 1189.
Lavender-leav'd Wormwood.
Nat. of the South of Europe.
Cult. 1656, by Mr. John Tradescant, Jun. *Mus. Trad.* 74.
Fl. August——October. H. ♄.

16. A.

16. A. foliis lanceolatis glabris integerrimis. *Sp. pl.* 1189. *Dracunculus.*
Tarragon.
Nat. of Siberia.
Cult. 1596, by Mr. John Gerard. *Hort. Ger.*
Fl. Auguſt. H. ♃.

17. A. foliis ſimplicibus lyrato-ſinuatis, caulibus procumbentibus, floribus pedunculatis ſolitariis globoſis oppoſitifoliis. *Sp. pl.* 1190. *maderaſpatana.*
Tanacetum ægyptiacum. *Jacqu. hort.* 3. *p.* 46. *t.* 88.
Madras Wormwood.
Nat. of the Eaſt Indies.
Introd. 1780, by Monſ. Thouin.
Fl. July and Auguſt. S. ☉.

18. A. foliis cuneiformibus repandis, caule procumbente, floribus axillaribus ſeſſilibus. *Sp. pl.* 1190. *minima.*
Leaſt Wormwood.
Nat. of China.
Introd. 1778, by Monſ. Thouin.
Fl. July. S. ☉.

GNAPHALIUM. *Gen. pl.* 946.

Recept. nudum. *Pappus* plumoſus. *Cal.* imbricatus: ſquamis marginalibus rotundatis, ſcarioſis, coloratis.

* *Fruticoſa Argyrocoma.*

1. G. fruticoſum, foliis ſeſſilibus linearibus ſupra glabris margine revolutis, floribus ſubcapitatis, pedunculis elongatis. *Sp. pl.* 1191. *arboreum.*
Tree Everlaſting.
Nat. of the Cape of Good Hope.
Introd. 1770, by Mr. William Malcolm.
Fl. Moſt part of the Year. G. H. ♄.

2. G.

grandiflorum. 2. G. fruticosum, foliis amplexicaulibus ovatis trinerviis utrinque lanuginosis. *Sp. pl.* 1191.
Great-flower'd Everlasting.
Nat. of the Cape of Good Hope.
Introd. 1787, by Mr. Francis Masson.
Fl. G. H. ♄.

ericoides. 3. G. fruticosum, foliis sessilibus linearibus, calycibus exterioribus rudibus; interioribus incarnatis. *Sp. pl.* 1193.
Heath-leav'd Everlasting.
Nat. of the Cape of Good Hope.
Introd. 1774, by Mr. Francis Masson.
Fl. March——August. G. H. ♄.

** *Fruticosa Chrysocoma.*

Stœchas. 4. G. fruticosum, foliis linearibus, ramis virgatis, corymbo composito. *Sp. pl.* 1193.
Common shrubby Everlasting.
Nat. of Germany, France, and Spain.
Cult. 1629. *Park. parad.* 374. *n.* 4.
Fl. June——October. H. ♄.

patulum. 5. G. fruticosum, foliis amplexicaulibus spathulatis, ramis patentibus, corymbis aggregatis. *Sp. pl.* 1194.
Spreading Everlasting.
Nat. of the Cape of Good Hope.
Introd. 1774, by Mr. Francis Masson.
Fl. August——January. G. H. ♄.

crassifolium. 6. G. fruticosum, foliis lato-lanceolatis subpetiolatis coriaceis tomentosis, corymbo composito, caule prolifero. *Linn. mant.* 112.
Thick-leav'd Everlasting.
Nat. of the Cape of Good Hope.

Introd. 1774, by Monſ. Richard.
Fl. July——September. G. H. ♄.

7. G. fruticoſum ramoſiſſimum, foliis lanceolatis ſeſſilibus acutiuſculis, calycinis intimis ſquamis aureis. *Linn. mant.* 283. *maritimum.*
Sea Everlaſting.
Nat. of the Cape of Good Hope.
Introd. about 1772.
Fl. June——Auguſt. G. H. ♄.

*** *Herbacea Chryſocoma.*

8. G. ſubherbaceum, foliis lineari-lanceolatis ſeſſilibus, corymbo compoſito, pedunculis elongatis. *Sp. pl.* 1195. *orientale.*
α Elichryſum orientale. *Bauh. pin.* 264.
Broad-leav'd Eaſtern Everlaſting.
β Gnaphalium *(fruticoſum)* fruteſcens, foliis inferne lanceolatis caulinis lineari-lanceolatis utrinque tomentoſis, corymbo compoſito terminali. *Mill. dict.*
Narrow-leav'd Eaſtern Everlaſting.
Nat. of Africa.
Cult. 1629. *Park. parad.* 374. *n.* 3.
Fl. April——Auguſt. G. H. ♄.

9. G. herbaceum, foliis lanceolatis, corymbo decompoſito, caule inferne ramoſo. *Syſt. veget.* 747. *rutilans.*
Shining-flower'd Everlaſting.
Nat. of the Cape of Good Hope.
Cult. 1732, by James Sherard, M.D. *Dill. elth.* 127. *t.* 107. *f.* 127.
Fl. June and July. G. H. ♄.

10. G. herbaceum, foliis lanceolatis trinerviis ſupra glabris, racemo terminali, caule inferne ramoſo. *Sp. pl.* 1195. *cymoſum.*

Branching

Branching Everlaſting.
Nat. of Africa.
Cult. 1732, by James Sherard, M.D. *Dill. elth.* 128. *t.* 107. *f.* 128.
Fl. April——Auguſt. G. H. ♃.

luteo-album. 11. G. herbaceum, foliis ſemiamplexicaulibus enſiformibus repandis obtuſis utrinque pubeſcentibus, floribus conglomeratis. *Sp. pl.* 1196.
Jerſey Everlaſting, or Cudweed.
Nat. of England.
Fl. July and Auguſt. H. ⊙.

odoratiſſimum. 12. G. herbaceum, foliis decurrentibus mucronatis, utrinque tomentoſis planis. *Sp. pl.* 1196.
Sweet-ſcented Everlaſting.
Nat. of the Cape of Good Hope.
Cult. 1691, in the Royal Garden at Hampton-court. *Pluk. phyt. t.* 173. *f.* 6.
Fl. April——Auguſt. G. H. ♃.

**** *Herbacea Argyrocoma.*

fœtidum. 13. G. herbaceum, foliis amplexicaulibus integerrimis acutis ſubtus tomentoſis, caule ramoſo. *Sp. pl.* 1197.
Strong-ſcented Everlaſting.
Nat. of the Cape of Good Hope.
Cult. 1692, by Mr. George London. *Pluk. phyt. t.* 243. *f.* 1.
Fl. June——September. H. ♂.

undulatum. 14. G. herbaceum, foliis ſubdecurrentibus lanceolatis acutis undulatis ſubtus tomentoſis, caule ramoſo. *Syſt. veget.* 748.
Wav'd Everlaſting.
Nat. of Africa.

Cult.

Cult. 1732, by James Sherard, M. D. *Dill. elth.* 130. *t.* 108. *f.* 130.
Fl. June——September. G. H. ☉.

15. G. herbaceum, foliis amplexicaulibus lanceolatis, corymbis compositis, calycum squamis subplicatis. *Syst. veget.* 748. *helianthemifolium.*
Dwarf Cistus-leav'd Everlasting.
Nat. of the Cape of Good Hope.
Introd. 1774, by Mr. Francis Masson.
Fl. G. H. ♃.

16. G. herbaceum, foliis lanceolatis, caule tomentoso paniculato, floribus glomeratis conicis terminalibus. *Sp. pl.* 1198. *obtusifolium.*
Blunt-leav'd Everlasting.
Nat. of Virginia and Pensylvania.
Cult. 1699, by Mr. Jacob Bobart. *Moris. hist.* 3. *p.* 88. *n.* 21. *f.* 7. *t.* 10. *f.* 19.
Fl. July——September. H. ☉.

17. G. herbaceum, foliis lineari-lanceolatis acuminatis alternis, caule superne ramoso, corymbis fastigiatis. *Sp. pl.* 1198. *margaritaceum.*
American Everlasting, or Cudweed.
Nat. of England.
Fl. July——September. H. ♃.

18. G. sarmentis procumbentibus, caule simplicissimo, foliis radicalibus ovatis maximis. *Sp. pl.* 1199. *plantagineum.*
Plantain-leav'd Everlasting.
Nat. of Virginia.
Cult. 1759, by Mr. Ph. Miller. *Mill. dict. edit.* 7. *n.* 22.
Fl. June and July. H. ♃.

dioicum. 19. G. ſarmentis procumbentibus, caule ſimpliciſſimo, corymbo ſimplici terminali, floribus dioicis. *Sp. pl.* 1199.
Mountain Everlaſting, or Cudweed.
Nat. of Britain.
Fl. May and June. H. ♃.

alpinum. 20. G. ſarmentis procumbentibus, caule ſimpliciſſimo, capitulo aphyllo, floribus oblongis. *Syſt. veget.* 748.
Alpine Everlaſting.
Nat. of Switzerland.
Introd. 1775, by the Doctors Pitcairn and Fothergill.
Fl. June and July. H. ♃.

***** *Filaginoidea.*

ſylvaticum. 21. G. caule herbaceo ſimpliciſſimo erecto, floribus ſparſis. *Syſt. veget.* 749.
Wood Everlaſting, or Upright Cudweed.
Nat. of Britain.
Fl. Auguſt. H. ♂.

declinatum. 22. G. caule herbaceo, foliis lineari-lanceolatis, calycibus radio albo lanceolato. *Syſt. veget.* 749.
Creeping Everlaſting.
Nat. of the Cape of Good Hope.
Introd. 1787, by Mr. Francis Maſſon.
Fl. July. G. H. ♃.

uliginoſum. 23. G. caule herbaceo ramoſo diffuſo, floribus confertis terminalibus. *Syſt. veget.* 749.
Marſh Everlaſting.
Nat. of Britain.
Fl. Auguſt. H. ⊙.

24. G.

24. G. caule herbaceo diffuso, calycibus squamis interioribus subulatis nudis, foliis subamplexicaulibus. *Syst. veget.* 749. *glomeratum.*
Cluster-flower'd Everlasting.
Nat. of the Cape of Good Hope.
Introd. 1774, by Mr. Francis Masson.
Fl. March——September. G. H. ♃.

XERANTHEMUM. *Gen. pl.* 947.

Recept. paleaceum. *Pappus* setaceus. *Cal.* imbricatus, radiatus: radio colorato.

1. X. herbaceum, foliis lanceolatis patentibus. *Syst. veget.* 749. *Jacqu. austr.* 4. *p.* 46. *t.* 388. *annuum.*
α Jacea oleæ folio, capitulis simplicibus. *Bauh. pin.* 272.
Purple and White Annual Xeranthemum.
β Xeranthemum oleæ folio, capitulis simplicibus, incanum fœtens, flore purpurascente minore. *Moris. hist.* 3. *p.* 43. *s.* 6. *t.* 12. *f.* 1. inapertum.
Small-flower'd Annual Xeranthemum.
Nat. of the South of Europe.
Cult. 1658, in Oxford Garden. *Hort. oxon. edit.* 2. *p.* 150.
Fl. July and August. H. ⊙.

2. X. fruticosum erectum, foliis adnatis lanceolato-linearibus apice calloso-mucronatis, ramis unifloris foliosis. *Sp. pl.* 1201. *vestitum.*
Upright Xeranthemum.
Nat. of the Cape of Good Hope.
Introd. 1774, by Mr. Francis Masson.
Fl. July——September. G. H. ♄.

fulgidum. 3. X. fruticosum erectum, foliis oblongis margine tomentosis.
Xeranthemum fulgidum. *Linn. suppl.* 365. *Jacqu. ic.*
Gnaphalium aureum. *Houtt. nat. hist.* 10. *p.* 590. *t.* 67. *f.* 3.
Great yellow-flower'd Xeranthemum.
Nat. of the Cape of Good Hope. Mr. *Fr. Masson.*
Introd. 1774.
Fl. February——October. G. H. ♃.

speciosissimum. 4. X. fruticosum erectum, foliis amplexicaulibus lanceolatis trinerviis, floribus pedunculatis. *Syst. veget.* 750.
Shewy Xeranthemum.
Nat. of the Cape of Good Hope.
Introd. 1787, by Mr. Francis Masson.
Fl. G. H. ♄.

retortum. 5. X. caulibus frutescentibus provolutis, foliis tomentosis recurvatis. *Sp. pl.* 1202.
Trailing Xeranthemum.
Nat. of the Cape of Good Hope.
Cult. 1732, by James Sherard, M.D. *Dill. elth.* 433. *t.* 322. *f.* 415.
Fl. July and August. G. H. ♄.

CARPESIUM. *Gen. pl.* 948.

Recept. nudum. *Pappus* nullus. *Cal.* imbricatus: squamis exterioribus reflexis.

cernuum. 1. C. floribus terminalibus. *Sp. pl.* 1203. *Jacqu. austr.* 3. *p.* 2. *t.* 204.
Drooping Carpesium.
Nat. of Italy, Spain, and Austria.
Cult. 1768, by Mr. Philip Miller. *Mill. dict. edit.* 8.
Fl. July and August. G. H. ♃.

BACCHA-

BACCHARIS. *Gen. pl.* 949.

Recept. nudum. *Pappus* pilosus. *Cal.* imbricatus, cylindricus. *Flosculi* feminei hermaphroditis immixti.

1. B. foliis lanceolatis longitudinaliter dentato-serratis. *Sp. pl.* 1204. *ivæfolia.*
Notch'd-leav'd Baccharis.
Nat. of America.
Cult. 1696, in Chelsea Garden. *Pluk. alm.* 400. *t.* 328. *f.* 2.
Fl. July and August. G. H. ♃.

2. B. foliis lanceolatis superne uno alterove denticulo serratis. *Sp. pl.* 1204. *neriifolia.*
Oleander-leav'd Baccharis.
Nat. of the Cape of Good Hope.
Cult. 1768, by Mr. Philip Miller. *Mill. dict. edit.* 8.
Fl. August——November. G. H. ♄.

3. B. foliis obovatis superne emarginato-crenatis. *Sp. pl.* 1204. *halimifolia.*
Sea-purslane-leav'd Baccharis, or Virginian Groundsel-tree.
Nat. of Virginia.
Cult. 1688, by Bishop Compton. *Raj. hist.* 2. *p.* 1799.
Fl. October and November. H. ♄.

CONYZA. *Gen. pl.* 950.

Recept. nudum. *Pappus* simplex. *Cal.* imbricatus, subrotundus. *Cor.* radii trifidæ.

1. C. foliis lanceolatis acutis, caule herbaceo corymboso, calycibus squarrosis. *Syst. veget.* 752. *squarrosa.*

 Great

Great Flea-bane, or Plowman's-ſpiknard.
Nat. of Britain.
Fl. July and Auguſt. H. ♂.

Inuloides. 2. C. foliis cuneiformi-linearibus obtuſis crenato-denticulatis glabris, caule fruticoſo, antheris biſetis.
Chryſocoma dichotoma. *Linn. ſuppl.* 359. *Jacqu. ic. collect.* 1. *p.* 44.
Cluſter-flower'd Flea-bane.
Nat. of the Iſland of Teneriffe. Mr. *Fr. Maſſon.*
Introd. 1780.
Fl. July and Auguſt. G. H. ♄.

linifolia. 3. C. foliis lineari-lanceolatis integerrimis, corollis radiatis. *Sp. pl.* 1205.
Flax-leav'd Flea-bane.
Nat. of North America.
Cult. 1699, in Chelſea Garden. *Moriſ. hiſt.* 3. *p.* 122. *n.* 45.
Fl. Auguſt and September. H. ♃.

ſordida. 4. C. foliis linearibus integerrimis, pedunculis longis trifloris, caule ſuffruticoſo. *Syſt. veget.* 752.
Gnaphalium ſordidum. *Sp. pl.* 1193.
Small-flower'd Flea-bane.
Nat. of the South of Europe.
Cult. 1570, by Matthias de l'Obel. *Lobel. adv.* 203.
Fl. July——September. G. H. ♄.

ſaxatilis. 5. C. foliis linearibus ſubintegris ſubtus tomentoſis, pedunculis longiſſimis unifloris, ſquamis calycinis ſubulatis.
Conyza ſaxatilis. *Sp. pl.* 1206.
Rock Flea-bane.
Nat. of the South of Europe.

Introd.

Introd. 1774, by Monſ. Richard.
Fl. July and Auguſt. G. H. ♄.

6. C. foliis lineari-filiformibus caulibuſque tomentoſo-ſericeis, floribus paniculatis. *ſericea.*
Chryſocoma ſericea. *Linn. ſuppl.* 360.
Snowy Flea-bane.
Nat. of the Canary Iſlands. Mr. *Francis Maſſon.*
Introd. 1779.
Fl. G. H. ♄.

7. C. foliis lato-lanceolatis ſubſerratis, corollis radiatis, calycibus ſquarroſis. *Syſt. veget.* 752. *Aſteroides.*
Starwort Flea-bane.
Nat. of North America.
Introd. 1773, by Mr. William Young.
Fl. Auguſt and September. H. ♃.

8. C. foliis ovato-oblongis amplexicaulibus. *Sp. pl.* 1207. *bifrons.*
Oval-leav'd Flea-bane.
Nat. of Canada.
Cult. 1759, by Mr. Ph. Miller. *Mill. dict. edit.* 7. *n.* 2.
Fl. Auguſt and September. H. ♃.

9. C. foliis ovatis tomentoſis, floribus confertis, pedunculis lateralibus terminalibuſque. *Sp. pl.* 1208. *candida.*
Woolly Flea-bane.
Nat. of Candia.
Cult. 1714, by the Dutcheſs of Beaufort. *Br. Muſ. H. S.* 131. *fol.* 8.
Fl. June and July. G. H. ♄.

10. C. foliis oblongo-ſpathulatis dentatis piloſis, floribus ſubpaniculatis globoſis, foliolis calycinis ſubulatis mollibus. *ægyptiaca.*

Erigeron ægyptiacum. *Linn. mant.* 112. *Jacqui hort.* 3. *p.* 14. *t.* 19.
Erigeron ſerratum. *Forſk. deſcr. p.* 148.
Egyptian Flea-bane.
Nat. of Egypt.
Introd. 1778, by Monſ. Thouin.
Fl. July. S. ⊙.

anthelmintica. 11. C. foliis lanceolato-ovatis ſcabris, pedunculis unifloris, calycibus ſquarroſis. *Syſt. veget.* 753.
Purple Flea-bane.
Nat. of the Eaſt Indies.
Introd. 1770, by Monſ. Richard.
Fl. Auguſt and September. S. ♂.

rugoſa. 12. C. fruticoſa, foliis cuneiformibus ſerratis villoſis rugoſis reticulatis, pedunculis villoſis unifloris, floribus radiatis.
St. Helena Flea-bane.
Nat. of the Iſland of St. Helena.
Introd. 1772, by Sir Joſeph Banks, Bart.
Fl. November. G. H. ♄.

inciſa. 13. C. foliis ovatis ſubcordatis piloſo-viſcoſis dentatis baſi auriculatis, diſco receptaculi favoſo.
Ear-leav'd Flea-bane.
Nat. of the Cape of Good Hope. Mr. *Fr. Maſſon.*
Introd. 1774.
Fl. June——Auguſt. G. H. ♄.

patula. 14. C. foliis ellipticis ſerratis ſubtus villoſis, calycibus ſubgloboſis: foliolis lanceolato-ſubulatis, ramis patulis.
Serratula foliis oblongo-ovatis obtuſe dentatis, caule ramoſo patulo, calycibus ſubrotundis mollibus. *Mill. ic.* 165. *t.* 247.

Spreading

Spreading Flea-bane.
Nat. of China.
Cult. 1758, by Mr. Philip Miller. *Mill. ic. loc. cit.*
Fl. August and September. S. ☉.

15. C. foliis decurrentibus lanceolatis ſerrulatis, caule virgato, floribus ſpicatis ſparſis congeſtis. *Sp. pl.* 1206. *virgata.*
Winged ſtalk'd Flea-bane.
Nat. of Carolina and the Weſt Indies.
Introd. 1783, by Mr. William Young.
Fl. Auguſt and September. G. H. ♃.

ERIGERON. *Gen. pl.* 951.

Recept. nudum. *Pappus* piloſus. *Cor.* radii lineares, anguſtiſſimæ.

1. E. ſquamis calycinis inferioribus laxis florem ſuperantibus, pedunculis folioſis. *Syſt. veget.* 754. *ſiculum.*
Red-ſtalk'd Erigeron.
Nat. of Sicily.
Introd. 1779, by John Fothergill, M. D.
Fl. Auguſt and September. H. ☉.

2. E. caule paniculato, floribus ſubſolitariis terminalibus, foliis linearibus integerrimis. *Sp. pl.* 1210. *carolinianum.*
Carolina Erigeron.
Nat. of North America.
Cult. 1727, by James Sherard, M. D. *Dill. elth.* 412. *t.* 306. *f.* 394.
Fl. July and Auguſt. H. ♃.

3. E. caule floribuſque paniculatis hirtis, foliis lanceolatis ciliatis. *Syſt. veget.* 754. *canadenſe.*
Canadian Erigeron.

Nat.

Nat. of England.
Fl. Auguſt. H. ⊙.

bonari-enſe. 4. E. foliis baſi revolutis. *Sp. pl.* 1211.
Buck's-horn Erigeron.
Nat. of South America.
Cult. 1732, by James Sherard, M. D. *Dill. elth.* 344. *t.* 257. *f.* 334.
Fl. July and Auguſt. H. ⊙.

philadel-phicum. 5. E. caule multifloro, ſoliis lanceolatis ſubſerratis: caulinis ſemiamplexicaulibus, floſculis radii capillaceis longitudine diſci. *Syſt. veget.* 754.
Spreading Erigeron.
Nat. of North America.
Introd. about 1778.
Fl. June——Auguſt. H. ♃.

purpure-um. 6. E. caule multifloro piloſo, foliis oblongis ſubdentatis amplexicaulibus, corollis radii capillaceis diſco longioribus.
Purple Erigeron.
Nat. of Hudſon's-bay.
Introd. 1776, by Meſſrs. Gordon and Græffer.
Fl. July and Auguſt. H. ♃.
Descr. *Caulis* herbaceus, ramoſus, pedalis et ultra. *Folia* alterna, acuta, nunc integra, nunc dentibus paucis notata. *Flores* paniculati. *Calycis* foliola lanceolata, acuta, æqualia, glabra, ſubcarinata: carina pilis raris adſperſa. *Corollulæ* radii numeroſiſſimæ, purpureæ; diſci flavæ.

Gouani. 7. E. floribus congeſtis, calycibus ſcarioſis, foliis lanceolatis ſubdentatis margine ſcabris. *Syſt. veget.* 754. *Jacqu. hort.* 3. *p.* 43. *t.* 79.
Cluſter-flower'd Erigeron.

Nat. of the Canary Islands.
Introd. 1772, by Monf. Richard.
Fl. July and August. H. ☉.

8. E. pedunculis alternis unifloris. *Sp. pl.* 1211. *Curtis lond.* *acre.*
Blue Erigeron.
Nat. of Britain.
Fl. July and August. H. ♃.

9. E. caule fubbifloro, calyce fubhirfuto, foliis obtufis fubtus villofis. *Syft. veget.* 754. *alpinum.*
Alpine Erigeron.
Nat. of Switzerland.
Cult. 1759, by Mr. Ph. Miller. *Mill. dict. edit.* 7. *n.* 5.
Fl. July. H. ♃.

10. E. caule unifloro, calyce pilofo. *Sp. pl.* 1211. *uniflorum.*
Dwarf Erigeron.
Nat. of Lapland and Switzerland.
Introd. 1775, by the Doctors Pitcairn and Fothergill.
Fl. August and September. H. ♃.

11. E. foliis lanceolato-linearibus retufis, floribus corymbofis. *Sp. pl.* 1213. *fœtidum.*
Stinking Erigeron.
Nat. of Africa.
Cult. 1722, in Chelfea Garden. *R. S. n.* 11.
Fl. August——November. G. H. ♃.

TUSSILAGO. *Gen. pl.* 952.

Recept. nudum. *Pappus* fimplex. *Cal.* fquamæ æquales, difcum æquantes, fubmembranaceæ.

1. T. fcapo fubnudo unifloro, foliis cordato-orbiculatis crenatis. *Sp. pl.* 1213. *Jacqu. auftr.* 3. *p.* 26. *t.* 246. *alpina.*

Alpine

Alpine Colt's-foot.
Nat. of Switzerland, Auſtria, and Siberia.
Cult. 1731, by Mr. Ph. Miller. *Mill. dict. edit.* 1. *n.* 2.
Fl. May. H. ♃.

Farfara. 2. T. ſcapo imbricato unifloro, foliis ſubcordatis angulatis denticulatis. *Sp. pl.* 1214. *Curtis lond.*
Common Colt's-foot.
Nat. of Britain.
Fl. March and April.

palmata. 3. T. thyrſo faſtigiato, foliis palmatis dentatis. TAB. 11.
Palmated Colt's-foot.
Nat. of Newfoundland and Labrador.
Introd. 1777, by John Fothergill, M.D.
Fl. April. H. ♃.

alba. 4. T. thyrſo faſtigiato, floſculis femineis nudis paucis. *Sp. pl.* 1214.
White Colt's-foot, or Butter-bur.
Nat. of Europe.
Cult. 1683, by Mr. James Sutherland. *Sutherl. hort. edin.* 267. *n.* 1.
Fl. January——April. H. ♃.

hybrida. 5. T. thyrſo oblongo, floſculis femineis nudis plurimis. *Sp. pl.* 1214.
Long-ſtalk'd Colt's-foot, or Butter-bur.
Nat. of Britain.
Fl. March and April. H. ♃.

paradoxa. 6. T. thyrſo ſubovato, floſculis femineis nudis multis; hermaphroditis ternis, antheris liberis, foliis triangulari-cordatis denticulatis. *Retzii obſ. bot.* 2. *p.* 24. *n.* 81. *tab.* 3.

Petaſites

The material originally positioned here is too large for reproduction in this reissue. A PDF can be downloaded from the web address given on page iv of this book, by clicking on 'Resources Available'.

Petaſites. floribus ſpicatis, erectis, foliis calycinis lanceolatis, floſculis pauciſſimis androgynis. *Hall. hiſt.* 141.
Downy-leav'd Colt's-foot.
Nat. of Switzerland.
Cult. 1758, by Mr. Philip Miller.
Fl. April. H. ♃.

7. T. thyrſo ovato, floſculis femineis nudis paucis. *Syſt. veget.* 756. *Curtis lond.* *Petaſites.*
Great Colt's-foot, or Butter-bur.
Nat. of Britain.
Fl. March and April. H. ♃.

SENECIO. *Gen. pl.* 953.

Recept. nudum. *Pappus* ſimplex. *Calyc.* cylindricus, calyculatus: ſquamis apice ſphacelatis.

* *Floribus floſculoſis.*

1. S. corollis nudis, foliis amplexicaulibus laceris, caule herbaceo erecto. *Sp. pl.* 1215. *hieracifolius.*
Hieracium-leav'd Groundſel.
Nat. of North America.
Cult. 1752, by Mr. Ph. Miller. *Mill. dict. edit.* 6. *n.* 2.
Fl. Auguſt. H. ⊙.

2. S. corollis nudis, calycibus ventricoſis ſubimbricatis, foliis filiformi-linearibus integerrimis glabris. *reclinatus.*
Senecio reclinatus. *Linn. ſuppl.* 369. *L'Herit. ſtirp. nov. p.* 9. *tab.* 5.
Senecio Chryſocoma. *Meerb. ic.* 39.
Senecio graminifolius. *Jacqu. ic. miſc.* 2. *p.* 322.
Graſs-leav'd Groundſel.
Nat. of the Cape of Good Hope.

Introd.

Introd. 1774, by Mr. Francis Maſſon.
Fl. June——Auguſt. G. H. ♂.

cernuus. 3. S. corollis nudis, foliis ellipticis dentato-ſerratis piloſiuſculis, pedunculis elongatis unifloris.
Senecio cernuus. *Linn. ſuppl.* 370.
Senecio rubens. *Jacqu. hort.* 3. *p.* 50. *t.* 98.
Drooping Groundſel.
Nat. of the Eaſt Indies.
Introd. 1780, by Monſ. Thouin.
Fl. July and Auguſt. S. ☉.

erubeſcens. 4. S. corollis nudis, foliis lyratis utrinque piloſis viſcoſis; ſuperioribus oblongo-lanceolatis dentatis, caulibus adſcendentibus.
Bluſh-colour'd Groundſel.
Nat. of the Cape of Good Hope. Mr. *Fr. Maſſon.*
Introd. 1774.
Fl. June——October. G. H. ☉.

purpureus. 5. S. corollis nudis, foliis lyratis hirtis; ſuperioribus lanceolatis dentatis. *Sp. pl.* 1215.
Purple Groundſel.
Nat. of the Cape of Good Hope.
Introd. 1774, by Mr. Francis Maſſon.
Fl. July——September. G. H. ♃.

japonicus. 6. S. corollis nudis, foliis pinnatifidis: laciniis lanceolatis acutis inciſis, ſtipulis foliaceis ſubpalmatis.
Senecio japonicus. *Thunb. japon.* 315.
Jagged-leav'd Groundſel.
Nat. of Japan.
Cult. 1774, by Mr. James Gordon.
Fl. Auguſt. H. ♃.

7. S.

7. S. corollis nudis, ſcapo ſubnudo longiſſimo. *Sp. pl.* 1216. *Pſeudo-China.*
Chineſe Groundſel.
Nat. of the Eaſt Indies.
Cult. 1732, by James Sherard, M. D. *Dill. elth.* 345. *t.* 258. *f.* 335.
Fl. S. ♃.

8. S. corollis nudis, foliis pinnato-ſinuatis amplexicaulibus, floribus ſparſis. *Sp. pl.* 1216. *Curtis lond.* *vulgaris.*
Common Groundſel.
Nat. of Britain.
Fl. April——October. H. ☉.

** *Floribus radiatis: radio revoluto.*

9. S. corollis revolutis, foliis ſeſſilibus ſinuatis, calycibus conicis: ſquamis minimis intactis. *Sp. pl.* 1216. *triflorus.*
Three-flower'd Groundſel.
Nat. of Egypt.
Introd. 1776, by Monſ. Thouin.
Fl. July——September. H. ☉.

10. S. corollis revolutis, foliis amplexicaulibus ſinuatis, ſquamis calycinis brevioribus integris ſphacelatis. *Sp. pl.* 1216. *ægyptius.*
Egyptian Groundſel.
Nat. of Egypt.
Introd. 1771, by John Earl of Bute.
Fl. July and Auguſt. H. ☉.

11. S. corollis revolutis, foliis pinnatifidis tomentoſis margine revolutis, panicula patula, ſquamis calycinis exterioribus patentibus. *cineraſcens.*
Gray Groundſel.
Nat. of the Cape of Good Hope. Mr. *Fr. Maſſon.*
Introd.

Introd. 1774.
Fl. May——July. G. H. ♄.

viscosus. 12. S. corollis revolutis, foliis pinnatifidis viscidis, squamis calycinis laxis longitudine perianthii. *Sp. pl.* 1217.
Stinking Groundsel.
Nat. of Britain.
Fl. June——August. H. ⊙.

sylvaticus. 13. S. corollis revolutis, foliis pinnatifidis denticulatis, caule corymboso erecto. *Sp. pl.* 1217.
Mountain Groundsel.
Nat. of Britain.
Fl. July. H. ⊙.

*** *Floribus radiatis: radio patente, foliis pinnatifidis.*

hastatus. 14. S. corollis radiantibus, petiolis amplexicaulibus, pedunculis folio triplo longioribus, foliis pinnato-sinuatis. *Sp. pl.* 1218.
Spleen-wort-leav'd Groundsel.
Nat. of the Cape of Good Hope.
Cult. 1732, by James Sherard, M.D. *Dill. elth.* 183. *t.* 152. *f.* 184.
Fl. Most part of the Summer. G. H. ♃.

venustus. 15. S. corollis radiantibus, caule, calyce foliisque glabris, foliis pinnatifidis: laciniis linearibus acutis dentatis.
Wing-leav'd Groundsel.
Nat. of the Cape of Good Hope. Mr. *Fr. Masson.*
Introd. 1774.
Fl. July——September. G. H. ♂.
Obs. Costa foliorum, præcipue inferiorum, sæpe aculeis mollibus obsita.

16. S.

16. S. corollis radiantibus, foliis piloſo-viſcidis pinnatifidis æqualibus patentiſſimis, rachi inferne anguſtata, calycibus hirtis. *elegans.*
Senecio elegans. *Sp. pl.* 1218.
Elegant Groundſel, or Purple Jacobea.
Nat. of the Cape of Good Hope.
Cult. 1700, by Charles Dubois, Eſq. *Pluk. mant.* 106.
Fl. June——Auguſt. H. ⊙.

17. S. corollis radiantibus, foliis pinnatifidis dentatis ſubhirtis, caule erecto. *Sp. pl.* 1218. *Curtis lond.* *erucifolius.*
Hoary perennial Groundſel.
Nat. of Britain.
Fl. July. H. ♃.

18. S. corollis radiantibus, foliis utrinque tomentoſis ſubpinnatis obtuſis, corymbo ſubrotundo. *Sp. pl.* 1219. *incanus.*
Downy Groundſel.
Nat. of the Alps of Switzerland and Auſtria.
Cult. 1759, by Mr. Philip Miller. *Mill. dict. edit.* 7. *n.* 12.
Fl. July and Auguſt. H. ♃.

19. S. corollis radiantibus, foliis pinnato-multifidis linearibus nudis acutis, pedunculis ſubbifloris. *Syſt. veget.* 757. *Jacqu. auſtr.* 1. *p.* 50. *t.* 79. *abrotanifolius.*
Southern-wood-leav'd Groundſel.
Nat. of the Alps of Auſtria.
Cult. 1759, by Mr. Ph. Miller. *Mill. dict. edit.* 7. *n.* 6.
Fl. July——October. H. ♃.

20. S. corollis radiantibus, foliis pinnato-lyratis: laciniis lacinulatis, caule erecto. *Sp. pl.* 1219. *Jacobæa.*

Rag-wort Groundſel.
Nat. of Britain.
Fl. July. H. ♃.

aureus. 21. S. corollis radiantibus, foliis crenatis: infimis cordatis petiolatis; ſuperioribus pinnatifidis lyratis. *Sp. pl.* 1220.
Golden Groundſel.
Nat. of Virginia and Canada.
Cult. 1759, by Mr. Ph. Miller. *Mill. dict. edit.* 7. *n.* 3.
Fl. May and June. H. ♃.

**** *Floribus radiatis: radio patente, foliis indiviſis.*

lanceus. 22. S. corollis radiantibus, foliis lanceolatis baſi cordatis amplexicaulibus lævibus argute ſerratis, caule fruteſcente.
Spear-leav'd Groundſel.
Nat. of the Cape of Good Hope. Mr. *Fr. Maſſon.*
Introd. 1774.
Fl. July——October. G. H. ♄.

paludoſus. 23. S. corollis radiantibus, foliis enſiformibus acute ſerratis ſubtus ſubvilloſis, caule ſtricto. *Sp. pl.* 1220.
Marſh Groundſel, or Bird's-tongue.
Nat. of England.
Fl. July and Auguſt. H. ♃.

nemorenſis. 24. S. corollis radiantibus octonis, foliis lanceolatis biſerratis ſubtus villoſis, caule ramoſo. *Sp. pl.* 1221. *Jacqu. auſtr.* 2. *p.* 50. *t.* 184.
Branching Groundſel.
Nat. of Auſtria and Switzerland.
Introd. 1775, by the Doctors Pitcairn and Fothergill.
Fl. July. H. ♃.

25. S.

25. S. corollis radiantibus, ſquamis calycinis adpreſſis, foliis ſubdecurrentibus ſubtus villoſiuſculis lanceolatis ſerratis. *coriaceus.*
Senecio orientalis. *Mill. dict.*
Doria, quæ Jacobæa orientalis Limonii folio Tourn. *Dill. elth.* 125. *t.* 105. *f.* 125.
Thick-leav'd Groundſel.
Nat. of the Levant.
Cult. 1732, by James Sherard, M. D. *Dill. elth. loc. cit.*
Fl. July and Auguſt. H. ♃.

26. S. corollis radiantibus, floribus corymboſis, foliis lanceolatis ſerratis glabriuſculis. *Sp. pl.* 1221. *Jacqu. auſtr.* 2. *p.* 52. *t.* 186. *ſarracenicus.*
Creeping Groundſel.
Nat. of England.
Fl. July——October. H. ♃.

27. S. corollis radiantibus, floribus corymboſis, foliis ſubdecurrentibus nudis lanceolatis denticulatis: ſuperioribus ſenſim minoribus. *Syſt. veget.* 758. *Jacqu. auſtr.* 2. *p.* 51. *t.* 185. *Doria.*
Broad-leav'd Groundſel.
Nat. of Auſtria.
Cult. 1570, by Mr. Hugh Morgan. *Lobel. adv.* 124.
Fl. July——September. H. ♃.

28. S. corollis radiantibus, caule indiviſo ſubunifloro, foliis indiviſis ſerratis: radicalibus ovatis ſubtus villoſis. *Sp. pl.* 1222. *Jacqu. auſtr.* 5. *p.* 53. *t. app.* 45. *Doronicum.*
Alpine Groundſel.
Nat. of the South of Europe.
Cult. 1705, by Dr. Uvedale. *Pluk. amalth.* 71.
Fl. July——September. H. ♃.

longifolius. 29. S. corollis radiantibus, foliis linearibus sparsis, caule fruticoso. *Sp. pl.* 1222.
Long-leav'd Groundsel.
Nat. of the Cape of Good Hope.
Introd. 1775, by Monf. Thouin.
Fl. August——November. G. H. ♄.

halimifolius. 30. S. corollis radiantibus, foliis obovatis carnosis subdentatis, caule fruticoso. *Sp. pl.* 1223.
Succulent-leav'd Groundsel.
Nat. of the Cape of Good Hope.
Cult. 1732, by James Sherard, M. D. *Dill. elth.* 124. *t.* 104. *f.* 124.
Fl. July. G. H. ♄.

ilicifolius. 31. S. corollis radiantibus, foliis sagittatis amplexicaulibus dentatis, caule herbaceo. *Syst. veget.* 759.
Ilex-leav'd Groundsel.
Nat. of the Cape of Good Hope.
Cult. 1731, by Mr. Philip Miller. *Mill. dict. edit.* 1. Jacobæa 3.
Fl. June and July. G. H. ♄.

asper. 32. S. corollis radiantibus, foliis lanceolato-linearibus dentatis rigidis, calycibus sublanuginosis.
Rough-leav'd Groundsel.
Nat. of the Cape of Good Hope.
Introd. 1774, by Mr. Francis Masson.
Fl. July and August. G. H. ♄.

rigidus. 33. S. corollis radiantibus, foliis amplexicaulibus spathulatis repandis erosis scabris, caule fruticoso. *Sp. pl.* 1224.
Hard-leav'd Groundsel.
Nat. of the Cape of Good Hope.

Cult.

Cult. 1759, by Mr. Philip Miller. *Mill. dict. edit.* 7. *n.* 13.
Fl. June——September. G. H. ♄.

BOLTONIA. *L'Herit. sert. angl.*

Recept. favosum, hemisphæricum. *Pappus* aristatus, subbicornis. *Cor.* radii plurimæ. *Cal.* imbricatus.

1. B. foliis integerrimis. *L'Herit. sert. angl. tab.* 36. *asteroides.*
Matricaria asteroides. *Linn. mant.* 116.
Star-wort-flower'd Boltonia.
Nat. of North America.
Cult. 1758, by Mr. Philip Miller.
Fl. August——October. H. ♃.

2. B. foliis inferioribus serratis. *L'Herit. sert. angl. tab.* 35. *glastifolia.*
Glaucous-leav'd Boltonia.
Nat. of North America.
Cult. 1758, by Mr. Philip Miller.
Fl. September. H. ♃.

ASTER. *Gen. pl.* 954.

Recept. nudum. *Pappus* simplex. *Cor.* radii plures 10. *Cal.* imbricati squamæ inferiores patulæ.

1. A. fruticosus, foliis linearibus punctatis, pedunculis unifloris nudis. *Sp. pl.* 1225. *fruticosus.*
Shrubby Star-wort.
Nat. of the Cape of Good Hope.
Cult. 1759, by Mr. Ph. Miller. *Mill. dict. ed.* 7. *n.* 32.
Fl. March——July. G. H. ♄.

2. A. fruticosus, foliis ovatis sinuatis hirtis, calycibus imbricatis hirtis. *Cymbalariæ.*

 Cymbalaria-

Cymbalaria-leav'd Star-wort.

Nat. of the Cape of Good Hope. Mr. *Fr. Masson.*

Introd. 1786.

Fl. Most part of the Summer. G. H. ♄.

alpinus. 3. A. foliis subspathulatis hirtis integerrimis, caulibus simplicibus unifloris.

Aster alpinus. *Sp. pl.* 1226. *Jacqu. austr.* 1. *p.* 55. *t.* 88.

Alpine Star-wort.

Nat. of the Alps of Europe.

Cult. 1759, by Mr. Philip Miller. *Mill. dict. edit.* 7. *n.* 1.

Fl. July and August. H. ♃.

tenellus. 4. A. foliis filiformibus aculeato-ciliatis, calycibus hemisphæricis: foliolis æqualibus.

Aster tenellus. *Sp. pl.* 1225. *Curtis magaz.* 33.

Dwarf Star-wort.

Nat. of the Cape of Good Hope.

Introd. 1774, by Mr. Francis Masson.

Fl. April——July. G. H. ♂.

nemoralis. 5. A. foliis lineari-lanceolatis basi attenuatis scabriusculis, ramis filiformibus unifloris, calycibus laxis imbricatis: foliolis acutis.

Wood Star-wort.

Nat. of Nova Scotia.

Introd. 1778, by Mr. William Malcolm.

Fl. August and September. H. ♃.

Obs. *Caulis* pedalis. *Radius* cæruleus; *Discus* albus.

linarifolius. 6. A. foliis linearibus integerrimis mucronatis scabris strictis: superioribus laxis remotis, calycibus imbricatis, ramis fastigiatis.

§ Aster

Aster linarifolius. *Sp. pl.* 1227.
Savory-leav'd Star-wort.
Nat. of North America.
Cult. 1712. *Philosoph. transact. n.* 333. *p.* 422. *n.* 79.
Fl. September and October. H. ♃.

7. A. foliis lanceolatis basi attenuatis integerrimis margine scabris, ramis corymbosis fastigiatis. *umbellatus.*
Aster umbellatus. *Mill. dict.*
Umbel'd Star-wort.
Nat. of Nova Scotia.
Cult. 1759, by Mr. Philip Miller. *Mill. dict. edit.* 7. *n.* 22.
Fl. July and August. H. ♃.
Obs. *Caulis* sexpedalis, strictus, glaber, superne tantummodo ramosus. *Radius* albus.

8. A. foliis oblongo-lanceolatis integris scabris, ramis corymbosis, calycibus imbricatis subsquarrosis: foliolis obtusis: interioribus membranaceis apice coloratis. *Amellus.*
Aster Amellus. *Sp. pl.* 1226. *Jacqu. austr.* 5. *p.* 12. *t.* 425.
α foliis obovato-lanceolatis planis, corollis radii numerosissimis subimbricatis.
Plain-leav'd Italian Star-wort.
β foliis lanceolatis rugosiusculis subundulatis, corollis radii divaricatis.
Wrinkl'd-leav'd Italian Star-wort.
Nat. of the South of Europe.
Cult. 1596, by Mr. John Gerard. *Hort. Ger.*
Fl. August and September. H. ♃.

9. A. foliis lineari-lanceolatis integerrimis carnosis glabris trinerviis, foliolis calycinis submembranaceis obtusis. *Tripolium.*

 Aster

After Tripolium. *Sp. pl.* 1226.
α flore radiato.
Radiated Sea Star-wort.
β flore discoideo.
Naked-flower'd Sea Star-wort.
Nat. of England; β. of the Isle of Wight.
Fl. August and September. H. ♃.

hyssopifolius. 10. A. foliis lineari-lanceolatis basi attenuatis integerrimis strictis, ramulis corymbosis fastigiatis: foliolis crebris linearibus imbricatis, calycibus imbricatis.
After hyssopifolius. *Linn. mant.* 114.
Hyssop-leav'd Star-wort.
Nat. of North America.
Cult. 1760, by Mr. Philip Miller.
Fl. September and October. H. ♃.

linifolius. 11. A. foliis linearibus integerrimis scabriusculis, ramis corymbosis fastigiatis foliolosis, calycibus imbricatis, radiis disco subæqualibus.
After linifolius. *Sp. pl.* 1228.
Flax-leav'd Star-wort.
Nat. of North America.
Cult. 1739, by Mr. Philip Miller. *Mill. dict. vol.* 2. *n.* 14.
Fl. July and August. H. ♃.

grandiflorus. 12. A. foliis linearibus integerrimis subamplexicaulibus hispidis ciliatis: ramorum calycisque reflexis.
After grandiflorus. *Sp. pl.* 1231.
Great-flower'd Star-wort.
Nat. of North America.
Cult. about 1720, by Mr. Thomas Fairchild. *Mill. ic.* 188. *t.* 282.
Fl. October. H. ♃.

13. A.

13. A. foliis linearibus amplexicaulibus erectis integerrimis glaberrimis margine ſcabris, pedunculis fere nudis, calycibus ſquarroſis. *paludoſus.*

Marſh Star-wort.

Nat. of the Swamps of Carolina.

Introd. 1784, by Mr. John Fairbairne.

Fl. September and October. H. ♃.

Obs. *Folia* tri-vel quadriuncialia, lineas duas vel tres lata, remota. *Pedunculi* nudi præter foliola, plerumque duo, ſub calyce. *Radius* cæruleus, magnus. *Diſcus* luteus.

14. A. foliis lanceolatis integerrimis cordatis amplexicaulibus piloſis, calycibus diſcum ſuperantibus laxis: foliolis lineari-lanceolatis ſubæqualibus, caule hiſpido. *novæ angliæ.*

Aſter novæ angliæ. *Sp. pl.* 1229.

α ramis breviſſimis ſubconfertis.

New England cluſter'd Star-wort.

β ramis numeroſis paniculatis.

New England panicl'd Star-wort.

Nat. of North America.

Cult. 1731, by Mr. Philip Miller. *Mill. dict. edit.* 1. *n.* 3.

Fl. September and October. H. ♃.

15. A. foliis oblongis integris acutis cordatis ſubamplexicaulibus ſcabris, ramis patulis elongatis paucifloris, calycibus imbricatis ſubſquarroſis, caule hirto. *patens.*

Spreading hairy-ſtalk'd Star-wort.

Nat. of Virginia.

Introd. about 1773, by George Aufrere, Eſq.

Fl. September and October. H. ♃.

Descr. *Caulis* tripedalis. *Rami* remoti, divaricati, pubeſcentes. *Folia* baſi oblique flexa: ſuprema minuta,

minuta, ſubimbricata. *Floris* radius dilute cæruleus; diſcus fulvus.

dumoſus. 16. A. foliis linearibus integerrimis glabris: ramulorum breviſſimis, ramis paniculatis, calycibus cylindraceis arcte imbricatis.

Aſter dumoſus. *Sp. pl.* 1227.

α radio pallide violaceo, caule pubeſcente.

Purple-flower'd buſhy Star-wort.

β radio albo, caule glabro.

White-flower'd buſhy Star-wort.

Nat. of North America.

Cult. 1725, in Chelſea Garden. *R. S. n.* 153.

Fl. September and October. H. ♃.

folioloſus. 17. A. foliis lanceolato-linearibus integerrimis glabris: ramulorum patentiſſimis, calycibus imbricatis: foliolis acutis, caule pubeſcente.

Aſter ericoides, meliloti agrariæ umbone. *Dill. elth.* 39. *t.* 35. *f.* 39.

Leafy Star-wort.

Nat. of North America.

Cult. 1732, by James Sherard, M. D. *Dill. elth. loc. cit.*

Fl. October. H. ♃.

ericoides. 18. A. foliis linearibus integerrimis glaberrimis: ramulorum ſubulatis approximatis; caulinis elongatis, calycibus ſubſquarroſis: foliolis acutis, caule glabro.

Aſter ericoides. *Sp. pl.* 1227. (excluſo ſynonymo Dillenii.)

Heath-leav'd Star-wort.

Nat. of North America.

Cult.

Cult. 1758, by Mr. Philip Miller.
Fl. September. H. ♃.

19. A. foliis linearibus integerrimis glabriuſculis, ramulis ſecundis, calycibus imbricatis ſquarroſis: foliolis ſubfoliaceis acutis, caule pubeſcente. *multiflorus.*

Aſter ericoides dumoſus. *Dill. elth.* 40. *t.* 36. *f.* 40.

α ramulis ſubmultifloris.
Early flowering ſmall-leav'd Star-wort,

β ramulis unifloris.
Late flowering ſmall-leav'd Star-wort.

Nat. of North America.
Cult. 1732, by James Sherard, M.D. *Dill. elth. loc. cit.*
Fl. September and October. H. ♃.

20. A. foliis lineari-lanceolatis integerrimis glabris, calycibus imbricatis laxis, caule lævi. *ſalicifolius.*

Willow-leav'd Star-wort.
Nat. of North America.
Cult. 1760, by Mr. Philip Miller.
Fl. September and October. H. ♃.
Obs. *Caulis* humanæ altitudinis. *Calycis* foliola acuta, apice patula. *Corollæ* radius e cæruleo incarnatus.

21. A. foliis lanceolatis ſubamplexicaulibus integerrimis glabris margine ſcabris, calycibus laxis: foliolis æqualibus. *æſtivus.*

Labrador Star-wort.
Nat. of North America.
Introd. 1776, by Meſſrs. Gordon and Græffer.
Fl. July and Auguſt. H. ♃.
Obs. *Caulis* bipedalis, hiſpidus. *Radius* cæruleus.

22. A.

tradeſ-canti. 22. A. foliis lanceolatis ſerratis ſeſſilibus glabris, ramis virgatis, calycibus arcte imbricatis, caule tereti glabro.

Aſter tradeſcanti. *Sp. pl.* 1230.

α floribus cæruleis.

Tradeſcant's dwarf Star-wort.

β floribus albis.

Tradeſcant's tall Star-wort.

Nat. of North America.

Cult. 1656, by Mr. John Tradeſcant, Jun. *Mill. dict. edit.* 8.

Fl. July——September. H. ♃.

junceus. 23. A. foliis lanceolato-linearibus ſeſſilibus glabris: infimis ſubſerratis; ramulorum lanceolatis, ramis virgatis, calycibus imbricatis, caule glabriuſculo.

Slender-ſtalk'd Star-wort.

Nat. of North America.

Cult. 1758, by Mr. Philip Miller.

Fl. October. H. ♃.

Obs. *Caulis* quadripedalis. *Calycis* foliola acuta, apice ſubpatula. *Radius* leviter incarnatus; diſcus elevatus, pallide flavus.

pendulus. 24. A. foliis elliptico-lanceolatis ſerratis glabris: ramulorum remotiuſculis, ramis divaricatiſſimis pendulis, caule pubeſcente.

Pendulous Star-wort.

Nat. of North America.

Cult. 1758, by Mr. Philip Miller.

Fl. October. H. ♃.

Obs. *Radius* albus; *diſcus* luteus, tandem ferrugineus.

25. A.

25. A. foliis elliptico-lanceolatis ſerratis glabris proportionatis, ramis patentibus, calycibus imbricatis, caule pubeſcente. *diffuſus.*

α caule tomentoſo, ramis patentiſſimis, diſco rubro.
Diffuſe red-flower'd Star-wort.

β caule pubeſcente, ramis virgatis, diſco ſtramineo.
Diffuſe white-flower'd Star-wort.

Nat. of North America.

Introd. 1777, by Meſſrs. Kennedy and Lee.

Fl. September. H. ♃.

Obs. *Radius* albus.

26. A. foliis elliptico-lanceolatis ſerratis glabris: caulinis lineari-lanceolatis elongatis, ramis patentibus, calycibus imbricatis, caule pubeſcente. *divergens.*

Spreading downy-ſtalk'd Star-wort.

Nat. of North America.

Cult. 1758, by Mr. Philip Miller.

Fl. October. H. ♃.

Descr. *Caulis* quinquepedalis et ultra, debilis. *Calyx* cylindraceus: foliola numeroſa, acuta. *Radius* albus, calyce brevior. *Diſcus* rubens.

27. A. foliis ſeſſilibus lanceolatis ſubſerratis glabris, calycibus imbricatis: foliolis acutis, diſco radiis æquali, caule villoſiuculo. *miſer.*

Aſter miſer. *Sp. pl.* 1232. (excluſo ſynonymo Dillenii.)

Small white-flower'd Star-wort.

Nat. of North America.

Introd. 1776, by Monſ. Thouin.

Fl. September and October. H. ♃.

28. A. foliis ſubamplexicaulibus lanceolatis ſerratis lævibus inferne attenuatis, ramulis virgatis, calycibus ſubfoliaceis laxis, caule glabro. *mutabilis.*

Aſter

Aster mutabilis. *Sp. pl.* 1230.
Variable Star-wort.
Cult. 1731, by Mr. Ph. Miller. *Mill. dict. edit.* 1. *n.* 4.
Fl. October. H. ♃.

novi belgii. 29. A. foliis subamplexicaulibus lanceolatis glabris margine scabris: inferioribus serratis, ramis subdivisis, calycibus laxe imbricatis: foliolis lineari-lanceolatis, caule tereti glabro.
Aster novi belgii. *Sp. pl.* 1231.
α foliis viridibus.
New Holland green Star-wort.
β foliis glaucis.
New Holland glaucous Star-wort.
Nat. of North America.
Cult. 1759, by Mr. Ph. Miller. *Mill. dict. edit.* 7. *n.* 9.
Fl. September and October. H. ♃.

lævis. 30. A. foliis amplexicaulibus oblongis integerrimis lucidis: radicalibus subserratis, ramis simplicibus subunifloris, calycibus imbricatis: foliolis subcuneiformibus acutis apice incrassatis, caule glabro.
Aster lævis. *Sp. pl.* 1230.
Smooth Star-wort.
Nat. of North America.
Cult. 1758, by Mr. Philip Miller.
Fl. September and October. H. ♃.

undulatus. 31. A. foliis serratis pilosis undulatis: inferioribus cordatis: petiolis alatis basi dilatatis, ramulis virgatis, calycibus imbricatis, caule hispido.
Aster undulatus. *Sp. pl.* 1228.
Wave-leav'd Star-wort.
Nat. of North America.

Cult.

Cult. 1699, by Mr. Jacob Bobart. *Moriſ. hiſt.* 3. *p.* 120. *n.* 26.
Fl. October. H. ♃.

32. A. foliis glabris acutis ſerratis petiolatis: radicalibus cordatis; caulinis ovatis: ſupremis lanceolatis integris, ramis paniculatis, calycibus laxis ſubimbricatis, caule glabro. *paniculatus.*
Smooth-ſtalk'd panicl'd Star-wort.
Nat. of North America.
Cult. 1774, by Mr. James Gordon.
Fl. October. H. ♃.

33. A. foliis cordatis glabris acuminatis omnibus argute ſerratis petiolatis: petiolis ſimplicibus, ramis faſtigiatis, caule glabro. *corymboſus.*
α caule purpureo.
Heart-leav'd purple-ſtalk'd Star-wort.
β caule viridi.
Heart leav'd green-ſtalk'd Star-wort.
Nat. of North America.
Cult. 1765, by Peter Collinſon, Eſq.
Fl. September. H. ♃.

34. A. foliis cordatis acutis argute ſerratis ſubtus piloſis petiolatis: petiolis ſubſimplicibus, ramis paniculatis, caule hirto. *cordifolius.*
Aſter cordifolius. *Sp. pl.* 1229.
Heart-leav'd Star-wort.
Nat. of North America.
Cult. 1759, by Mr. Ph. Miller. *Mill. dict. edit.* 7. *n.* 13.
Fl. July and Auguſt. H. ♃.

35. A. foliis ſerratis oblongis: ſupremis ovatis ſeſſilibus; caulinis cordatis petiolatis: petiolis ſuperioribus alatis. *macrophyllus.*

Aſter

Aster macrophyllus. *Sp. pl.* 1232. *L'Herit. stirp. nov. tom.* 2. *tab.* 66.

α foliis radicalibus oblongis cordatis.
Broad-leav'd white Star-wort.

β foliis radicalibus ovatis cordatis.
Broad-leav'd blue Star-wort.

Nat. of North America.
Cult. 1739, by Mr. Philip Miller. *Mill. dict. vol.* 2. *n.* 23.
Fl. July——September. H. ♃.

chinensis. 36. A. foliis ovato-oblongis integris sinuatisve ciliatis, calycibus foliaceis patentibus ciliatis.
Aster chinensis. *Sp. pl.* 1232.
Chinese Star-wort, or Aster.
Nat. of China.
Introd. 1731, by Mr. Philip Miller. *Mill. dict. edit.* 8.
Fl. July and August. H. ☉.

sibiricus. 37. A. foliis lanceolatis subamplexicaulibus serratis piloso-scabris, calycibus laxis: foliolis lanceolatis acuminatis foliaceis hispidis.
Aster sibiricus. *Sp. pl.* 1226.

α foliis basi latioribus, floribus numerosis.
Siberian broad-leav'd Star-wort.

β foliis basi angustioribus, floribus paucis.
Siberian narrow-leav'd Star-wort.

Nat. of Siberia.
Cult. 1768, by Mr. Philip Miller. *Mill. dict. edit.* 8.
Fl. July——October. H. ♃.

puniceus. 38. A. foliis amplexicaulibus lanceolatis serratis scabriusculis, ramis paniculatis, calycibus laxis discum superantibus: foliolis lineari-lanceolatis subæqualibus, caule hispido.
Aster puniceus. *Sp. pl.* 1229.

α sep-

α ſeptempedalis, caulibus ſaturate purpureis.
Tall purple-ſtalk'd Star-wort.
β octopedalis, caulibus e viridi-rufeſcentibus.
Aſter altiſſimus. *Mill. dict.*
Giant Star-wort.
γ tripedalis.
Dwarf purple-ſtalk'd Star-wort.
Nat. of North America.
Cult. 1739, in Chelſea Garden. *Rand. chel. n.* 6.
Fl. July——October. H. ♃.

39. A. foliis lanceolatis ſcabriuſculis: inferioribus ſerratis, ramis corymboſis, foliolis calycinis laxis foliaceis ſubcuneiformibus acutiuſculis ſquarroſis. *ſpectabilis.*
Shewy Star-wort.
Nat. of North America.
Introd. 1777, by William Pitcairn, M. D.
Fl. Auguſt and September. H. ♃.
Obs. *Caulis* bipedalis. *Radius* cæruleus.

40. A. foliis ſeſſilibus lanceolatis baſi attenuatis ſerratis glabris, calycibus laxis: foliolis lanceolato-linearibus ſubæqualibus glabris. *tardiflorus.*
Aſter tardiflorus. *Sp. pl.* 1231.
Spear-leav'd Star-wort.
Nat. of North America.
Introd. 1775, by Mr. John Cree.
Fl. July——September. H. ♃.

41. A. foliis piloſiuſculis: inferioribus ſubovatis ſerratis; ſuperioribus lanceolatis, calycibus hemiſphæricis: foliolis ſubæqualibus ſtrigoſis. *annuus.*
Aſter annuus. *Sp. pl.* 1229.
American annual Star-wort.
Nat. of North America.

Cult. 1640. *Park. theat.* 528. *f.* 4.
Fl. July——September. H. ☉.

Radula. 42. A. foliis lanceolatis ſerratis acuminatis rugoſis ſcaberrimis, calycibus imbricatis: foliolis lanceolatis obtuſis.
Rough Star-wort.
Nat. of Nova Scotia.
Introd. 1785, by William Pitcairn, M. D.
Fl. September. H. ♃.

SOLIDAGO. *Gen. pl.* 955.

Recept. nudum. *Pappus* ſimplex. *Cor.* radii circiter 5
Cal. ſquamæ imbricatæ, clauſæ.

* *Racemis ſecundis.*

canadenſis. 1. S. caule villoſo erecto, foliis lanceolatis ſerratis triplinerviis ſcabris, racemis paniculatis ſecundis recurvis, ligulis abbreviatis.
α foliis nudiuſculis: ſerraturis pauciſſimis, racemis terminalibus elongatis.
Solidago canadenſis. *Sp. pl.* 1233.
Common Canadian Golden-rod.
β foliis ſcabris: ſerraturis paucis, racemis ſubæqualibus patentiſſimis.
Solidago humilis. *Mill. dict.*
Dwarf Canadian Golden-rod.
γ foliis ſubtus villoſis: ſuperioribus integerrimis, racemis elongatis patentibus.
Intire-leav'd Canadian Golden-rod.
Nat. of North America.
Cult. 1656, by Mr. John Tradeſcant, Jun. *Muſ. Trad.* 176.
Fl. July——September. H. ♃.

2. S.

2. S. caule villoso erecto, foliis lanceolatis serratis triplinerviis scabris subtus villosis, racemis spiciformibus erectis; innuptis nutantibus, ligulis abbreviatis. *procera.*

Great Golden-rod.

Nat. of North America.

Cult. 1758, by Mr. Philip Miller.

Fl. September and October. H. ♃.

3. S. caule erecto tereti lævi, foliis lineari-lanceolatis glabris margine asperis serratis triplinerviis, racemis paniculatis secundis. *serotina.*

Upright smooth Golden-rod.

Nat. of North America.

Cult. 1758, by Mr. Philip Miller.

Fl. July and August. H. ♃.

4. S. caule erecto glabro, foliis lanceolatis glabris serratis margine scabris, racemis paniculatis secundis, pedunculis hirtis, ligulis abbreviatis. *gigantea.*

Gigantic Golden-rod.

Nat. of North America.

Cult. 1758, by Mr. Philip Miller.

Fl. August and September. H. ♃.

5. S. caule erecto villoso, foliis lanceolatis subserratis triplinerviis scabris reflexis, racemis paniculatis subsecundis. *reflexa.*

Reflex'd Golden-rod.

Nat. of North America.

Cult. 1758, by Mr. Philip Miller.

Fl. August and September. H. ♃.

6. S. caule erecto pilosiusculo, foliis lanceolatis subtriplinerviis glabris margine scabris: inferioribus subserratis, racemis paniculatis subrecurvis secundis. *lateriflora.*

Solidago lateriflora. *Sp. pl.* 1234.

α caulibus rubentibus vix pilosis.

Lateral-flower'd red-stalk'd Golden-rod.

β caulibus viridibus pilosis.

Lateral-flower'd green-stalk'd Golden-rod.

Nat. of North America.

Cult. 1758, by Mr. Philip Miller.

Fl. August and September. H. ♃.

aspera. 7. S. caule erecto tereti piloso, foliis ovatis subellipticis scaberrimis rugosis serratis enerviis, racemis paniculatis secundis.

Virga aurea americana aspera foliis brevioribus serratis. *Dill. elth.* 411. *tab.* 305. *fig.* 392.

Rough-leav'd Golden-rod.

Nat. of North America.

Cult. 1732, by James Sherard, M.D. *Dill. elth. loc. cit.*

Fl. September. H. ♃.

altissima. 8. S. caule erecto hirto, foliis lanceolatis scaberrimis rugosis serratis enerviis, paniculis secundis.

Solidago altissima. *Sp. pl.* 1233.

Virga aurea novæ angliæ, rugosis foliis crenatis. *Dill. elth.* 416. *tab.* 308. *fig.* 396.

α caule quinquepedali piloso, serraturis profundis inæqualibus, racemis divaricatis.

Solidago altissima. *Mill. dict.*

Tall Golden-rod.

β caule tripedali hirsuto, serraturis profundis subæqualibus, racemis adscendentibus.

Solidago pilosa. *Mill. dict.*

Hairy Golden-rod.

γ caule tripedali villoso, serraturis profundis subæqualibus.

Solidago

Solidago recurvata. *Mill. dict.*
Recurv'd Golden-rod.

δ caule quinquepedali villosissimo, serraturis magnis, racemis vix divergentibus.
Solidago virginiana. *Mill. dict.*
Virginian Golden-rod.

ε caule tripedali villosiusculo, serraturis parvis subæqualibus, racemis divaricatis.
Solidago rugosa. *Mill. dict.*
Wrinkled-leav'd Golden-rod.
Nat. of North America.
Cult. 1732, by James Sherard, M. D. *Dill. elth. loc. cit.*
Fl. August and September. H. ♃.

9. S. caule erecto tomentoso, foliis caulinis lanceolatis hispidis integerrimis; radicalibus subcuneiformibus serratis, racemis paniculatis secundis. *nemoralis.*
Woolly-stalk'd Golden-rod.
Nat. of North America.
Introd. 1769, by Samuel Martin, M. D.
Fl. September. H. ♃.

10. S. caule erecto glabro, foliis glabris argute inæqualiter serratis: caulinis ellipticis; radicalibus ovato-oblongis, racemis paniculatis secundis, ligulis elongatis. *arguta.*
Sharp-notch'd Golden-rod.
Nat. of North America.
Cult. 1758, by Mr. Philip Miller.
Fl. July and August. H. ♃.

11. S. caule erecto glabro, foliis lanceolatis glabris margine scabris: inferioribus serratis, racemis paniculatis secundis. *juncea.*

 Rush-

Rush-stalk'd Golden-rod.
Nat. of North America.
Introd. 1769, by Samuel Martin, M. D.
Fl. August and September. H. ♃.

elliptica. 12. S. caule erecto glabro, foliis ellipticis lævibus serratis, racemis paniculatis secundis, ligulis mediocribus.
Solidago latissimifolia. *Mill. dict.*
Oval-leav'd Golden-rod.
Nat. of North America.
Cult. 1759, by Mr. Ph. Miller. *Mill. dict. edit.* 7. *n.* 16.
Fl. August. H. ♃.

sempervirens. 13. S. caule erecto glabro, foliis lineari-lanceolatis subcarnosis lævibus integerrimis margine scabris, racemis paniculatis secundis, pedunculis pilosis.
Solidago sempervirens. *Sp. pl.* 1232.
Narrow-leav'd evergreen Golden-rod.
Nat. of North America.
Cult. 1699, by the Dutchess of Beaufort. *Br. Mus. Sloan. mss.* 525 & 3349.
Fl. September and October. H. ♃.

odora. 14. S. caule erecto pubescente, foliis lineari-lanceolatis integerrimis glabris margine scabris, racemis paniculatis secundis.
Virga aurea americana Tarraconis facie et sapore, panicula speciosissima. *Pluk. alm.* 389. *t.* 116. *f.* 6.
Sweet-scented Golden-rod.
Nat. of North America.
Cult. 1763, by Mr. James Gordon.
Fl. July and August. H. ♃.

** *Racemis erectis.*

lanceolata. 15. S. caule glabro ramosissimo, foliis lanceolato-linearibus

bus integerrimis trinerviis glabris, corymbis terminalibus, ligulis altitudine disci.
Solidago lanceolata. *Linn. mant.* 114.
Chrysocoma graminifolia. *Sp. pl.* 1178.
Grass-leav'd Golden-rod.
Nat. of North America.
Cult. 1758, by Mr. Philip Miller.
Fl. October. H. ♃.

16. S. caule erecto lævi, foliis lanceolatis carnosis integerrimis undique lævibus, racemis paniculatis erectis, pedunculis squamosis villosis, ligulis elongatis. *lævigata.*
Fleshy-leav'd Golden-rod.
Nat. of North America.
Cult. 1699, in Chelsea Garden. *Moris. hist.* 3. *p.* 124. *n.* 15.
Fl. October and November. H. ♃.

17. S. caule obliquo glabro, foliis lanceolatis subcarnosis integerrimis undique lævibus, racemis paniculatis erectis, pedunculis squamosis glabris, ligulis elongatis. *mexicana.*
Solidago mexicana. *Sp. pl.* 1234.
Mexican Golden-rod.
Nat. of North America.
Cult. 1697, by the Dutchess of Beaufort. *Br. Mus. Sloan. mss.* 3357. *fol.* 71.
Fl. July and August. H. ♃.

18. S. caule erecto subpubescente, foliis lineari-lanceolatis membranaceis basi attenuatis glabris margine scabris: infimis subserratis, racemis erectis, ligulis elongatis. *viminea.*
Solidago integerrima. *Mill. dict.*
Twiggy Golden-rod.

Nat. of North America.
Cult. 1759, by Mr. Philip Miller. *Mill. dict. edit.* 7. *n.* 26.
Fl. September. H. ♃.

stricta. 19. S. caule erecto glabro, foliis caulinis lanceolatis integerrimis glabris margine scabris; radicalibus serratis, racemis paniculatis erectis, pedunculis glabris.
Willow-leav'd Golden-rod.
Nat. of North America.
Cult. 1758, by Mr. Philip Miller.
Fl. September. H. ♃.

petiolaris. 20. S. caule erecto villoso, foliis ellipticis scabriusculis petiolatis, racemis erectis, ligulis elongatis.
Late flowering Golden-rod.
Nat. of North America.
Cult. 1758, by Mr. Philip Miller.
Fl. October——December. H. ♃.

bicolor. 21. S. caule foliisque ellipticis pilosis: inferioribus serratis, ramis foliolosis, racemis erectis, foliolis calycinis obtusis.
Solidago bicolor. *Linn. mant.* 114.
Solidago alba. *Mill. dict.*
Two-colour'd Golden-rod.
Nat. of North America.
Cult. 1759, by Mr. Philip Miller. *Mill. dict. edit.* 7. *n.* 28.
Fl. September. H. ♃.

rigida. 22. S. caule foliisque ovato-oblongis pilosis scabris: caulinis integerrimis; infimis serratis, ramis floriferis paniculatis, racemis compactis erectis, ligulis elongatis.

Solidago

Solidago rigida. *Sp. pl.* 1235.
Hard-leav'd Golden-rod.
Nat. of North America.
Cult. 1759, by Mr. Philip Miller. *Mill dict. edit.* 7. *n.* 19.
Fl. September. H. ♃.

23. S. caule lævi recto, foliis lanceolatis ſerratis glabris, racemis erectis, ligulis mediocribus. *cæſia.*
Solidago cæſia. *Sp. pl.* 1234.
Maryland Golden-rod.
Nat. of North America.
Cult. 1732, by James Sherard, M. D. *Dill. elth.* 414. *t.* 307. *f.* 395.
Fl. September. H. ♃.

24. S. caule flexuoſo glabro angulato, foliis ovatis acuminatis ſerratis glabris, racemis erectis, ligulis mediocribus. *flexicaulis.*
Solidago flexicaulis. *Sp. pl.* 1234.
Solidago latifolia. *Sp. pl.* 1234.
Crooked-ſtalk'd Golden-rod.
Nat. of North America.
Cult. 1731, by Mr. Philip Miller. *Mill. dict. edit.* 1. Virga aurea 11.
Fl. September. H. ♃.

25. S. caule ſubflexuoſo glabro angulato ramoſo, foliis oblongo-lanceolatis denſe ſerratis ſubtus piloſiuſculis, racemis erectis, ligulis elongatis. *ambigua.*
Angular-ſtalk'd Golden-rod.
Nat.
Cult. 1759, by Mr. Philip Miller.
Fl. July and Auguſt. H. ♃.

26. S.

Virga aurea. 26. S. caule erecto tereti pubescente superne ramoso, foliis inferioribus ellipticis pilosiusculis serratis, racemis erectis, ligulis elongatis.
Solidago Virgaurea. *Sp. pl.* 1235.
Common Golden-rod.
Nat. of Britain.
Fl. July and August. H. ♃.

cambrica. 27. S. caule simplicissimo pubescente, foliis cuneiformi-lanceolatis serratis pilosiusculis, racemis erectis, ligulis elongatis.
Solidago cambrica. *Hudf. angl.* 367.
Solidago minuta. *Mill. dict.*
Virga aurea cambrica, floribus conglobatis. *Dill. elth.* 413. *tab.* 306. *fig.* 393.
Welch Golden-rod.
Nat. of Wales.
Fl. July. H. ♃.

multiradiata. 28. S. caule villosiusculo, foliis sessilibus lanceolatis glabris ciliatis: inferioribus apice serratis, racemo terminali erecto, ligulis elongatis numerosis.
Labrador Golden-rod.
Nat. of Labrador.
Introd. 1776, by the Hudson's-bay Company.
Fl. July. H. ♃.

minuta. 29. S. caule simplicissimo piloso, foliis lanceolatis acutis serratis glabris, racemo terminali simplici erecto, ligulis elongatis.
Solidago minuta. *Sp. pl.* 1235.
Least Golden-rod.
Nat. of the Pyrenees.
Introd. 1772, by Monf. Richard.
Fl. July. H. ♃.

CINERA-

CINERARIA. *Gen. pl.* 957.

Recept. nudum. *Pappus* ſimplex. *Cal.* ſimplex, polyphyllus, æqualis.

1. C. pedunculis unifloris, foliis oppoſitis ovatis nudis, caule ſuffruticoſo. *Syſt. veget.* 765. *Amelloides.*

Blue-flower'd Cineraria, or Cape-aſter.

Nat. of the Cape of Good Hope.

Introd. 1753, by Mr. Ph. Miller. *Mill. ic.* 51. *t.* 76. *f.* 2.

Fl. February——September. G. H. ♄.

2. C. pedunculis unifloris, foliis reniformibus ſubangulatis, petiolis apice auriculatis nudiſve. *L'Herit. ſert. angl. n.* 1. *humifuſa.*

Aſter flore luteo, folio cymbalariæ. *Raj hiſt.* 3. *p.* 158.

Aſter africanus minimus monanthes luteus, foliolis anguloſis minimis, aceris forma vel cymbalariæ. *Raj. hiſt.* 3. *p.* 161.

Trailing Cineraria.

Nat. of the Cape of Good Hope.

Introd. 1774, by Mr. Francis Maſſon.

Fl. July and Auguſt. G. H. ♃.

3. C. pedunculis unifloris, foliis pinnatifido-lobatis acutis viſcidis carnuloſis. *L'Herit. ſert. angl. n.* 2. *viſcoſa.*

Clammy Cin raria.

Nat. of the Cape of Good Hope. Mr. *Fr. Maſſon.*

Introd. 1774.

Fl. June——Auguſt. G. H. ♂.

4. C. pedunculis unifloris, foliis cordato-ſubrotundis ſeptangulis ſubtus lanuginoſis. *L'Herit. ſert. angl. n.* 5. *tab.* 30. *lanata.*

Cineraria lanata. *Curtis magaz.* 53.

Woolly Cineraria.

Nat. of the Canary Iſlands. Mr. *Francis Maſſon.*

Introd.

Introd. 1780.
Fl. May——September. G. H. ♄.

geifolia. 5. C. pedunculis ramosis, foliis reniformibus angulatis sublobatis pubescentibus, petiolis superne auritis.
Cineraria geifolia. *Sp. pl.* 1242.
Kidney-leav'd Cineraria.
Nat. of the Cape of Good Hope.
Cult. 1759, by Mr. Philip Miller. *Mill. dict. edit.* 7. Othonna 3.
Fl. April——August. G. H. ♄.

populifolia. 6. C. floribus corymbosis, foliis cordatis subangulatis subtus tomentosis, petiolis apice multijugo-appendiculatis. *L'Herit. sert. angl. n.* 6.
Cacalia appendiculata. *Linn. suppl.* 352.
Poplar-leav'd Cineraria.
Nat. of the Canary Islands. Mr. *Francis Masson.*
Introd. 1780.
Fl. G. H. ♄.

aurita. 7. C. floribus corymbosis, foliis cordatis subangulatis subtus tomentosis, petiolis basi biauritis. *L'Herit. sert. angl. n.* 7. *tab.* 31.
Purple-flower'd Cineraria.
Nat. of Madeira. Mr. *Francis Masson.*
Introd. 1777.
Fl. June and July. G. H. ♄.

malvæfolia. 8. C. floribus cymosis, foliis cordatis angulatis infra subtomentosis, petiolis simplicibus. *L'Herit. sert. angl. n.* 9. *tab.* 32.
Mallow-leav'd Cineraria.
Nat. of the Azores. Mr. *Francis Masson.*
Introd. 1777.
Fl. August. G. H. ♃.

9. C.

9. C. floribus cymosis, foliis cordatis angulatis subtus purpurascentibus, petiolis basi auritis. *L'Herit. sert. angl. n.* 11. *tab.* 33. *cruenta.*

Purple-leav'd Cineraria.

Nat. of the Canary Islands.

Introd. 1777, by Mr. Francis Masson.

Fl. February and March. G. H. ♃.

10. C. floribus subcorymbosis, foliis subrotundis multilobatis glabris, petiolis basi auritis, calycibus subcalyculatis. *L'Herit. sert. angl. n.* 13. *tab.* 34. *lobata.*

Lobed Cineraria.

Nat. of the Cape of Good Hope. Mr. *Fr. Masson.*

Introd. 1774.

Fl. June——August. G. H. ♄.

11. C. racemo simplici, foliis cordatis obtusis denticulatis lævibus, caule simplicissimo monophyllo. *Sp. pl.* 1242. *sibirica.*

Siberian Cineraria.

Nat. of Siberia.

Introd. 1784, by Mr. John Bell.

Fl. H. ♃.

12. C. panicula pauciflora, foliis caulinis petiolatis cordatis acute serratis glabris, caule angulato. *Linn. suppl.* 375. *cordifolia.*

Cineraria cordifolia. *Jacqu. austr.* 2. *p.* 47. *t.* 176.

Senecio alpinus. *Linn. suppl.* 371. (exclusis synonymis.)

Senecio foliis cordato-lanceolatis serratis. *Hall. hist.* 63.

Heart-leav'd Cineraria.

Nat. of Austria and Switzerland.

Introd. 1775, by the Doctors Pitcairn and Fothergill.

Fl. July and August. H. ♃.

13. C.

alpina. 13. C. umbella involucrata, pedunculo communi nudiuſculo, foliis oblongis villoſis. *Sp. pl.* 1243. *Relhan cantabr.* 320. *cum fig.*
Cineraria integrifolia pratenſis. *Jacqu. auſtr.* 2. *p.* 48. *t.* 180.
Mountain Cineraria.
Nat. of England.
Fl. June. H. ♃.

paluſtris. 14. C. floribus corymboſis, foliis lato-lanceolatis dentato-ſinuatis, caule villoſo. *Sp. pl.* 1243.
Marſh Cineraria.
Nat. of England.
Fl. June and July. H. ♃.

maritima. 15. C. floribus paniculatis, foliis pinnatifidis tomentoſis: laciniis ſinuatis, caule fruteſcente. *Sp. pl.* 1244.
Sea Cineraria, or Rag-wort.
Nat. of Italy, and the South of France.
Cult. 1633, by Mr. Ralph Tuggy. *Ger. emac.* 280. *f.* 4.
Fl. July——September. H. ♄.

I N U L A. *Gen. pl.* 956.

Recept. nudum. *Pappus* ſimplex. *Antheræ* baſi in ſetas duas deſinentes.

Helenium. 1. I. foliis amplexicaulibus ovatis rugoſis ſubtus tomentoſis, calycum ſquamis ovatis. *Sp. pl.* 1236.
Common Inula, or Elecampane.
Nat. of Britain.
Fl. July and Auguſt. H. ♃.

odora. 2. I. foliis amplexicaulibus dentatis hirſutiſſimis: radicalibus ovatis; caulinis lanceolatis, caule pauciflo-ro. *Sp. pl.* 1236.
Sweet-rooted Inula.
Nat. of the South of Europe.

Cult. 1759, by Mr. Ph. Miller. *Mill. dict. edit.* 7. *n.* 2.
Fl. July and August. H. ♃.

3. I. foliis amplexicaulibus lanceolatis distinctis serratis subtus villosis, caule ramoso villoso erecto. *Sp. pl.* 1237. *Britannica.*
Creeping-rooted Inula.
Nat. of Germany.
Cult. 1759, by Mr. Ph. Miller. *Mill. dict. edit.* 7. *n.* 8.
Fl. July——September. H. ♃.

4. I. foliis amplexicaulibus cordato-oblongis subtomentosis, caule villoso paniculato, squamis calycinis setaceis. *Sp. pl.* 1237. *Curtis lond.* *dysenterica.*
Meadow Inula, or Middle Fleabane.
Nat. of England.
Fl. August and September. H. ♃.

5. I. foliis lanceolatis denticulatis sessilibus basi reflexis, pedunculis lateralibus unifloris foliolosis. *viscosa.*
Erigeron viscosum. *Sp. pl.* 1209.
Clammy Inula.
Nat. of the South of Europe.
Cult. 1633. *Ger. emac.* 481. *f.* 1.

6. I. foliis amplexicaulibus cordato-lanceolatis undulatis. *Linn. mant.* 115. *undulata.*
Wave-leav'd Inula.
Nat. of Egypt.
Introd. 1777, by Mons. Thouin.
Fl. July. H. ☉.

7. I. foliis amplexicaulibus undulatis, caule prostrato, floribus subglobosis: radio brevissimo. *Syst. veget.* 766. *Curtis lond.* *Pulicaria.*

Small

Small Inula.
Nat. of England.
Fl. Auguſt and September. H. ☉.

ſquarroſa. 8. I. foliis ſeſſilibus ovalibus lævibus reticulato-venoſis ſubcrenatis, calycibus ſquarroſis. *Syſt. veget.* 766.
Nét-leav'd Inula.
Nat. of Italy and the South of France.
Cult. 1768, by Mr. Philip Miller. *Mill. dict. edit.* 8.
Fl. July——September. H. ♃.

ſolicina. 9. I. foliis lanceolatis recurvis ſerrato-ſcabris, ramis angulatis, floribus inferioribus altioribus. *Syſt. veget.* 767.
Willow-leav'd Inula.
Nat. of the North of Europe.
Cult. 1683, by Mr. James Sutherland. *Sutherl. hort. edin.* 40. *n.* 2.
Fl. Auguſt and September. H. ♃.

hirta. 10. I. foliis ſeſſilibus lanceolatis recurvatis ſubſerrato-ſcabris, floribus inferioribus altioribus, caule teretiuſculo ſubpiloſo. *Sp. pl.* 1239. *Jacqu. auſtr.* 4. *p.* 30. *t.* 358.
Hairy Inula.
Nat. of Siberia, Auſtria, and France.
Cult. 1759, by Mr. Philip Miller. *Mill. dict. edit.* 7. *n.* 9.
Fl. June——September. H. ♃.

ſuaveolens. 11. I. foliis ellipticis baſi attenuatis ſubpetiolatis piloſis: inferioribus dentatis, caule multifloro.
Inula ſuaveolens. *Jacqu. hort.* 3. *p.* 29. *t.* 51.
Woolly-leav'd Inula.
Nat. of the South of Europe.

Cult.

Cult. 1758, by Mr. Philip Miller.
Fl. June——August. H. ♃.

12. I. foliis sessilibus lanceolatis subserratis pilosis, pedunculis subunifloris subviscosis: foliolis linearibus. *Sp. pl.* 1240. *mariana.*
American Inula.
Nat. of Maryland and Carolina.
Introd. 1742, by Thomas Dale, M. D. *Mill. ic.* 38. *t.* 57. *f.* 1.
Fl. July. H. ♃.

13. I. foliis linearibus carnosis tricuspidatis. *Sp. pl.* 1240. *crithmoides.*
Trifid Inula, or Golden Samphire.
Nat. of England.
Fl. August and September. H. ♃.

14. I. foliis subserratis subtus tomentosis radicalibus petiolatis ovatis, caule erecto unifloro. *Sp. pl.* 1241. *provincialis.*
Oval-leav'd Inula.
Nat. of the South of France.
Introd. 1778, by Monf. Thouin.
Fl. July and August. H. ♃.

15. I. foliis lanceolatis hirsutis integerrimis, caule unifloro, calyce brevi imbricato. *Sp. pl.* 1241. *montana.*
Mountain Inula.
Nat. of the South of Europe.
Cult. 1759, by Mr. Ph. Miller. *Mill. dict. edit.* 7. *n.* 6.
Fl. July and August. H. ♃.

ARNICA. *Gen. pl.* 958.

Recept. nudum. *Pappus* ſimplex. *Corollulæ* radii filamentis 5 abſque antheris.

montana. 1. A. foliis ovatis integris: caulinis geminis oppoſitis. *Sp. pl.* 1245.
Mountain Arnica.
Nat. of Europe.
Cult. 1759, by Mr. Ph. Miller, *Mill. dict. edit.* 7. *n.* 1.
Fl. July. H. ♃.

ſcorpioides. 2. A. foliis alternis ſerratis. *Sp. pl.* 1246. *Jacqu. auſtr.* 4. *p.* 26. *t.* 349.
Alternate-leav'd Arnica.
Nat. of Switzerland and Auſtria.
Cult. 1759, by Mr. Philip Miller. *Mill. dict. edit.* 7. *n.* 2.
Fl. July and Auguſt. H. ♃.

DORONICUM. *Gen. pl.* 959.

Recept. nudum. *Pappus* ſimplex. *Calycis* ſquamæ duplicis ordinis æquales, diſco longiores. *Sem.* radii nuda pappoque deſtituta.

Pardalianches. 1. D. foliis cordatis obtuſis denticulatis: radicalibus petiolatis; caulinis amplexicaulibus. *Sp. pl.* 1247. *Jacqu. auſtr.* 4. *p.* 26. *t.* 350.
Great Leopard's-bane.
Nat. of Scotland.
Fl. May. H. ♃.

plantagineum. 2. D. foliis ovatis acutis ſudentatis, ramis alternis. *Sp. pl.* 1247.
Plantain-leav'd Leopard's-bane.

Nat.

Nat. of France, Spain, and Portugal.
Cult. 1597, by Mr. John Gerard. *Ger. herb.* 620. *f.* 2.
Fl. May. H. ♃.

3. D. caule nudo ſimpliciſſimo unifloro. *Sp. pl.* 1247. *Jacqu. auſtr.* 4. *p.* 53. *t.* 400. *Bellidiaſtrum.*
Daiſy-leav'd Leopard's-bane.
Nat. of Germany, Switzerland, and Italy.
Cult. 1759, by Mr. Ph. Miller. *Mill. dict. edit.* 7. *n.* 4.
Fl. June——Auguſt. H. ♃.

HELENIUM. *Gen. pl.* 961.

Recept. nudum: *Radii* paleaceum. *Pappus* 5-ariſtatus. *Cal.* ſimplex, multipartitus. *Corollulæ* radii ſemitrifidæ.

1. H. foliis glaberrimis. *autumnale.*
Helenium autumnale. *Sp. pl.* 1248.
Smooth Helenium.
Nat. of North America.
Cult. 1729, in Chelſea Garden. *R. S. n.* 371.
Fl. Auguſt——October. H. ♃.

2. H. foliis pubeſcentibus. *pubeſcens.*
Downy Helenium.
Nat. of North America.
Introd. 1776, by Mr. William Malcolm.
Fl. Auguſt and September. H. ♃.

BELLIS. *Gen. pl.* 962.

Recept. nudum, conicum. *Pappus* nullus. *Cal.* hemiſphæricus: ſquamis æqualibus. *Sem.* obovata.

1. B. ſcapo nudo. *Sp. pl.* 1248. *Curtis lond.* *perennis.*

 α Bellis

α Bellis ſylveſtris minor. *Bauh. pin.* 261.
Common Daiſy.

hortenſis. β Bellis hortenſis, flore pleno. *Bauh pin.* 261.
Double Daiſy.

fiſtuloſa. γ Bellis hortenſis rubra, flore multiplici fiſtuloſo. *Tournef. inſt.* 491.
Quill'd Daiſy.

prolifera. δ Bellis hortenſis prolifera. *Bauh. pin.* 262.
Proliferous, or Hen and Chicken Daiſy.
Nat. of Britain.
Fl. March——July. H. ♃.

annua. 2. B. caule ſubfolioſo. *Sp. pl.* 1249.
Annual Daiſy.
Nat. of Spain and the South of France.
Cult. 1759, by Mr. Ph. Miller. *Mill. dict. edit.* 7. *n.* 2.
Fl. H. ⊙.

BELLIUM. *Linn. mant.* 157.

Recept. nudum. *Sem.* conica corona paleacea 8-phylla, *Pappoque* ariſtato. *Cal.* foliolis æqualibus.

minutum. 1. B. caule folioſo. *Linn. mant.* 286.
Pectis minuta. *Sp. pl.* 1250.
Dwarf Bellium, or Baſtard Daiſy.
Nat. of the Levant.
Introd. 1772, by Monſ. Richard.
Fl. June——October. H. ⊙.

TAGETES. *Gen. pl.* 964.

Recept. nudum. *Pappus* ariſtis 5, erectis. *Cal.* 1-phyllus, 5-dentatus, tubuloſus. *Floſculi* radii 4, perſiſtentes.

patula. 1. T. caule ſubdiviſo patulo. *Sp. pl.* 1249.

α flore

α flore ſimplici.
Single-flower'd French Marygold.
β flore pleno.
Double-flower'd French Marygold.
Nat. of Mexi o.
Cult. 1596, by Mr. John Gerard. *Hort. Ger.*
Fl. July——October. H. ⊙.

2. T. caule ſimplici erecto, pedunculis nudis unifloris. *Sp. pl.* 1249. *erecta.*
α Tanacetum africanum majus, ſimplici flore. *Bauh. pin.* 133.
Single-flower'd African Marygold.
β Tagetes maximus rectus, flore maximo multiplicato. *Bauh. hiſt.* 3. *p.* 100.
Double-flower'd African Marygold.
γ Caryophyllus mexicanus fiſtuloſo flore ſimplex. *Col. ecphr.* 2. *p.* 47. *t.* 46. *f.* 1.
Quill'd-flower'd African Marygold.
Nat. of Mexico.
Cult. 1596, by Mr. John Gerard. *Hort. Ger.*
Fl. June——September. H. ⊙.

LEYSERA. *Gen. pl.* 965.

Recept. ſubpaleaceum. *Pappus* paleaceus; diſci etiam plumoſus. *Cal.* ſcarioſus.

1. L. foliis ſparſis, floribus pedunculatis. *Syſt. veget.* 770. *gnaphalodes.*
Woolly Leyſera.
Nat. of the Cape of Good Hope.
Introd. 1774, by Mr. Francis Maſſon.
Fl. July——September. G. H. ♄.

RELHANIA. *L'Herit. ſert. angl.*

Recept. paleaceum. *Pappus* membranaceus, cylindraceus, brevis. *Cal.* imbricatus, ſcarioſus. *Corollulæ* radii plurimæ.

ſquarroſa. 1. R. foliis oblongis acuminatis enervibus apice recurvis. *L'Herit. ſert. angl. n.* 1. *tab.* 29.
Athanaſia ſquarroſa. *Sp. pl.* 1180.
Croſs-leav'd Relhania.
Nat. of the Cape of Good Hope.
Introd. 1774, by Mr. Francis Maſſon.
Fl. May and June. G. H. ♄.

ZINNIA. *Gen. pl.* 974.

Recept. paleaceum. *Pappus* ariſtis 2 erectis. *Cal.* ovato-cylindricus, imbricatus. *Floſculi* radii 5, perſiſtentes, integri.

pauciflora. 1. Z. floribus ſeſſilibus. *Syſt. veget.* 771.
Yellow Zinnia.
Nat. of Peru.
Introd. 1753, by Mr. Ph. Miller. *Mill. ic.* 43. *t.* 64.
Fl. July and Auguſt. H. ☉.

multiflora. 2. Z. floribus pedunculatis. *Syſt. veget.* 771.
Red Zinnia.
Nat. of North America.
Introd. about 1771.
Fl. June——October. H. ☉.

CHRYSAN-

CHRYSANTHEMUM. *Gen. pl.* 966.

Recept. nudum. *Pappus* marginatus. *Cal.* hemiſphæricus, imbricatus: ſquamis marginalibus membranaceis.

* *Leucanthema.*

1. C. fruticoſum, foliis carnoſis linearibus pinnatis dentatis apice trifidis. *Syſt. veget.* 771. *fruteſcens.*
Shrubby Chryſanthemum.
Nat. of the Canary Iſlands.
Cult. 1699, in Oxford Garden. *Moriſ. hiſt.* 3. *p.* 35. *n.* 7.
Fl. Moſt part of the Year. G. H. ♄.

2. C. fruticoſum, foliis glabris baſi attenuatis pinnatifidis: laciniis inciſis. *pinnatifidum.*
Chryſanthemum pinnatifidum. *Linn. ſuppl.* 377.
Cut-leav'd Chryſanthemum.
Nat. of Madeira. Mr. *Francis Maſſon.*
Introd. 1777.
Fl. May——Auguſt. G. H. ♄.

3. C. foliis lanceolatis ſuperne ſerratis utrinque acuminatis. *Sp. pl.* 1251. *ſerotinum.*
Creeping-rooted Chryſanthemum.
Nat.
Cult. 1731, by Mr. Philip Miller. *Mill. dict. edit.* 1. Leucanthemum 3.
Fl. September and October. H. ♃.

4. C. foliis omnibus cuneiformibus oblongis inciſis carnoſis. *Syſt. veget.* 772. *atratum.*
Fleſhy-leav'd Chryſanthemum.
Nat. of Auſtria and Switzerland.

Introd. 1775, by the Doctors Pitcairn and Fothergill.
Fl. July and August. H. ♃.

alpinum. 5. C. foliis cuneiformibus pinnatifidis: laciniis integris, caulibus unifloris. *Sp. pl.* 1253.
Alpine Chrysanthemum.
Nat. of Switzerland.
Cult. 1759, by Mr. Ph. Miller. *Mill. dict. edit.* 7. *n.* 6.
Fl. July and August. H. ♃.

Leucanthemum. 6. C. foliis amplexicaulibus oblongis: superne serratis; inferne dentatis. *Sp. pl.* 1251. *Curtis lond.*
Ox-eye Daisy, or Chrysanthemum.
Nat. of Britain.
Fl. June and July. H. ♃.

Achilleæ. 7. C. foliis bipinnatis: pinnis imbricatis, caule stricto multifloro. *Syst. veget.* 772.
Milfoil-leav'd Chrysanthemum.
Nat. of Italy.
Introd. 1775, by Monf. Thouin.
Fl. June—August. H. ♃.

corymbosum. 8. C. foliis pinnatis inciso serratis, caule multifloro. *Sp. pl.* 1251. *Jacqu. austr.* 4. *p.* 41. *t.* 379.
Mountain Chrysanthemum.
Nat. of Germany and Siberia.
Cult. 1759, by Mr. Ph. Miller. *Mill. dict. edit.* 7. *n.* 7.
Fl. June—August. H. ♃.

** *Chrysanthema.*

segetum. 9. C. foliis amplexicaulibus: superne laciniatis; inferne dentato-serratis. *Sp. pl.* 1254.
Common Corn Marygold, or Chrysanthemum.

Nat.

Nat. of Britain.

Fl. June and July. H. ☉.

10. C. foliis lingulatis obtuſis ſerratis, calycum ſquamis æqualibus. *Sp. pl.* 1254. *myconis.*

Tongue-leav'd Chryſanthemum.

Nat. of Italy and Portugal.

Introd. 1775, by Monſ. Thouin.

Fl. July and Auguſt. H. ☉.

11. C. foliis pinnatifidis inciſis extrorſum latioribus. *Sp. pl.* 1254. *coronarium.*

α flore ſimplici.

Common Garden Chryſanthemum.

β flore pleno.

Double-flower'd Garden Chryſanthemum.

Nat. of Candia and Sicily.

Cult. 1629. *Park. parad.* 297. *f.* 1.

Fl. July——September. H. ☉.

12. C. floſculis omnibus uniformibus hermaphroditis. *Sp. pl.* 1255. *floſculoſum.*

Baſtard Chryſanthemum.

Nat. of Candia and Africa.

Introd. before 1605, by Mr. John Parkinſon. *Lobel. adv. p. alt. pag.* 508.

Fl. June——October. G. H. ♄.

MATRICARIA. *Gen. pl.* 967.

Recept. nudum. *Pappus* nullus. *Cal.* hemiſphæricus, imbricatus: marginalibus ſolidis, acutiuſculis.

1. M. foliis compoſitis planis: foliolis ovatis inciſis, pedunculis ramoſis. *Sp. pl.* 1255. *Parthenium.*

α floribus ſimplicibus.

Common

Common Feverfew.

β floribus plenis.

Double Feverfew.

Nat. of Britain.

Fl. June——September. H. ♃.

maritima. 2. M. receptaculis hemisphæricis, foliis bipinnatis subcarnosis: supra convexis; subtus carinatis. *Sp. pl.* 1256.

Sea Feverfew.

Nat. of Britain.

Fl. June——October. H. ♃.

suaveolens. 3. M. receptaculis conicis, radiis deflexis, squamis calycinis margine æqualibus. *Sp. pl.* 1256.

Sweet Feverfew.

Nat. of Britain.

Fl. June——August. H. ☉.

Chamomilla. 4. M. receptaculis conicis, radiis patentibus, squamis calycinis margine æqualibus. *Sp. pl.* 1256. *Curtis lond.*

Corn Feverfew.

Nat. of Britain.

Fl. June and July. H. ☉.

COTULA. *Gen. pl.* 968.

Recept. subnudum. *Pappus* marginatus. *Corolulæ* disci 4-fidæ; radii fere nullæ.

Anthemoides. 1. C. foliis pinnato-multifidis dilatatis, floribus flosculosis. *Syst. veget.* 774.

Dwarf Cotula.

Nat. of the Island of St. Helena.

Cult.

Cult. 1732, by James Sherard, M. D. *Dill. elth.* 26. *t.* 23. *f.* 25.
Fl. July. S. ⊙.

2. C. foliis pinnatifidis planis nudis punctatis, caule erecto ſtricto, floribus radiatis. *Syſt. veget.* 774. *ſtricta.*
Silvery Cotula.
Nat. of the Cape of Good Hope.
Introd. 1774, by Mr. Francis Maſſon.
Fl. May and June. G. H. ♄.

3. C. foliis lanceolato-linearibus amplexicaulibus dentatis, floribus floſculoſis. *Syſt. veget.* 774. *coronopifolia.*
Buck's-horn Cotula.
Nat. of the Cape of Good Hope.
Cult. 1714. *Philoſoph. tranſact. n.* 346. *p.* 354. *n.* 96.
Fl. July and Auguſt. G. H. ⊙.

4. C. receptaculis ſubtus inflatis turbinatis, floribus radiatis. *Syſt. veget.* 775. *turbinata.*
Radiated Cotula.
Nat. of the Cape of Good Hope.
Cult. 1713, in Chelſea Garden. *Philoſoph. tranſact. n.* 337. *p.* 59. *n.* 89.
Fl. July and Auguſt. G. H. ⊙.

ANACYCLUS. *Gen. pl.* 969.

Recept. paleaceum. *Pappus* emarginatus. *Semin.* lateribus membranaceis.

1. A. foliis decompoſitis linearibus: laciniis diviſis planis. *Sp. pl.* 1258. *creticus.*
Trailing Anacyclus.
Nat. of the Iſland of Candia.

Introd.

Introd. 1782, by Monſ. Thouin.
Fl. H. ⊙.

aureus. 2. A. foliis bipinnatis teretiuſculis incanis excavato-punctatis. *Linn. mant.* 287.
Golden-flower'd Anacyclus.
Nat. of the South of Europe and the Levant.
Cult. 1570. *Lobel. adv.* 343.
Fl. H. ⊙.

valentinus. 3. A. foliis decompoſitis linearibus: laciniis diviſis teretiuſculis acutis, floribus floſculoſis. *Sp. pl.* 1258.
Fine-leav'd Anacyclus.
Nat. of Spain.
Cult. 1656, by Mr. John Tradeſcant. *Muſ. Trad.* 93.
Fl. June and July. G. H. ⊙.

ANTHEMIS. *Gen. pl.* 970.

Recept. paleaceum. *Pappus* nullus. *Cal.* hemiſphæricus, ſubæqualis. *Floſculi* radii plures quam 5.

* *Radio diſcolore ſ. albo.*

altiſſima. 1. A. erecta, foliis pinnatis: pinnarum baſibus denticulo reflexo aſperis. *Syſt. veget.* 776.
Tall Camomile.
Nat. of France, Spain, and Italy.
Cult. 1748, by Mr. Philip Miller. *Mill. dict. edit.* 5. Chamæmelum 6.
Fl. Auguſt. H. ⊙.

maritima. 2. A. foliis pinnatis dentatis carnoſis nudis punctatis, caule proſtrato, calycibus ſubtomentoſis. *Sp. pl.* 1259.
Sea Camomile.

Nat.

Nat. of England.
Fl. August. H. ♃.

3. A. foliis simplicibus dentato-laciniatis. *Sp. pl.* 1260. *mixta.*
Simple-leav'd Camomile.
Nat. of France and Italy.
Cult. 1731, by Mr. Philip Miller. *Mill. dict. edit.* 1. Chamæmelum 9.
Fl. July. H. ⊙.

4. A. foliis pinnatifidis laciniatis, pedunculis nudis subvillosis. *Sp. pl.* 1260. *chia.*
Cut-leav'd Camomile.
Nat. of Chio.
Cult. 1731, by Mr. Philip Miller. *Mill. dict. edit.* 1. Chamæmelum 7.
Fl. June——October. H. ⊙.

5. A. foliis pinnato-compositis linearibus acutis subvillosis. *Sp. pl.* 1260. *nobilis.*
Common Camomile.
Nat. of Britain.
Fl. July and August. H. ♃.

6. A. receptaculis conicis: paleis setaceis, seminibus coronato-marginatis. *Sp. pl.* 1261. *arvensis.*
Corn Camomile.
Nat. of Britain.
Fl. July. H. ♂.

7. A. receptaculis conicis: paleis setaceis, seminibus nudis. *Sp. pl.* 1261. *Curtis lond.* *Cotula.*
Stinking Camomile, or May-weed.
Nat. of Britain.
Fl. June and July. H. ⊙.

8. A.

Pyrethrum. 8. A. caulibus ſimplicibus unifloris decumbentibus, foliis pinnato-multifidis. *Sp. pl.* 1262.
Spaniſh Camomile, or Pellitory of Spain.
Nat. of the South of Europe and the Levant.
Cult. 1570. *Lobel. adv.* 346.
Fl. June and July. H. ♃.

** *Radio concolore ſ. luteo.*

valentina. 9 A. caule ramoſo, foliis pubeſcentibus tripinnatis ſetaceis, calycibus villoſis pedunculatis. *Syſt. veget.* 776.
Purple-ſtalk'd Camomile.
Nat. of the South of Europe.
Cult. 1621, by Mr. John Goodyer. *Ger. emac.* 745. *n.* 7.
Fl. July and Auguſt. H. ☉.

odorata. 10. A. foliis apice pinnatifidis, pedunculis elongatis, calycibus membranaceis, radio ſterili.
Shrubby Camomile.
Nat. of the Cape of Good Hope. Mr. *Fr. Maſſon.*
Introd. 1774.
Fl. April——June. G. H. ♄.

tinctoria. 11. A. foliis bipinnatis ſerratis ſubtus tomentoſis, caule corymboſo. *Sp. pl.* 1263.
Yellow Camomile.
Nat. of England.
Fl. June——November. H. ♃.

ACHILLEA. *Gen. pl.* 971.

Recept. paleaceum. *Pappus* nullus. *Cal.* ovatus, imbricatus. *Floſculi* radii circiter 4.

* *Corollis flavis.*

Santolina. 1. A. foliis ſetaceis dentatis: denticulis ſubintegris ſubulatis reflexis. *Sp. pl.* 1264.

§ Lavender-

Lavender-cotton-leav'd Milfoil.
Nat. of the Levant.
Cult. 1759, by Mr. Philip Miller. *Mill. dict. edit.* 7. *n.* 2.
Fl. June——August. H. ♃.

2\. A. foliis lanceolatis obtusis acute serratis. *Sp. pl.* 1264. *Ageratum.*
Sweet Milfoil, or Maudlin.
Nat. of the South of Europe.
Cult. 1570. *Lobel. adv.* 208.
Fl. August——October. H. ♃.

3\. A. foliis pinnatis hirsutis: pinnis linearibus dentatis. *Sp. pl.* 1264. *tomentosa.*
Woolly Milfoil.
Nat. of the South of Europe.
Cult. 1658, in Oxford Garden. *Hort. oxon. edit.* 2. *p.* 110.
Fl. May——October. H. ♃.

4\. A. foliis pinnatis: foliolis lanceolatis incisis serratis subtus lanigeris. *Sp. pl.* 1264. *pubescens.*
Downy Milfoil.
Nat. of the Levant.
Cult. 1739, in Chelsea Garden. *Rand. chel.* Millefolium 11.
Fl. June——September. H. ♃.

5\. A. foliis pinnatis supradecompositis: laciniis linearibus distantibus. *Sp. pl.* 1265. *abrotanifolia.*
Southernwood-leav'd Milfoil.
Nat. of the Levant.
Cult. 1739, by Mr. Philip Miller. *Rand. chel.* Millefolium 10.
Fl. June and July. H. ♃.

ægyptiaca. 6. A. foliis pinnatis: foliolis obtuſe lanceolatis ſerrato-dentatis. *Sp. pl.* 1265.
Egyptian Milfoil.
Nat. of the Levant.
Cult. 1712, in Chelſea Garden. *Br. Muſ. H. S.* 134. *fol.* 36.
Fl. July——September. G. H. ♃.

** *Corollis radio albis.*

macrophylla. 7. A. foliis pinnatis: pinnis inciſo-ſerratis: extimis majoribus coadunatis. *Syſt. veget.* 777.
Feverfew-leav'd Milfoil.
Nat. of Italy and Switzerland.
Cult. 1759, by Mr. Ph. Miller. *Mill. dict. edit.* 7. *n.* 11.
Fl. July and Auguſt. H. ♃.

Clavennæ. 8. A. foliis laciniatis planis obtuſis tomentoſis. *Sp. pl.* 1266. *Jacqu. auſtr.* 1. *p.* 49. *t.* 76.
Silvery-leav'd Milfoil.
Nat. of Auſtria.
Cult. 1683, by Mr. James Sutherland. *Sutherl. hort. edin.* 3. *n.* 2.
Fl. June and July. H. ♃.

Ptarmica. 9. A. foliis lanceolatis acuminatis argute ſerratis. *Sp. pl.* 1266. *Curtis lond.*
α flore ſimplici.
Common Sneeze-wort, or Milfoil.
β flore pleno.
Double-flower'd Sneeze-wort, or Milfoil.
Nat. of Britain.
Fl. July——November. H. ♃.

alpina. 10. A. foliis lanceolatis dentato-ſerratis: denticulis tenuiſſime ſerratis. *Sp. pl.* 1266.

Alpine

Alpine Milfoil.

Nat. of Siberia and Switzerland.

Cult. 1731, by Mr. Philip Miller. *Mill. dict. edit.* 1: Ptarmica 4.

Fl. July——November. H. ♃.

11. A. foliis lineari-lanceolatis ſeſſilibus tomentoſis, profunde ſerratis, baſi laciniatis. *Retz. obſ. bot.* 2. *p.* 25. *n.* 83. *ſerrata.*

Notch-leav'd Milfoil.

Nat.

Introd. 1784, by Mr. John Græffer.

Fl. Auguſt and September. H. ♃.

12. A. foliis linearibus ſerratis, ſerraturis tranſverſis criſtatis, caule ramoſo debili. *Retz. obſ. bot.* 2. *p.* 25. *n.* 84. *criſtata.*

Slender-branch'd Milfoil.

Nat.

Introd. 1784, by Mr. John Græffer.

Fl. July and Auguſt. H. ♃.

13. A. foliis pinnulis pectinatis integriuſculis, pedunculis villoſis. *Sp. pl.* 1267. *Jacqu. auſtr.* 1. *p.* 49. *t.* 77. *atrata.*

Black-cupp'd Milfoil.

Nat. of Auſtria and Switzerland.

Introd. 1774, by the Doctors Pitcairn and Fothergill.

Fl. July——September. H. ♃.

14. A. foliis pinnatis punctatis: pinnis remotis linearibus ſubulatis ſubintegris, radiis longitudine calycis. *moſchata.*

Achillea moſchata. *Jacqu. auſtr.* 5. *p.* 45. *tab. app.* 33. *Allion. pedem.* 1. *p.* 182.

Achillea livia. *Scop. inſubr.* 1. *p.* 6. *tab.* 3.

Achillea foliis pinnatis: pinnis ſimplicibus glabris punctatis. *Hall. hiſt.* 112.
Muſk-ſmelling Milfoil.
Nat. of Switzerland and Italy.
Introd. 1775, by the Doctors Pitcairn and Fothergill.
Fl. June and July. H. ♃.

magna. 15. A. foliis bipinnatis ſubpiloſis: laciniis linearibus dentatis: auriculis decuſſatis. *Sp. pl.* 1267.
Great Milfoil, or Yarrow.
Nat. of the South of Europe.
Cult. 1683, by Mr. James Sutherland. *Sutherl. hort. edin.* 231. *n.* 3.
Fl. June——November. H. ♃.

Millefolium. 16. A. foliis bipinnatis nudis: laciniis linearibus dentatis, caulibus ſuperne ſulcatis. *Syſt. veget.* 778.
α Millefolium vulgare album. *Bauh. pin.* 140.
Common white Milfoil, or Yarrow.
β Millefolium purpureum majus. *Bauh. pin.* 140.
Common purple Milfoil, or Yarrow.
Nat. of Britain.
Fl. June——October. H. ♃.

nobilis. 17. A. foliis bipinnatis: inferioribus nudis planis; ſuperioribus obtuſis tomentoſis, corymbis convexis confertiſſimis. *Sp. pl.* 1268.
Shewy Milfoil.
Nat. of Germany and France.
Cult. 1640, by Mr. John Parkinſon. *Park. theat.* 695. *n.* 9.
Fl. June——Auguſt. H. ♃.

ſquarroſa. 18. A. foliis lanceolato-linearibus pinnatifidis: pinnis ovato-cuneiformibus inciſo-acuminatis verticalibus, caule villoſiuſculo.

§ Rough-

Rough-headed Milfoil.
Nat.
Introd. 1775, by Monf. Thouin.
Fl. July and Auguft. H. ♃.

AMELLUS. *Gen. pl.* 978.

Recept. paleaceum. *Pappus* fimplex. *Cal.* imbricatus. *Corollulæ* radii indivifæ.

1. A. foliis oppofitis lanceolatis obtufis tomentofis, pedunculis unifloris. *Syft. veget.* 778. *lychnitis.*
Trailing Amellus.
Nat. of the Cape of Good Hope.
Cult. 1768, by Mr. Philip Miller. *Mill. dict. edit.* 8.
Fl. June and July. G. H. ♄.

ECLIPTA. *Linn. mant.* 157.

Recept. paleaceum. *Pappus* nullus. *Corollulæ* difci 4-fidæ.

1. E. caule erecto, foliis bafi deflexis feffilibus. *Linn. mant.* 286. *erecta.*
Verbefina alba. *Sp. pl.* 1272.
Upright Eclipta.
Nat. of the Weft Indies.
Cult. 1690, in the Royal Garden at Hampton-court. *Catal. mff.*
Fl. July——September. S. ⊙.

2. E. caule erecto, foliis ovatis petiolatis. *Linn. fuppl.* 378. *latifolia.*
Oval-leav'd Eclipta.
Nat. of the Eaft Indies.
Introd. 1777, by Sir Jofeph Banks, Bart.
Fl. September and October. S. ⊙.

prostrata. 3. E. caule prostrato, foliis subundulatis subpetiolatis. *Linn. mant.* 286.
Verbesina prostrata. *Sp. pl.* 1272.
Trailing Eclipta.
Nat. of the East Indies.
Cult. 1759, by Mr. Philip Miller. *Mill. dict. edit.* 7. Verbesina 4.
Fl. August. S. ⊙.

SIGESBECKIA. *Gen. pl.* 973.

Recept. paleaceum. *Pappus* nullus. *Cal.* exterior 5-phyllus, proprius, patens. *Radius* dimidiatus.

orientalis. 1. S. petiolis sessilibus, calycibus exterioribus linearibus majoribus patentibus. *Sp. pl.* 1269.
Oriental Sigesbeckia.
Nat. of India and China.
Cult. 1730, by Mr. Philip Miller. *R. S. n.* 424.
Fl. July and August. S. ⊙.

occidentalis. 2. S. petiolis decurrentibus, calycibus nudis. *Sp. pl.* 1269.
American Sigesbeckia.
Nat. of Virginia.
Cult. 1731, by Mr. Philip Miller. *Mill. dict. edit.* 1. Corona Solis 16.
Fl. October and November. H. ♃.

flosculosa. 3. S. flosculis tridentatis, hermaphroditis triandris. *L'Herit. stirp. nov. p.* 37. *tab.* 19.
Small-flower'd Sigesbeckia.
Nat. of Peru.
Introd. 1784, by Chev. Thunberg.
Fl. June and July. H. ⊙.

VERBE-

VERBESINA. *Gen. pl.* 975.

Recept. paleaceum. *Pappus* ariſtatus. *Cal.* 2-plici ordine. *Floſculi* radii circiter 5.

1. V. foliis alternis decurrentibus undulatis obtuſis. *Sp. pl.* 1270. *alata.*
Wing-ſtalk'd Verbeſina.
Nat. of South America.
Cult. 1714, by the Dutcheſs of Beaufort. *Br. Muſ. H. S.* 139. *fol.* 57.
Fl. Moſt part of the Summer. S. ♃.

2. V. foliis alternis profunde pinnatifidis, caule fruticoſo. *Jacqu. ic. collect.* 1. *p.* 53. *giganica.*
Bidens fruteſcens, ſphondylii folio et facie. *Plum. ic.* 41. *t.* 51.
Tree Verbeſina.
Nat. of the Weſt Indies.
Cult. 1758, by Mr. Philip Miller.
Fl. S. ♄.

3. V. foliis oppoſitis ovatis ſerratis, calycibus oblongis ſeſſilibus: caulinis lateralibus. *Syſt. veget.* 779. *nodiflora.*
Seſſile-flower'd Verbeſina.
Nat. of the Weſt Indies.
Cult. 1732, by James Sherard, M.D. *Dill. elth.* 53. *t.* 45. *f.* 53.
Fl. June and July. S. ⊙.

BUPHTHALMUM. *Gen. pl.* 977.

Recept. paleaceum. *Pappus:* margo obſoletus. *Seminum* latera, præſertim radii, marginata. *Stigma* floſcul. hermaphroditicorum indiviſum.

1. B. foliis ſparſis cuneiformibus acutis integerrimis villoſo-ſericeis. *ſericeum.*

Buphthalmum ſericeum. *Linn. ſuppl.* 379. *L'Herit. ſert. angl. tab.* 37.
Silky Ox-eye.
Nat. of Fuertaventura, one of the Canary Iſlands. Mr. *Francis Maſſon.*
Introd. 1779.
Fl. May——July. G. H. ♄.

fruteſcens. 2. B. foliis oppoſitis lanceolatis, petiolis bidentatis, caule fruticoſo. *Sp. pl.* 1273.
Shrubby Ox-eye.
Nat. of Virginia and Jamaica.
Introd. 1696, by Biſhop Compton. *Mill. dict. edit.* 8.
Fl. June——Auguſt. G. H. ♄.

arboreſcens. 3. B. foliis oppoſitis lanceolatis utrinque tomentoſis edentulis integerrimis, caule fruticoſo. *Sp. pl.* 1273.
Tree Ox-eye.
Nat. of Bermudas.
Cult. 1699, in the Royal Garden at Hampton-court. *Moriſ. hiſt.* 3. *p.* 25. *n.* 90.
Fl. Moſt part of the Summer. G. H. ♄.

ſpinoſum. 4. B. calycibus acute foliosis, foliis alternis lanceolatis amplexicaulibus integerrimis, caule herbaceo. *Sp. pl.* 1274.
Prickly Ox-eye.
Nat. of Spain and Italy.
Cult. 1570. *Lobel. adv.* 147.
Fl. June——September. H. ⊙.

aquaticum. 5. B. calycibus obtuſe folioſis ſeſſilibus axillaribus, foliis alternis oblongis obtuſis, caule herbaceo. *Sp. pl.* 1274.
Sweet-ſcented Ox-eye.

Nat.

Nat. of the South of Europe.
Cult. 1731, by Mr. Philip Miller. *Mill. dict. edit.* 1. Asteriscus 2.
Fl. July and August. H. ⊙.

6. B. calycibus obtuse foliosis pedunculatis, foliis alternis spathulatis, caule herbaceo. *Sp. pl.* 1274. *maritimum.*
Sea Ox-eye.
Nat. of Sicily.
Cult. 1640. *Park. theat.* 129. *f.* 3.
Fl. July——September. H. ♃.

7. B. foliis alternis lanceolatis subserratis villosis, calycibus nudis, caule herbaceo. *Sp. pl.* 1275. *Jacqu. austr.* 4. *p.* 36. *t.* 370. *salicifolium.*
Willow-leav'd Ox-eye.
Nat. of Austria, Switzerland, and the South of France.
Cult. 1759, by Mr. Ph. Miller. *Mill. dict. edit.* 7. *n.* 3.
Fl. June——October. H. ♃.

8. B. foliis alternis lanceolatis subdenticulatis glabris, calycibus nudis, caule herbaceo. *Sp. pl.* 1275. *grandiflorum.*
Great-flower'd Ox-eye.
Nat. of Austria and France.
Cult. 1722, in Chelsea Garden. *R. S. n.* 6.
Fl. June——October. H. ♃.

9. B. foliis oppositis ovatis serratis triplinerviis, calycibus foliosis, caule herbaceo. *Sp. pl.* 1275. *L'Herit. stirp. nov. p.* 93. *tab.* 45. *Helianthoides.*
Helianthus lævis. *Sp. pl.* 1278.
Silphium solidaginoides. *Sp. pl.* 1302.
Sun-flower-leav'd Ox-eye.
Nat. of North America.

Cult. 1714, by the Dutchess of Beaufort. *Br. Mus.* *H. S.* 139. *fol.* 56.
Fl. July——October. H. ♃.

POLYGAMIA FRUSTRANEA.

HELIANTHUS. *Gen. pl.* 979.

Recept. paleaceum, planum. *Pappus* 2-phyllus. *Cal.* imbricatus, subsquarrosus.

annuus. 1. H. foliis omnibus cordatis trinervatis, pedunculis incrassatis, floribus cernuis. *Syst. veget.* 781.
α floribus simplicibus.
Common annual Sun-flower.
β floribus plenis.
Double-flower'd annual Sun-flower.
Nat. of Mexico and Peru.
Cult. 1596, by Mr. John Gerard. *Hort. Ger.*
Fl. June——October. H. ☉.

indicus. 2. H. foliis omnibus cordatis trinervatis, pedunculis æqualibus, calycibus foliosis. *Linn. mant.* 117.
Dwarf annual Sun-flower.
Nat.
Introd. 1785, by Chevalier Thunberg.
Fl. June——October. H. ☉.

multiflorus. 3. H. foliis inferioribus cordatis trinervatis; superioribus ovatis. *Sp. pl.* 1277.
α floribus simplicibus.
Many-flower'd perennial Sun-flower,
β floribus plenis.

Double-

Double-flower'd perennial Sun-flower.
Nat. of Virginia.
Introd. before 1699, by Lord Lemſter. *Moriſ. hiſt.* 3. *p.* 23. *n.* 61.
Fl. Auguſt——October. H. ♃.

4. H. foliis ovato-cordatis triplinerviis. *Sp. pl.* 1277. *Jacqu. hort.* 2. *p.* 75. *t.* 161. *tuberoſus.*
Tuberous-rooted Sun-flower, or Jeruſalem Artichoke.
Nat. of Brazil.
Cult. 1617, by Mr. Franqueuill. *Ger. emac.* 753.
Fl. September and October. H. ♃.

5. H. caule inferne lævi, foliis triplinerviis lanceolato-cordatis, radiis decapetalis, pedunculis ſcabris. *Sp. pl.* 1277. *decapetalus.*
Ten-petal'd Sun-flower.
Nat. of Canada.
Cult. 1759, by Mr. Ph. Miller. *Mill. dict. edit.* 7. *n.* 10.
Fl. Auguſt——November. H. ♃.

6. H. radice fuſiformi. *Sp. pl.* 1277. *ſtrumoſus.*
Carrot-rooted Sun-flower.
Nat. of Canada.
Cult. 1759, by Mr. Ph. Miller. *Mill. dict. edit.* 7. *n.* 4.
Fl. July——September. H. ♃.

7. H. foliis alternis lanceolatis ſcabris baſi ciliatis, caule ſtricto ſcabro. *Sp. pl.* 1278. *giganteus.*
Gigantic Sun-flower.
Nat. of Virginia and Canada.
Cult. 1714, by the Dutcheſs of Beaufort. *Br. Muſ. H. S.* 139. *fol.* 60.
Fl. September and October. H. ♃.

8. H.

altissimus. 8. H. foliis alternis latiuscule lanceolatis scabris, petiolis ciliatis, caule stricto glabro. *Sp. pl.* 1278. *Jacqu. hort.* 2. *p.* 75. *t.* 160.
Tall Sun-flower.
Nat. of Pensylvania.
Cult. 1739, by Mr. Philip Miller. *Mill. dict. edit.* 1. Corona Solis 13.
Fl. August and September. H. ♃.

divaricatus. 9. H. foliis oppositis sessilibus ovato-oblongis trinerviis, panicula dichotoma. *Sp. pl.* 1279.
Rough-leav'd Sun-flower.
Nat. of North America.
Cult. 1759, by Mr. Ph. Miller. *Mill. dict. edit.* 7. *n.* 6.
Fl. August——October. H. ♃.

atrorubens. 10. H. foliis oppositis spathulatis crenatis triplinerviis scabris, squamis calycinis erectis longitudine disci. *Sp. pl.* 1279.
Dark-red Sun-flower.
Nat. of Carolina and Virginia.
Cult. 1732, by James Sherard, M. D. *Dill. elth.* 111. *t.* 94. *f.* 110.
Fl. July——September. H. ♃.

RUDBECKIA. *Gen. pl.* 980.

Recept. paleaceum, conicum. *Pappus* margine 4-dentato. *Cal.* 2-plici ordine squamarum.

laciniata. 1. R. foliis compositis laciniatis. *Sp. pl.* 1279.
Broad jagged-leav'd Rudbeckia.
Nat. of Virginia and Canada.
Cult. 1640, by Mr. John Tradescant, Sen. *Park. theat.* 322. *f.* 10.
Fl. July. H. ♃.

2. R.

2. R. foliis inferioribus compofitis; caulinis quinatis ternatifque: fummis fimplicibus. *Mill. dict.* *digitata.*
Narrow jagged-leav'd Rudbeckia.
Nat. of North America.
Cult. 1759, by Mr. Ph. Miller. *Mill. dict. edit.* 7. *n.* 6.
Fl. Auguft and September. H. ♃.

3. R. foliis indivifis fpathulato-ovatis triplinerviis ferratis hirtis, receptaculo conico, paleis lanceolatis. *hirta.*
Rudbeckia hirta. *Sp. pl.* 1280.
Hairy Rudbeckia.
Nat of Virginia and Canada.
Cult. 1732, by James Sherard, M. D. *Dill. elth.* 295. *t.* 218. *f.* 285.
Fl. June——November. H. ♂.

4. R. foliis oblongo-lanceolatis denticulatis hifpidis bafi anguftatis fubcordatis, receptaculo hemifphærico, paleis lanceolatis. *fulgida.*
Bright Rudbeckia.
Nat. of North America.
Introd. 1760, by Meffrs. Kennedy and Lee.
Fl. July and Auguft. H. ♃.

5. R. foliis lanceolato-ovatis alternis indivifis, petalis radii bifidis. *Sp. pl.* 1280. *Curtis magaz.* 2. *purpurea.*
Purple Rudbeckia.
Nat. of Carolina and Virginia.
Introd. before 1699, by the Rev. John Banifter. *Morif. hift.* 3. *p.* 42. Dracunculus.
Fl. July——October. H. ♃.

6. R. foliis oppofitis linearibus integerrimis. *Sp. pl.* 1281. *anguftifolia.*
Narrow-leav'd Rudbeckia.

Nat.

Nat. of Virginia.
Cult. 1758, by Mr. Philip Miller. *Mill. ic.* 150. *t.* 224. *f.* 2.
Fl. August and September. H. ♃.

COREOPSIS. *Gen. pl.* 981.

Recept. paleaceum. *Pappus* bicornis. *Cal.* erectus, polyphyllus, basi radiis patentibus cinctus.

verticillata. 1. C. foliis decomposito-pinnatis linearibus. *Sp. pl.* 1281.
Whorl'd-leav'd Coreopsis, or Tick-seed Sun-flower.
Nat. of Virginia.
Cult. 1759, by Mr. Ph. Miller. *Mill. dict. edit.* 7. *n.* 4.
Fl. July——October. H. ♃.

aurea. 2. C. foliis serratis: radicalibus tripartitis; caulinis trifidis integrisve lanceolato-linearibus.
Hemp-leav'd Coreopsis.
Nat. of North America.
Introd. 1785, by Charles Earl of Tankerville.
Fl. August and September. H. ♃.

tripteris. 3. C. foliis subternatis integerrimis. *Sp. pl.* 1282.
Three-leav'd Coreopsis.
Nat. of Virginia and Canada.
Cult. 1759, by Mr. Ph. Miller. *Mill. dict. edit.* 7. *n.* 5.
Fl. August——October. H. ♃.

auriculata. 4. C. foliis integerrimis ovatis: inferioribus ternatis. *Sp. pl.* 1282.
Ear-leav'd Coreopsis.
Nat. of Virginia.

Cult.

Cult. 1699. *Moris. hist.* 3. *p.* 21. *n.* 46. *f.* 6. *t.* 3. *f.* 45.
Fl. August——October. H. ♃.

5. C. foliis obovato-oblongis obtusis integerrimis pubescentibus. *crassifolia.*
Thick-leav'd Coreopsis.
Nat. of Carolina.
Introd. 1786, by Mr. Adam Robson.
Fl. August——October. H. ♃.

6. C. foliis alternis lineari-lanceolatis integerrimis lævibus, petalis radii oblongis trifidis: lacínia media majore. *angustifolia.*
Narrow-leav'd Coreopsis.
Nat. of Carolina and Florida.
Cult. 1778, by John Fothergill, M. D.
Fl. June. G. H. ♃.

7. C. foliis lanceolatis serratis alternis petiolatis decurrentibus. *Sp. pl.* 1283. *Jacqu. hort.* 2. *p.* 50. *t.* 110. *alternifolia.*
Alternate-leav'd Coreopsis.
Nat. of Virginia and Canada.
Cult. 1759, by Mr. Philip Miller. *Mill. dict. edit.* 7. *n.* 1.
Fl. September——November. H. ♃.

8. C. foliis ellipticis acuminatis serratis petiolatis venosis decurrentibus: inferioribus verticillatis; superioribus alternis. *procera.*
Tall Coreopsis.
Nat. of North America. Mr. *John Cree.*
Introd. 1765.
Fl. September and October. H. ♃.

GORTE-

GORTERIA. *Gen. pl.* 982.

Recept. nudum. *Pappus* lanatus. *Cor.* radii ligulatæ. *Cal.* imbricatus ſquamis ſpinoſis.

perſonata. 1. G. foliis lanceolatis integris ſinuatiſque, caule erecto, floribus pedunculatis. *Sp. pl.* 1283.
Annual Gorteria.
Nat. of the Cape of Good Hope.
Introd. 1774, by Mr. Francis Maſſon.
Fl. July and Auguſt. G. H. ⊙.

rigens. 2. G. foliis lanceolatis pinnatifidis, caule depreſſo, ſcapis unifloris. *Sp. pl.* 1284.
Great-flower'd Gorteria.
Nat. of the Cape of Good Hope.
Cult. 1755, by Mr. Ph. Miller. *Mill. ic.* 33. *t.* 49.
Fl. Moſt part of the Summer. G. H. ♃.

echinata. 3. G. foliis oblongis ſinuato-inciſis ſpinuloſis, caulibus adſcendentibus, receptaculis paleaceis.
Prickly Gorteria.
Nat. of the Cape of Good Hope. Mr. *Fr. Maſſon.*
Introd. 1774.
Fl. July. G. H. ⊙.
Caules adſcendentes, pedales, anguloſi, rubicundi, læves, ſed interdum lanugine hinc inde veſtiti. *Folia* ſparſa, oblonga, ſeſſilia, amplexicaulia, ſubdecurrentia, inciſo-ſinuata, lævia, ſed lanugine hinc inde induta. *Flores* terminales, ſolitarii. *Calycis* ſquamæ exteriores breves, palmato-ſpinoſæ; interiores longiores, lanceolatæ, ſpina terminatæ, et baſi utrinque ſpinulis armatæ. *Corollulæ radii* omnino ſteriles: *Ligulæ* apice quadrifidæ, unciam longæ, patentes, luteæ,

luteæ, ſubtus prope apicem ſordide purpuraſcentes. *Corollulæ diſci* luteæ. *Semina* obovata. *Receptaculum* paleaceum: *paleæ* longæ, reticulatim ſubconnatæ.

4. G. foliis lanceolatis decurrentibus adnatis ciliato-ſpinuloſis, floribus ſeſſilibus. *Syſt. veget.* 783. *ſquarroſa.*
Cob-web Gorteria.
Nat. of the Cape of Good Hope.
Introd. 1786, by Baron Hake.
Fl. June——Auguſt. G. H. ♄.

5. G. foliis imbricatis bifariam ciliatis: ciliis exterioribus ſpinaque terminali reflexis. *Sp. pl.* 1284. *ciliaris.*
Ciliated Gorteria.
Nat. of the Cape of Good Hope.
Introd. 1774, by Mr. Francis Maſſon.
Fl. May and June. G. H. ♄.

6. G. foliis lanceolatis integris dentato-ſpinoſis ſubtus tomentoſis, caule fruticoſo. *Sp. pl.* 1284. *fruticoſa.*
Gorteria Aſteroides. *Linn. ſuppl.* 381.
Shrubby Gorteria.
Nat. of the Cape of Good Hope.
Cult. 1739, by Mr. Philip Miller. *Rand. chel.* Carthamus 2.
Fl. Auguſt and September. G. H. ♄.
Obs. Diverſa omnino a Gorteria fruticoſa. *Berg. cap.* 302, quæ eſt Atractylis oppoſitifolia *Syſt. veget.* 730.

7. G. foliis oblongis amplexicaulibus dentato-ſpinoſis patentibus glabris, calycibus ciliato-ſerratis, floribus cernuis. *Linn. ſuppl.* 382. *cernua.*
Drooping Gorteria.

Nat.

Nat. of the Cape of Good Hope.
Introd. 1785, by Baron Hake.
Fl. May. G. H. ♄.

DIDELTA. *L'Herit. ſtirp. nov.*

Recept. favoſum, diſcedens in partes ſemina retinentes. *Cal.* patens: exterior foliaceus. *Papp.* paleaceus, polyphyllus.

carnoſa. 1. D. foliis alternis lanceolato-oblongis carnoſis.
Didelta tetragoniæfolia. *L'Herit. ſtirp. nov. p.* 55. *tab.* 28.
Polymnia carnoſa. *Linn. ſuppl.* 384.
Succulent-leav'd Didelta.
Nat. of the Cape of Good Hope. Mr. *Fr. Maſſon.*
Introd. 1774.
Fl. July. G. H. ♄.

ſpinoſa. 2. D. foliis oppoſitis ſubamplexicaulibus ovatis.
Polymnia ſpinoſa. *Linn. ſuppl.* 384.
Oppoſite-leav'd Didelta.
Nat. of the Cape of Good Hope. Mr. *Fr. Maſſon.*
Introd. 1774.
Fl. June and July. G. H. ♄.

CENTAUREA. *Gen. pl.* 984.

Recept. ſetoſum. *Pappus* ſimplex. *Cor.* radii infundibuliformes, longiores, irregulares.

* Jaceæ: *calycibus lævibus inermibus.*

Crupina. 1. C. calycibus inermibus: ſquamis lanceolatis, foliis pinnatis ſerratis ſubciliatis. *Syſt. veget.* 784.
Black-ſeeded Centaury.
Nat. of Italy and France.

Cult.

Cult. 1640. *Park. theat.* 786. *f.* 4.
Fl. June and July. H. ⊙.

2. C. calycibus inermibus ſubrotundis glabris: ſquamis ovatis, foliis lyrato-dentatis. *Sp. pl.* 1286. *moſchata.*
α Cyanus orientalis major moſchatus flore purpureo & albo. *Moriſ. hiſt.* 3. *p.* 135. *ſ.* 7. *t.* 25. *f.* 5.
Purple ſweet Centaury, or Sultan.
β Cyanus orientalis major, foliis minus diſſectis, flore luteo. *Moriſ. hiſt.* 3. *p.* 135. *ſ.* 7. *t.* 25. *f.* 9. Amberboi.
Yellow ſweet Centaury, or Sultan.
Nat. of Perſia.
Cult. 1629. *Park. parad.* 327. *n.* 2.
Fl. July——October. H. ⊙.

3. C. calycibus inermibus: ſquamis mucronatis, foliis ſubdecurrentibus lyrato-dentatis. *Syſt. veget.* 784. *Lippii.*
Egyptian Centaury.
Nat. of Egypt.
Cult. 1759, by Mr. Ph. Miller. *Mill. dict. edit.* 7. *n.* 12.
Fl. June and July. H. ⊙.

4. C. calycibus inermibus: ſquamis ovatis obtuſis, foliis pinnatis glabris integerrimis: impari ſerrato. *Sp. pl.* 1286. *alpina.*
Alpine Centaury.
Nat. of Italy.
Cult. 1640. *Park. theat.* 466. *f.* 4.
Fl. July and Auguſt. H. ♃.

5. C. calycibus inermibus: ſquamis ovatis, foliis pinnatis: foliolis decurrentibus ſerratis. *Sp. pl.* 1287. *Centaurium.*
Great Centuary.
Nat. of Italy.

Cult. 1596, by Mr. John Gerard. *Hort. Ger.*
Fl. July and Auguſt. H. ♃.

** Cyani: *calycinis ſquamis ſerrato-ciliatis.*

phrygia. 6. C. calycibus recurvato-plumoſis, foliis indiviſis oblongis ſcabris. *Syſt. veget.* 785.
Auſtrian Centaury.
Nat. of Auſtria and Switzerland.
Cult. 1727, by Mr. Philip Miller. *R. S. n.* 276.
Fl. June——October. H. ♃.

nigra. 7. C. calycibus ciliatis: ſquamula ovata: ciliis capillaribus erectis, foliis lyrato-angulatis, floribus floſculoſis. *Sp. pl.* 1288.
Black Centaury, or Knapweed.
Nat. of Britain.
Fl. May——July. H. ♂.

montana. 8. C. calycibus ſerratis, foliis lanceolatis decurrentibus, caule ſimpliciſſimo. *Sp. pl.* 1289. *Jacqu. auſtr.* 4. *p.* 37. *t.* 371. *Curtis magaz.* 77.
α Cyanus montanus latifolius vel Verbaſculum cyanoides. *Bauh. pin.* 273.
Broad leav'd Mountain Centaury, or Blue-bottle.
β Jacea integrifolia humilis. *Bauh. pin.* 271.
Dwarf Mountain Centaury, or Blue-bottle.
Nat. of Auſtria and Switzerland.
Cult. 1596, by Mr. John Gerard. *Hort. Ger.*
Fl. June——Auguſt. H. ♃.

Cyanus. 9. C. calycibus ſerratis: foliis linearibus integerrimis: infimis dentatis. *Sp. pl.* 1289.
Corn Centaury, or Blue-bottle.
Nat. of Britain.
Fl. Moſt part of the Summer. H. ⊙.

10. C. calycibus ciliatis: ſquamis planis, foliis bipinnatifidis; rameis pinnatifidis linearibus, caule paniculato. *Sp. pl.* 1289. *Jacqu. auſtr.* 4. *p.* 10. *t.* 320. *paniculata.*
Panicl'd Centaury.
Nat. of Europe.
Cult. 1640. *Park. theat.* 476. *n.* 4.
Fl. July and Auguſt. H. ♂.

11. C. calycibus ciliatis: ſquamis planis adpreſſis apice ciliatis, foliis caulinis oblongis decurrentibus. *alata.*
Centaurea alata. *De Lamarck. encycl.* 1. *p.* 665.
Centaurea ſemialata. *L'Herit. ſtirp. nov. tom.* 2. *tab.* 87.
Upright Wing-ſtalk'd Centaury.
Nat.
Introd. 1781, by Monſ. Thouin.
Fl. Auguſt and September. H. ♃.

12. C. calycibus ciliatis: ſquamis planis obtuſis apice ciliatis, foliis pinnatifidis, diſco radio æquali. *intybacea.*
Centaurea intybacea. *De Lamarck encycl.* 1. *p.* 671.
Succory-leav'd Centaury.
Nat. of the South of Europe.
Introd. 1778, by Monſ. Thouin.
Fl. July——September. H. ♃.

13. C. calyce ſubciliato, ramis ſpinoſis. *Sp. pl.* 1290. *ſpinoſa.*
Prickly-branch'd Centaury.
Nat. of the Iſland of Candia.
Introd. 1788, by John Sibthorp, M.D.
Fl. G. H. ♃.

14. C. calycibus ciliatis, foliis tomentoſis pinnatifidis: foliolis obtuſis ovatis integerrimis: exterioribus majoribus. *Sp. pl.* 1290. *raguſina*

Cretan Centaury.
Nat. of the Island of Candia.
Cult. 1714, by the Dutchess of Beaufort. *Br. Mus.* *H. S.* 133. *fol.* 69.
Fl. June and July. G. H. ♃.

Cineraria. 15. C. calycibus ciliatis terminali-sessilibus, foliis tomentosis bipinnatifidis: lobis acutis. *Sp. pl.* 1290.
White-leav'd Mountain Centaury.
Nat. of Italy.
Cult. 1731, by Mr. Philip Miller. *Mill. dict. edit.* 1. Jacea 3.
Fl. July and August. G. H. ♃.

argentea. 16. C. calycibus serratis, foliis tomentosis: radicalibus pinnatifidis: foliolis inauritis. *Syst. veget.* 786.
Silvery Centaury.
Nat. of the Island of Candia.
Cult. 1768, by Mr. Philip Miller. *Mill. dict. edit.* 8.
Fl. July and August. G. H. ♄.

sibirica. 17. C. calycibus ciliatis, foliis tomentosis indivisis pinnatifidisque integerrimis, caule declinato. *Sp. pl.* 1291.
Siberian Centaury.
Nat. of Siberia.
Introd. 1782, by Mr. John Bush.
Fl. H. ♃.

sempervirens. 18. C. calycibus ciliatis, foliis lanceolatis serratis: infimo dente substipulatis; inferioribus hastatis. *Syst. veget.* 786.
Evergreen Centaury.
Nat. of Spain and Portugal.

Cult.

Cult. 1683, by Mr. James Sutherland. *Sutherl. hort. edin.* 166. *n.* 3.

Fl. July and Auguſt. G. H. ♃.

19. C. calycibus ciliatis, foliis pinnatifidis; pinnis lanceolatis. *Sp. pl.* 1291. *Scabioſa.*

Scabious Centaury, or Greater Knapweed.

Nat. of Britain.

Fl. June and July. H. ♃.

20. C. calycibus ciliatis: ſquamis acuminatis ſubſpinoſis, foliis oblongis pinnatifidiſque, floſculis radii diſco longioribus. *diluta.*

Pale-flower'd Centaury.

Nat. of the South of Europe.

Introd. 1781, by Monſ. Thouin.

Fl. July and Auguſt. H. ♃.

*** Rhapontica: *calycinis ſquamis aridis ſcarioſis.*

21. C. calycibus ſcarioſis ciliatis, foliis pinnatifidis: pinnis lanceolatis. *Syſt. veget.* 786. *orientalis.*

Oriental Centaury.

Nat. of Siberia.

Cult. 1759, by Mr. Philip Miller. *Mill. dict. edit.* 7. *n.* 18.

Fl. July and Auguſt. H. ♃.

22. C. calycibus ſcarioſis laceris, foliis lanceolatis: radicalibus ſinuato-dentatis, ramis angulatis. *Sp. pl.* 1293. *Jacea.*

Common Centaury, or Knapweed.

Nat. of the North of Europe.

Cult. 1748, by Mr. Philip Miller. *Mill. dict. edit.* 5. Jacea 1.

Fl. June and July. H. ♃.

alba. 23. C. calycibus ſcarioſis integris mucronatis, foliis pinnato-dentatis: caulinis linearibus baſi dentatis. *Sp. pl.* 1293.
White-flower'd Centaury.
Nat. of Spain.
Introd. 1773, by Monſ. Richard.
Fl. June——September. H. ♃.

ſplendens. 24. C. calycibus ſcarioſis obtuſis, foliis radicalibus bipinnatifidis; caulinis pinnatis: dentibus lanceolatis. *Sp. pl.* 1293.
Shining Centaury.
Nat. of Spain and Siberia.
Cult. 1597, by Mr. John Gerard. *Ger. herb.* 590. *f.* 1.
Fl. July and Auguſt. H. ♂.

Rhapontica. 25. C. calycibus ſcarioſis, foliis ovato-oblongis denticulatis integris petiolatis ſubtus tomentoſis. *Sp. pl.* 1294.
Swiſs Centaury.
Nat. of Switzerland.
Cult. 1656, by Mr. John Tradeſcant, Jun. *Muſ. Trad.* 160.
Fl. July and Auguſt. H. ♃.

babylonica. 26. C. calycibus ſcarioſis, foliis ſubtomentoſis decurrentibus indiviſis: radicalibus lyratis. *Syſt. veget.* 787.
Serratula babylonica. *Sp. pl.* 1148.
Babylonian Centaury.
Nat. of the Levant.
Cult. 1748, by Mr. Philip Miller. *Mill. dict. edit.* 5. Jacea 14.
Fl. June——September. H. ♃.

27. C.

27. C. calycibus ſcarioſis, foliis decurrentibus indiviſis integerrimis. *Sp. pl.* 1294. *Curtis magaz.* 62. *glaſtifolia.*
Woad-leav'd Centaury.
Nat. of Siberia.
Cult. 1731, by Mr. Philip Miller. *Mill. dict. edit.* 1. Centaurium 6.
Fl. June——Auguſt. H. ♃.

28. C. calycibus ſcarioſis, foliis tomentoſis: radicalibus lanceolatis; caulinis pinnatifidis, caule ſimplici. *Sp. pl.* 1294. *conifera.*
Cone Centaury.
Nat. of the South of Europe.
Cult. 1757, by Mr. Ph. Miller. *Mill. ic.* 102. *t.* 153.
Fl. June——September. H. ♃.

* * * * Stœbæ: *calycinis ſpinis palmatis.*

29. C. calycibus palmato-ſpinoſis, foliis decurrentibus ſpinuloſis repando-dentatis. *Sp. pl.* 1294. *ſonchifolia.*
Sow-thiſtle-leav'd Centaury.
Nat. of the Coaſts of the Mediterranean.
Introd. about 1780, by Monſ. Thouin.
Fl. Auguſt——October. H. ♂.

30. C. calycibus palmato-ſpinoſis, foliis decurrentibus ſinuatis ſpinuloſis: radicalibus lyratis. *Sp. pl.* 1295. *napifolia.*
Turnep-leav'd Centaury.
Nat. of the Iſland of Candia.
Cult. 1759, by Mr. Philip Miller. *Mill. dict. edit.* 7. *n.* 15.
Fl. July——September. H. ☉.

31. C. calycibus palmato-triſpinoſis, foliis lanceolatis dentatis. *Syſt. veget.* 787. *aſpera.*

Rough Centaury.
Nat. of the South of Europe.
Introd. 1772, by Monſ. Richard.
Fl. June——October. H. ☉.

***** Calcitrapæ: *calycinis ſpinis compoſitis.*

benedicta. 32. C. calycibus duplicato-ſpinoſis lanatis involucratis, foliis ſemidecurrentibus denticulato-ſpinoſis. *Sp. pl.* 1296.
Bleſſed Thiſtle, or Centaury.
Nat. of Spain and the Levant.
Cult. 1597. *Ger. herb.* 1008. *f.* 2.
Fl. June——September. H. ☉.

eriophora. 33. C. calycibus duplicato-ſpinoſis lanatis, foliis ſemidecurrentibus integris ſinuatiſque, caule prolifero, *Sp. pl.* 1296.
Woolly-headed Centaury.
Nat. of Portugal.
Cult. 1768, by Mr. Ph. Miller. *Mill. dict. edit.* 8.
Fl. June——September. H. ☉.

Calcitrapa. 34. C. calycibus ſubduplicato-ſpinoſis ſeſſilibus, foliis linearibus pinnatifidis dentatis, caule piloſo. *Sp. pl.* 1297.
Common Star Centaury, or Thiſtle.
Nat. of England.
Fl. July and Auguſt. H. ☉.

Calcitrapoides. 35. C. calycibus ſubduplicato-ſpinoſis, foliis amplexicaulibus lanceolatis indiviſis ſerratis. *Sp. pl.* 1297.
Phœnician Centaury.
Nat. of the Levant.
Cult. 1683, by Mr. James Sutherland. *Sutherl. hort. edin.* 68. *n.* 4.
Fl. June and July. H. ☉.

36. C. calycibus duplicato-ſpinoſis ſolitariis, foliis rameis decurrentibus inermibus lanceolatis; radicalibus lyrato-pinnatifidis. *Sp. pl.* 1297. *ſolſtitialis.*
St. Barnaby's Centaury, or Thiſtle.
Nat. of England.
Fl. July and Auguſt. H. ⊙.

37. C. calycibus duplicato-ſpinoſis terminalibus confertis; foliis decurrentibus lanceolatis ſinuoſis inermibus. *Sp. pl.* 1297. *melitenſis.*
Cluſter-headed Centaury.
Nat. of the Iſland of Malta.
Cult. 1739, by Mr. Ph. Miller. *Rand. chel.* Jacea 18.
Fl. July and Auguſt. H. ⊙.

****** Crocodiloidea: *ſpinis ſimplicibus.*

38. C. calycibus ſimpliciſſime ſpinoſis: dentibus duobus oppoſitis, foliis lanceolatis integris decurrentibus. *Sp. pl.* 1299. *Jacqu. ic. collect.* 1. *p.* 90. *Verutum.*
Dwarf Centaury.
Nat. of the Levant.
Introd. 1780, by Mr. William Hudſon.
Fl. Auguſt and September. H. ⊙.

39. C. calycibus ſetula ſubſpinoſa exſtante glabris, foliis lyrato-runcinatis ſerratis. *Syſt. veget.* 788. *Jacqu. hort.* 1. *p.* 26. *t.* 64. *ſalmantica.*
Lyre-leav'd Centaury.
Nat. of the South of Europe.
Cult. 1596, by Mr. John Gerard. *Hort. Ger.*
Fl. July and Auguſt. H. ♂.

40. C. calycibus ſimpliciſſime ſpinoſis: ſpinis patentibus, floſculis æqualibus, foliis hirtis: inferioribus pinnatifidis. *aurea.*

Great

Great Golden Centaury.
Nat. of the South of Europe.
Cult. 1758, by Mr. Philip Miller.
Fl. July and August. H. ♃.

Crocodilium. 41. C. calycibus scariosis simplicissime spinosis, foliis pinnatifidis integerrimis: lacinia extima majore dentata. *Sp. pl.* 1299.
Blush Centaury.
Nat. of the Levant.
Introd. 1777, by John Earl of Bute.
Fl. July and August. H. ⊙.

galactites. 42. C. calycibus setaceo-spinosis, foliis decurrentibus sinuatis spinosis subtus tomentosis. *Syst. veget.* 789.
White-vein'd Centaury.
Nat. of the South of Europe.
Cult. 1739, by Mr. Philip Miller. *Rand. chel.* Carduus 9.
Fl. July and August. H. ⊙.

POLYGAMIA NECESSARIA.

MILLERIA. *Gen. pl.* 985.

Recept. nullum. *Pappus* nullus. *Cal.* 3-valvis. *Radius* corollæ dimidiatus.

quinqueflora. 1. M. foliis cordatis, pedunculis dichotomis. *Sp. pl.* 1301.
Five-flower'd Milleria.
Nat. of Vera Cruz.

Introd. 1731, by William Houſtoun, M. D. *Mill. dict. edit.* 8.
Fl. Auguſt—Octoḃer. S. ⊙.

BALTIMORA. *Linn. mant.* 158.

Recept. paleaceum. *Pappus* nullus. *Cal.* cylindricus, polyphyllus. *Radius* corollæ 5-florus.

1. BALTIMORA. *Linn. mant.* 288. *recta.*
Upright Baltimora.
Nat. of Maryland.
Introd. 1781, by Monſ. Thouin.
Fl. June and July. H. ⊙.

SILPHIUM. *Gen. pl.* 986.

Recept. paleaceum. *Pappus* marginato-bicornis. *Cal.* ſquarroſus.

1. S. foliis alternis pinnato-ſinuatis. *Sp. pl.* 1301. *laciniatum.*
Jagged-leav'd Silphium.
Nat. of North America.
Introd. 1781, by Monſ. Thouin.
Fl. July—September. H. ♃.

2. S. foliis alternis ovatis ſerratis ſcabris; radicalibus cordatis. *Linn. ſuppl.* 383. *Jacqu. hort.* 1. *p.* 16. *t.* 43. *terebinthinum.*
Broad-leav'd Silphium.
Nat. of North America.
Introd. 1756, by Meſſrs. Kennedy and Lee.
Fl. Auguſt and September. H. ♃.

3. S. foliis oppoſitis deltoidibus petiolatis perfoliatis, caule tetragono lævi. *Syſt. veget.* 789. *perfoliatum.*

Square-

Square-ſtalk'd Silphium.
Nat. of North America.
Cult. 1766, by Peter Collinſon, Eſq.
Fl. July——October. H. ♃.

connatum. 4. S. foliis oppoſitis ſeſſilibus perfoliatis, caule tereti ſcabro. *Linn. mant.* 574.
Round-ſtalk'd Silphium.
Nat. of North America.
Introd. 1765, by Meſſrs. Kennedy and Lee.
Fl. July——October. H. ♃.

Aſteriſcus. 5. S. foliis indiviſis ſeſſilibus oppoſitis: inferioribus alternis. *Sp. pl.* 1302.
Hairy-ſtalk'd Silphium.
Nat. of North America.
Cult. 1732, by James Sherard, M.D. *Dill. elth.* 42. *t.* 37. *f.* 42.
Fl. July——September. H. ♃.

trifoliatum. 6. S. foliis ternis. *Sp. pl.* 1302.
Three-leav'd Silphium.
Nat. of North America.
Cult. 1759, by Mr. Ph. Miller. *Mill. dict. edit.* 7. *n.* 1.
Fl. July——October. H. ♃.

POLYMNIA. *Gen. pl.* 987.

Recept. paleaceum. *Pappus* nullus. *Cal. exterior* 4 ſ. 5-phyllus: *interior* 10-phyllus: foliolis concavis.

Uvedalia. 1. P. foliis oppoſitis haſtato-ſinuatis. *Sp. pl.* 1303.
Broad-leav'd Polymnia.
Nat. of Virginia.

Cult.

Cult. 1712. *Philosoph. transact. n.* 333. *p.* 423. *n.* 82.

Fl. August——October. H. ♃.

2. P. foliis oppositis, spathulatis subdentatis. *Syst. veget.* 790. *Tetragonotheca.*

Tetragonotheca helianthoides. *Sp. pl.* 1273.

Narrow-leav'd Polymnia.

Nat. of Virginia.

Cult. 1726, by James Sherard, M.D. *Dill. elth.* 378. *t.* 283. *f.* 365.

Fl. August——October. H. ♃.

3. P. foliis oppositis sessilibus oblongo-lanceolatis subdentatis, calycibus quinquepartitis, flosculis omnibus seminiferis. *abyssinica.*

Polymnia abyssinica. *Linn. suppl.* 383.

Upright Polymnia.

Nat. of Africa. *James Bruce,* Esq.

Introd. 1775.

Fl. April and May. S. ♂.

MELAMPODIUM. *Gen. plant.* 989.

Recept. paleaceum, conicum. *Pappus* 1-phyllus, vulviformis. *Cal.* 5-phyllus.

1. M. caule erecto, foliis lyrato-dentatis sessilibus. *Swartz prodr.* 114. *humile.*

Dwarf Melampodium.

Nat. of Jamaica.

Introd. 1782, by Mr. Gilbert Alexander.

Fl. June——October. S. ⊙.

CALEN-

CALENDULA. *Gen. pl.* 990.

Recept. nudum. *Pappus* nullus. *Cal.* polyphyllus, æqualis. *Sem.* disci membranacea.

arvensis. 1. C. seminibus cymbiformibus muricatis incurvatis: extimis erectis protensis. *Sp. pl.* 1303.
Field Marygold.
Nat. of Europe.
Cult. 1683, by Mr. James Sutherland. *Sutherl. hort. edin.* 62. *n.* 3.
Fl. Most part of the Summer. H. ☉.

officinalis. 2. C. seminibus cymbiformibus muricatis incurvatis omnibus. *Sp. pl.* 1304.
α flore simplici.
Common Marygold.
β flore pleno.
Double-flower'd Marygold.
Nat. of the South of Europe.
Cult. 1597. *Ger. herb.* 600.
Fl. June——September. H. ♃.

pluvialis. 3. C. foliis lanceolatis sinuato-denticulatis, caule folioso, pedunculis filiformibus. *Sp. pl.* 1304.
Small Cape Marygold.
Nat. of the Cape of Good Hope.
Cult. 1726, by Mr. Philip Miller. *R. S. n.* 214.
Fl. June——August. H. ☉.

hybrida. 4. C. foliis lanceolatis dentatis, caule folioso, pedunculis superne incrassatis. *Sp. pl.* 1304.
Great Cape Marygold.
Nat. of the Cape of Good Hope.

Cult.

Cult. 1756, by Mr. Ph. Miller. *Mill. ic.* 50. *t.* 75.
Fl. June and July. H. ⊙.

5. C. foliis lanceolatis ſinuato-dentatis, caule ſubnudo. *Sp. pl.* 1305. *nudicaulis.*
Naked-ſtalk'd Cape Marygold.
Nat. of the Cape of Good Hope.
Cult. 1731, by Mr. Philip Miller. *Mill. dict. edit.* 1. Caltha 8.
Fl. June——Auguſt. H. ⊙.

6. C. cauleſcens, foliis alternis linearibus ſubintegerrimis piloſiuſculis, ſeminibus ſuborbiculatis. *Tragus.*
Bending-ſtalk'd Marygold.
Nat. of the Cape of Good Hope. Mr. *Fr. Maſſon.*
Introd. 1774.
Fl. May and June. G. H. ♄.

7. C. foliis linearibus ſubintegerrimis, caulibus ſubnudis unifloris, ſeminibus obcordato-orbiculatis lævibus. *graminifolia.*
Calendula graminifolia. *Sp. pl.* 1305.
Graſs-leav'd Marygold.
Nat. of the Cape of Good Hope.
Cult. 1731, by Mr. Philip Miller. *Mill. dict. edit.* 1. Caltha 9.
Fl. Moſt part of the Year. G. H. ♄.

8. C. foliis ellipticis dentatis ſcabriuſculis, ſeminum alis ſemiorbiculatis. *rigida.*
Rough-leav'd Marygold.
Nat. of the Cape of Good Hope. Mr. *Fr. Maſſon.*
Introd. 1774.
Fl. December. G. H. ♄.

9. C.

fruticosa. 9. C. foliis obovatis ſubdentatis, caule fruticoſo decumbente. *Syſt. veget.* 791.
Shrubby Marygold.
Nat. of the Cape of Good Hope.
Cult. 1759, by Mr. Ph. Miller. *Mill. dict. edit.* 7. *n.* 8.
Fl. June and July. G. H. ♄.

oppoſitifolia. 10. C. foliis oppoſitis linearibus integerrimis ſubcarnoſis glabris.
Glaucous Marygold.
Nat. of the Cape of Good Hope. Mr. *Fr. Maſſon.*
Introd. 1774.
Fl. Auguſt. G. H. ♄.

ARCTOTIS. *Gen. pl.* 991.

Recept. villoſum ſ. paleaceum. *Pappus* corona pentaphylla. *Cal.* imbricatus: ſquamis apice ſcarioſis.

calendulacea. 1. A. floſculis radiantibus ſterilibus, foliis runcinatis ſubtomentoſis. *Syſt. veget.* 791.
Marygold-flower'd Arctotis.
Nat. of the Cape of Good Hope.
Cult. 1752, by Mr. Ph. Miller. *Mill. dict. edit.* 6. *n.* 4.
Fl. June——Auguſt. G. H. ⊙.

grandiflora. 2. A. floſculis radiantibus fertilibus, foliis pinnatifidis denticulatis arachnoideis trinerviis.
Great-flower'd Arctotis.
Nat. of the Cape of Good Hope. Mr. *Fr. Maſſon.*
Introd. 1774.
Fl. March——May. G. H. ♂.
Obs. *Radius* maximus. *Petala* ſtraminea, ſubtus rubore tincta, ſupra prope baſin luteſcentia, ore nigropurpureo.

3. A.

3. A. flosculis radiantibus fertilibus, foliis lanceolato-ovatis nervosis denticulatis amplexicaulibus. *Sp. pl.* 1306. *plantaginea.*

Plantain-leav'd Arctotis.

Nat. of the Cape of Good Hope.

Cult. 1768, by Mr. Philip Miller. *Mill. dict. edit.* 8.

Fl. June——August. G. H. ♃.

4. A. flosculis radiantibus fertilibus, foliis lanceolato-linearibus integerrimis tomentosis. *argentea.*

Silvery Arctotis.

Nat. of the Cape of Good Hope. Mr. *Fr. Masson.*

Introd. 1774.

Fl. August. G. H. ♂.

Obs. *Flores* flavi.

5. A. flosculis radiantibus fertilibus, foliis pinnato-sinuatis villosis: laciniis oblongis dentatis. *Sp. pl.* 1307. *aspera.*

α Anemonospermos africana, foliis cardui benedicti, florum radiis intus sulphureis. *Comm. hort.* 2. *p.* 43. *t.* 22.

Broad-leav'd rough Arctotis.

β Anemonospermos afra, folio Jacobææ tenuiter laciniato, flore aurantio pulcherrimo. *Boerh. lugdb.* 1. *p.* 100. *t.* 100.

Narrow-leav'd rough Arctotis.

Nat. of the Cape of Good Hope.

Cult. 1731, by Mr. Philip Miller. *Mill. dict. edit.* 1. Anemonospermos 3.

Fl. July——September. G. H. ♄.

6. A. flosculis radiantibus sterilibus, paleis disco longioribus coloratis, foliis bipinnatis linearibus. *Sp. pl.* 1307. *paradoxa.*

Camomile-leav'd Arctotis.

Nat. of the Cape of Good Hope.
Introd. 1774, by Mr. Francis Maſſon.
Fl. Auguſt. G. H. ♂.

ſcarioſa. 7. A. floſculis radiantibus ſterilibus, paleis floſculos diſci æquantibus, foliis decompoſitis.
Southernwood-leav'd Arctotis.
Nat. of the Cape of Good Hope. Mr. *Fr. Maſſon.*
Introd. 1774.
Fl. April——Auguſt. G. H. ♄.
Obs. In hac ſpecie foliola calycina omnia obtuſa ſcarioſa cochleata ſquarroſa patentia; in ſequenti interiora tantum ſcarioſa obtuſa, exteriora ſubſucculenta lanceolata.

paleacea. 8. A. floſculis radiantibus ſterilibus, paleis floſculos diſci æquantibus, foliis pinnatis linearibus. *Sp. pl.* 1307.
Chaffy Arctotis.
Nat. of the Cape of Good of Hope.
Cult. 1768, by Mr. Philip Miller. *Mill. dict. edit.* 8.
Fl. April——Auguſt. G. H. ♄.

dentata. 9. A. floſculis radiantibus ſterilibus, foliis pinnatis: pinnis pinnatifido-dentatis. *Sp. pl.* 1307.
Fine-leav'd Arctotis.
Nat. of the Cape of Good Hope.
Introd. 1787, by Mr. Haneman.
Fl. July. G. H. ☉.

acaulis. 10. A. pedunculis radicalibus, foliis lyratis. *Sp. pl.* 1306.
Dwarf Arctotis.
Nat. of the Cape of Good Hope.
Cult. 1759, by Mr. Ph. Miller. *Mill. dict. edit.* 7. *n.* 7.
Fl. April——July. G. H. ♃.

OSTEOSPERMUM. *Gen. pl.* 992.

Recept. nudum. *Pappus* nullus. *Cal.* polyphyllus. *Sem.* globosa, colorata, ossea.

1. O. spinis ramosis. *Sp. pl.* 1308. *spinosum.*
Prickly Osteospermum.
Nat. of the Cape of Good Hope.
Cult. 1700, by Dr. Uvedale. *Pet. mus.* 799.
Fl. February——October. G. H. ♄.

2. O. foliis lanceolatis mucronatis subpetiolatis glabris serratis, ramulis denticulato-angulatis. *Sp. pl.* 1308. *pisiferum.*
Smooth Osteospermum.
Nat. of the Cape of Good Hope.
Cult. 1757, by Mr. Philip Miller. *Mill. ic.* 129. *t.* 194. *f.* 1.
Fl. March——May. G. H. ♄.

3. O. foliis obovatis serratis petiolatis subdecurrentibus. *Syst. veget.* 792. *moniliferum.*
Poplar-leav'd Osteospermum.
Nat. of the Cape of Good Hope.
Cult. 1714. *Philosoph. transact. n.* 346. *p.* 357. *n.* 107.
Fl. July and August. G. H. ♄.

4. O. foliis dentato-pinnatifidis pilosis, ramis inermibus. *rigidum.*
Rigid Osteospermum.
Nat. of the Cape of Good Hope. Mr. *Fr. Masson.*
Introd. 1774.
Fl. April——July. G. H. ♄.

5. O. foliis pinnatis: pinnis dentatis. *cæruleum.*
Osteospermum cæruleum. *Jacqu. ic. collect.* 1. *p.* 78.
Osteospermum pinnatifidum. *L'Herit. stirp. nov. p.* 11. *tab.* 6.

Blue-flower'd Osteospermum.
Nat. of the Cape of Good Hope. Mr. *Fr. Masson.*
Introd. 1774.
Fl. June——September. G. H. ♄.

OTHONNA. *Gen. pl.* 993.

Recept. nudum. *Pappus* subnullus. *Cal.* 1-phyllus, multifidus, subcylindricus.

Cacalioides. 1. O. tubere denudato digitato-lobato plantigero, scapis unifloris, foliis obovatis denticulatis.
Othonna Cacalioides. *Linn. suppl.* 388.
Tuberous Othonna, or African Ragwort.
Nat. of the Cape of Good Hope. Mr. *Fr. Masson.*
Introd. 1774.
Fl. May——November. G. H. ♃.

bulbosa. 2. O. foliis oblongis nudis petiolatis, caule herbaceo, pedunculis unifloris longissimis. *Sp. pl.* 1309.
Bulbous African Ragwort.
Nat. of the Cape of Good Hope.
Introd. 1774, by Mr. Francis Masson.
Fl. May and June. D. S. ♃.

denticulata. 3. O. foliis oblongis denticulatis glabris basi attenuatis amplexicaulibus, floribus paniculatis.
Dentated African Ragwort.
Nat. of the Cape of Good Hope. Mr. *Fr. Masson.*
Introd. 1774.
Fl. April——July. G. H. ♄.

pectinata. 4. O. foliis pinnatifidis: laciniis linearibus parallelis. *Sp. pl.* 1309.
Wormwood-leav'd African Ragwort.
Nat. of the Cape of Good Hope.

Cult.

Cult. 1731, by Mr. Philip Miller. *Mill. dict. edit.* 1. Jacobæa 4.
Fl. May and June. G. H. ♄.

5. O. foliis multifido-pinnatis linearibus, caulis geniculis villosis. *Sp. pl.* 1310. *abrotanifolia.*
Southernwood-leav'd African Ragwort.
Nat. of the Cape of Good Hope.
Cult. 1759, by Mr. Ph. Miller. *Mill. dict. edit.* 7. *n.* 8.
Fl. January——March. G. H. ♄.

6. O. foliis infimis lanceolatis integerrimis; superioribus sinuato-dentatis. *Sp. pl.* 1310. *coronopifolia.*
Buck's-horn African Ragwort.
Nat. of the Cape of Good Hope.
Cult. 1731, by Mr. Philip Miller. *Mill. dict. edit.* 1. Doria 7.
Fl. July——September. G. H. ♄.

7. O. foliis lanceolatis trinerviis integerrimis, caule suffruticoso repente. *Sp. pl.* 1310. *cheirifolia.*
Stock-leav'd African Ragwort.
Nat. of Africa.
Cult. 1758, by Mr. Philip Miller. *Mill. ic.* 163. *t.* 245. *f.* 1.
Fl. April——June. H. ♄.

8. O. foliis filiformibus carnosis, caule fruticoso. *Linn. mant.* 118. *tenuissima.*
Fine-leav'd African Ragwort.
Nat. of the Cape of Good Hope.
Introd. 1774, by Mr. Francis Masson.
Fl. July. D. S. ♄.

9. O. foliis oblongis integerrimis, caule arborescente carnoso: cicatricibus lanatis. *Syst. veget.* 795. *arborescens.*

Tree African Ragwort.
Nat. of the Cape of Good Hope.
Cult. 1729, by James Sherard, M.D. *Dill. elth.* 123, *t.* 103. *f.* 123.
Fl. G. H. ♄.

HIPPIA. *Linn. mant.* 158.

Recept. nudum. *Pappus* nullus. *Sem.* latiſſimis marginibus, nuda. *Cal.* hemiſphæricus, ſubimbricatus. *Corollulæ radii* 10, obſoletæ, ſubtrifidæ.

integrifolia. 1. H. hiſpida erecta, foliis ovatis ſerratis quinquenerviis, racemis terminalibus. *Linn. ſuppl.* 389.
Sphæranthus africanus. *Burm. ind.* 185. *tab.* 60. *fig.* 2.
Ethulia paniculata. *Houtt. nat. hiſt.* 10. *p.* 551. *tab.* 67. *fig.* 2.
Annual Hippia.
Nat. of the Eaſt Indies.
Introd. 1777, by Monſ. Thouin.
Fl. July and Auguſt. S. ⊙.

fruteſcens. 2. H. fruticoſa villoſa, foliis pinnatifidis, floribus corymboſis. *Syſt. veget.* 795.
Tanacetum fruteſcens. *Sp. pl.* 1183.
Shrubby Hippia.
Nat. of the Cape of Good Hope.
Cult. 1731, by Mr. Philip Miller. *Mill. dict. edit.* 1. Tanacetum 5.
Fl. February——Auguſt. G. H. ♄.

ERIOCEPHALUS. *Gen. pl.* 994.

Recept. ſubvilloſum. *Pappus* nullus. *Cal.* 10-phyllus, æqualis. *Radii* floſculi 5.

africanus. 1. E. foliis integris diviſiſque, floribus corymboſis. *Sp. pl.* 1310.

Cluſter-

Cluſter-leav'd Eriocephalus.
Nat. of the Cape of Good Hope.
Cult. 1732, by James Sherard, M. D. *Dill. elth.* 132. *t.* 110. *f.* 134.
Fl. January——March. G. H. ♄.

2. E. foliis linearibus indiviſis, floribus racemoſis. *Sp. pl.* 1311. *racemoſus.*
Silvery-leav'd Eriocephalus.
Nat. of the Cape of Good Hope.
Cult. 1758, by Mr. Philip Miller.
Fl. March and April. G. H. ♄.

FILAGO. *Gen. pl.* 995.

Recept. nudum. *Pappus* nullus. *Cal.* imbricatus. *Floſculi feminei* inter ſquamas calycis locati.

1. F. panicula dichotoma, floribus rotundatis axillaribus hirſutis, foliis acutis. *Sp. pl.* 1311. *germanica.*
Common Cudweed.
Nat. of Britain.
Fl. June and July. H. ⊙.

2. F. caule ſubdichotomo erecto, floribus conicis terminalibus axillaribuſque. *Syſt. veget.* 796. *montana.*
Leaſt Cudweed.
Nat. of Britain.
Fl. June and July. H. ⊙.

3. F. caule erecto dichotomo, floribus ſubulatis axillaribus, foliis filiformibus. *Sp. pl.* 1312. *gallica.*
Corn Cudweed.
Nat. of England.
Fl. June and July. H. ⊙.

Leontopodium. 4. F. caule ſimpliciſſimo, capitulo terminali bracteis hirſutiſſimis radiato. *Sp. pl.* 1312.
Lion's-foot Cudweed.
Nat. of Auſtria and Switzerland.
Introd. 1776, by Profeſſor de Sauſſure.
Fl. June and July. H. ♃.

MICROPUS. *Gen. pl.* 996.

Recept. paleaceum. *Pappus* nullus. *Cal.* calyculatus. *Radius* corollæ nullus. *Floſculi feminei* a ſquamis calycinis involuti.

ſupinus. 1. M. caule procumbente, foliis geminis. *Syſt. veget.* 796.
Trailing Micropus.
Nat. of Portugal, Italy, and the Levant.
Cult. 1759, by Mr. Philip Miller. *Mill. dict. edit.* 7.
Fl. June——September. G. H. ⊙.

POLYGAMIA SEGREGATA.

ELEPHANTOPUS. *Gen. pl.* 997.

Calyculus 4-florus. *Corollulæ* ligulatæ, hermaphroditæ. *Recept.* nudum. *Pappus* ſetaceus.

ſcaber. 1. E. foliis oblongis ſcabris. *Sp. pl.* 1313.
Rough-leav'd Elephant's-foot.
Nat. of the Eaſt Indies.
Cult. 1695, in Chelſea Garden. *Pluk. alm.* 132. *t.* 388. *f.* 6.
Fl. June——September. S. ♃.

SPHÆRAN-

SPHÆRANTHUS. *Gen. pl.* 998.

Calyc. 8-flori. *Coroll.* tubulosæ hermaphroditæ et obsoletæ femineæ. *Recept.* squamosum. *Pappus* nullus.

1\. S. foliis decurrentibus lanceolatis serratis, pedunculis crispatis. *Syst. veget.* 797. *indicus.*
Indian Sphæranthus.
Nat. of the East Indies.
Cult. 1709, by the Dutchess of Beaufort. *Br. Mus. Sloan. mss.* 525 & 3349.
Fl. August——December. S. ♃.

ECHINOPS. *Gen. pl.* 999.

Calyc. 1-florus. *Coroll.* tubulosæ, hermaphroditæ. *Recept.* setosum. *Pappus* obsoletus.

1\. E. capitulis globosis, foliis sinuatis pubescentibus. *Syst. veget.* 797. *sphærocephalus.*
Great Globe Thistle.
Nat. of Austria and Italy.
Cult. 1596, by Mr. John Gerard. *Hort. Ger.*
Fl. July and August. H. ♃:

2\. E. capitulo globoso, foliis pinnatifidis supra glabris. *Syst. veget.* 797. *Ritro.*
Small Globe Thistle.
Nat. of Siberia, France, and Italy.
Cult. 1629. *Park. parad.* 331. *f.* 6.
Fl. July——September. H. ♃.

3\. E. capitulis fascicularibus, calycibus lateralibus sterilibus, foliis supra strigosis. *Sp. pl.* 1315. *strigosus.*

Annual

Annual Globe Thiſtle.
Nat. of Spain.
Cult. 1731, by Mr. Philip Miller. *Mill. dict. edit.* 1. Echinopus 3.
Fl. July and Auguſt. H. ☉.

MONOGAMIA.

SERIPHIUM. *Gen. pl.* 1003.

Cal. imbricatus. *Cor.* 1-petala, regularis. *Sem.* 1, oblongum, infra corollam.

cinereum. 1. S. floribus verticillato-ſpicatis unifloris, foliis patentibus. *Syſt. veget.* 798.
Heath-leav'd Seriphium.
Nat. of the Cape of Good Hope.
Introd. 1774, by Mr. Francis Maſſon.
Fl. July——September. G. H. ♄.

JASIÒNE. *Gen. pl.* 1005.

Cal. communis 10-phyllus. *Cor.* 5-petala, regularis. *Capſ.* infera, bilocularis.

montàna. 1. Jasione. *Sp. pl.* 1317. *Curtis lond.*
α Rapunculus ſcabioſæ capitulo cæruleo. *Bauh. pin.* 92.
Annual Jaſione, or Sheep's Scabious.
β radice perenni. *Linn. ſuppl.* 392.
Perennial Jaſione.
Nat. of Britain.
Fl. June——September. H. *α.* ☉. *β.* ♃.

LOBELIA.

LOBELIA. *Gen. pl.* 1006.

Cal. 5-fidus. *Corolla* 1-petala, irregularis. *Capsula* infera, 2-s. 3-locularis.

* *Foliis integerrimis.*

1. L. fruticosa, foliis linearibus confertis integerrimis. *Sp. pl.* 1318. *pinifolia.*
Pine-leav'd Lobelia.
Nat. of the Cape of Good Hope.
Introd. 1786, by Mr. Francis Masson.
Fl. G. H. ♄.

2. L. foliis linearibus bilocularibus integerrimis, caule subnudo. *Sp. pl.* 1318. *Dortmanna.*
Water Lobelia, or Gladiole.
Nat. of Britain.
Fl. July and August. H. ♃.

* *Caule erecto, foliis incisis.*

3. L. caule erecto, foliis lanceolatis dentatis, racemo terminali aphyllo. *Linn. mant.* 120. *triquetra.*
Tooth-leav'd Lobelia.
Nat. of the Cape of Good Hope.
Introd. 1774, by Mr. Francis Masson.
Fl. May——September. G. H. ♃.

4. L. foliis lanceolatis dentatis, pedunculis brevissimis lateralibus, tubo corollæ filiformi longissimo. *Sp. pl.* 1319. *Jacqu. hort.* 1. *p.* 10. *t.* 27. *longiflora.*
Long-flower'd Lobelia.
Nat. of Jamaica.
Cult. 1739, by Mr. Philip Miller. *Mill. dict. vol.* 2. Trachelium 6.
Fl. June——August. S. ♃.

5. L.

aſſurgens. 5. L. foliis lanceolatis ſerratis inferne dentatis, racemis compoſitis terminalibus. *Sp. pl.* 1321.
Tree Lobelia.
Nat. of the Weſt Indies.
Introd. 1787, by Mr. Alexander Anderſon.
Fl. S. ♄.

Cardinalis. 6. L. caule erecto, foliis lato-lanceolatis ſerratis, racemo terminali ſecundo. *Syſt. veget.* 801.
Scarlet Lobelia, or Cardinal's-flower.
Nat. of Virginia.
Cult. 1629. *Park. parad.* 355. *f.* 6.
Fl. July——October. H. ♃.

debilis. 7. L. erecta, foliis lanceolatis ſerratis glabris, pedunculis lateralibus folio longioribus. *Linn. ſuppl.* 395.
Slender Lobelia.
Nat. of the Cape of Good Hope.
Introd. 1774, by Mr. Francis Maſſon.
Fl. July. G. H. ⊙.

ſiphilitica. 8. L. caule erecto, foliis ovato-lanceolatis ſubſerratis, calycum ſinubus reflexis. *Syſt. veget.* 801.
Blue Lobelia, or Cardinal-flower.
Nat. of Virginia.
Cult. 1665. *Rea's flora*, 140.
Fl. Auguſt——October. H. ♃.

inflata. 9. L. caule erecto, foliis ovatis ſubſerratis pedunculo longioribus, capſulis inflatis. *Sp. pl.* 1320.
Bladder-podded Lobelia.
Nat. of Virginia and Canada.
Cult. 1759, by Mr. Philip Miller. *Mill. dict. edit.* 7. Rapuntium 6.
Fl. July and Auguſt. H. ⊙.

10. L.

10. L. caule erecto, foliis cordatis lævibus obsolete dentatis petiolatis, corymbo terminali. *Syst. veg.* 801. *Cliffortiana.*
Purple Lobelia.
Nat. of America.
Cult. 1739, by Mr. Philip Miller. *Rand. chel.* Rapuntium 4.
Fl. July——October. H. ☉.

11. L. caule erectiusculo, foliis inferioribus subrotundis crenatis; superioribus lanceolatis serratis, floribus racemosis. *Sp. pl.* 1321. *urens.*
Stinging Lobelia.
Nat. of England.
Fl. June. H. ☉.

12. L. foliis radicalibus ovatis, scapis capillaribus. *Linn. mant.* 292. *minuta.*
Least Lobelia.
Nat. of the Cape of Good Hope.
Introd. 1772, by Monf. Richard.
Fl. June——September. G. H. ♃.

*** *Caule prostrato, foliis incisis.*

13. L. caule prostrato, foliis lanceolato-ovalibus crenatis, caule ramoso, pedunculis solitariis unifloris longissimis. *Sp. pl.* 1321. *Laurentia.*
Italian annual Lobelia.
Nat. of Italy.
Introd. 1778, by Monf. Thouin.
Fl. July. G. H. ☉.

14. L. caule patulo, foliis lanceolatis subdentatis, pedunculis longissimis. *Syst. veget.* 802. *Erinus.*
Small spreading Lobelia.
Nat. of the Cape of Good Hope.

Introd.

Introd. 1774, by Mr. Francis Maſſon.
Fl. June——September. G. H. ♃.

erinoides. 15. L. caulibus proſtratis filiformibus, foliis petiolatis oblongis dentatis. *Syſt. veget.* 802.
Trailing Lobelia.
Nat. of the Cape of Good Hope.
Introd. 1786, by Mr. Francis Maſſon.
Fl. July and Auguſt. G. H. ⊙.

lutea. 16. L. caulibus procumbentibus, foliis lanceolatis ſerratis, floribus ſeſſilibus ſubſpicatis. *Sp. pl.* 1322.
Yellow Lobelia.
Nat. of the Cape of Good Hope.
Introd. 1774, by Mr. Francis Maſſon.
Fl. June and July. G. H. ⊙.

coronopifolia. 17. L. foliis lanceolatis dentatis, pedunculis longiſſimis. *Sp. pl.* 1322.
Buck's-horn Lobelia.
Nat. of the Cape of Good Hope.
Introd. 1787, by Mr. Francis Maſſon.
Fl. July and Auguſt. G. H. ♃.

VIOLA. *Gen. pl.* 1007.

Cal. 5-phyllus. *Cor.* 5-petala, irregularis, poſtice cornuta. *Capſ.* ſupera, 3 valvis, 1-locularis.

* *Acaules.*

palmata. 1. V. acaulis, foliis palmatis quinquelobis dentatis indiviſiſque. *Sp. pl.* 1323.
Palmated Violet.
Nat. of Virginia.

Cult.

Cult. 1739, by Mr. Philip Miller. *Mill. dict. vol.* 2. *n.* 27.
Fl. May and June. H. ♃.

2. V. acaulis, foliis pedatis ſeptempartitis. *Sp. pl.* 1323. *pedata.*
Multifid Violet.
Nat. of North America.
Cult. 1759, by Mr. Ph. Miller. *Mill. dict. edit.* 7. *n.* 5.
Fl. May. H. ♃.

3. V. acaulis, foliis oblongis acutis cordato-ſagittatis ſerratis baſi inciſis, floribus inverſis. *ſagittata.*
Arrow-leav'd Violet.
Nat. of Penſylvania.
Introd. 1775, by John Fothergill, M. D.
Fl. July. H. ♃.
Descr. *Folia* inæqualiter et remote ſerrata, infra medium inciſo-ſinuata, venoſa, ſubpubeſcentia, digitalia. *Petioli* foliis longiores, ſemiteretes, piloſiuſculi. *Scapi* filiformes, longi, uniflori, tetragoni, villis raris adſperſi, foliolo uno alterove lanceolato minuto inſtructi, apice reflexi. *Calycis* foliola lanceolata, acuta, glabra. *Petala* oblongo-ovata, cærulea: *ſupremum* infra medium album, venis cæruleis pictum, baſi intus villoſum, ſeſquiuncia paulo longius; *lateralia* ſupremo paulo longiora, baſi villoſa et albida; *infima* longitudine lateralium, baſi albida, imberbia.

4. V. acaulis, foliis lanceolatis crenatis. *Sp. pl.* 1323. *lanceolata.*
Spear-leav'd Violet.
Nat. of North America.
Introd. 1785, by Mr. Archibald Menzies.
Fl. June. H. ♃.

5. V.

obliqua. 5. V. acaulis, foliis cordatis acutis ſubplanis glabris, floribus erectis, petalis oblique flexis.

Oblique-flower'd Violet.

Nat. of Penſylvania and Virginia.

Cult. 1762, by Mr. James Gordon.

Fl. May and June. H. ♃.

Descr. *Folia* crenato-ſerrata, ſeſquiuncialia. *Petioli* foliis duplo vel triplo longiores, ſemiteretes, canaliculati, glabri. *Scapi* filiformes, ſemiteretes, glabri, erecti, plerumque longitudine petiolorum. *Calycis* foliola glabra. *Petala* oblongo-ovata, ſtraminea; baſi cærulea: *ſupremum* ſemunciale, ſtriis cæruleis pictum, imberbe; *lateralia* ſupremo paulo anguſtiora & longiora, infra medium barbata; *infima* longitudine lateralium, ſed paulo latiora, imberbia.

cucullata. 6. V. acaulis, foliis cordatis acutiuſculis glabris baſi cucullatis, floribus inverſis, petalis oblique flexis.

Hollow-leav'd Violet.

Nat. of North America.

Introd. 1772, by Samuel Martin, M. D.

Fl. July. H. ♃.

Descr. *Folia* ſerrata, apice attenuata, biuncialia et ultra. *Petioli* foliis duplo longiores, ſupra plani, glabri. *Scapi* filiformes, ſubteretes, in medio foliolo uno alterove minutiſſimo inſtructi, longitudine plerumque petiolorum, apice reflexi. *Calycis* foliola glabra. *Petala* ovato-oblonga, cærulea: *ſupremum* ſemunciale, infra medium albidum, venis violaceis, imberbe; *lateralia* ſupremo longiora, infra medium albida, barbata; *infima* baſi albida, imberbia.

primulifolia. 7. V. acaulis, foliis oblongis ſubcordatis, petiolis membranaceis. *Sp. pl.* 1324.

Primroſe-leav'd Violet.

Nat.

Nat. of North America.
Introd. 1783, by Mr. William Young.
Fl. June. H. ♃.

8. V. acaulis, foliis cordatis piloso-hispidis. *Sp. pl.* 1324. *Curtis lond.* *hirta.*
Hairy Violet.
Nat. of England.
Fl. April and May. H. ♃.

9. V. acaulis, foliis reniformibus. *Sp. pl.* 1324. *Curtis lond.* *palustris.*
Marsh Violet.
Nat. of Britain.
Fl. June. H. ♃.

10. V. acaulis, foliis cordatis, stolonibus reptantibus. *Sp. pl.* 1324. *Curtis lond.* *odorata.*
α Viola martia purpurea, flore simplici odoro. *Bauh. pin.* 199.
Purple-flower'd sweet Violet.
β Viola martia alba. *Bauh. pin.* 199.
White-flower'd sweet Violet.
γ Viola martia multiplici flore. *Bauh. pin.* 199.
Double-flower'd sweet Violet.
Nat. of Britain.
Fl. March and April. H. ♃.

** *Caulescentes.*

11. V. caule adultiore adscendente, foliis oblongo-cordatis. *Sp. pl.* 1324. *Curtis lond.* *canina.*
Dog's Violet.
Nat. of Britain.
Fl. April. H. ♃.

montana. 12. V. caulibus erectis, foliis cordatis oblongis. *Sp. pl.* 1325.
Mountain Violet.
Nat. of the Alps of Europe.
Cult. 1714. *Philosoph. transact. n.* 344. *p.* 270. *n.* 51.
Fl. May and June. H. ♃.

canadensis. 13. V. caule erecto teretiusculo, foliis cordatis acuminatis glabris, stipulis integris.
Viola canadensis. *Sp. pl.* 1326.
Canadian Violet.
Nat. of North America.
Introd. 1783, by Mr. William Young.
Fl. June. H. ♃.

striata. 14. V. caule erecto semitereti, foliis ovatis cordatis acutis serratis, stipulis lanceolatis serrato-ciliatis.
Striated Violet.
Nat. of North America.
Introd. 1772, by Mr. William Young.
Fl. June and July. H. ♃.

pubescens. 15. V. caule erecto villoso tereti hinc planiusculo, foliis cordatis acutis pubescentibus, stipulis oblongis apice serrulatis.
Downy Violet.
Nat. of North America.
Introd. 1772, by Mr. William Young.
Fl. June. H. ♃.

mirabilis. 16. V. caule triquetro, foliis reniformi-cordatis, floribus caulinis apetalis. *Sp. pl.* 1326. *Jacqu. austr.* 1. *p.* 14. *t.* 19.
Broad-leav'd Violet.
Nat. of Germany and Sweden.

Cult. 1732, by James Sherard, M. D. *Dill. elth.* 408. *t.* 303. *f.* 390.

Fl. July and August. H. ♃.

17. V. caule bifloro, foliis reniformibus serratis. *Sp. pl.* 1326. *biflora.*

Two-flower'd Violet.

Nat. of the Alps of Europe.

Cult. 1739, by Mr. Philip Miller. *Mill. dict. vol.* 2. *n.* 12.

Fl. April and May. H. ♃.

18. V. caule unifloro, foliis cordatis dentatis. *Sp. pl.* 1327. *uniflora.*

Siberian Violet.

Nat. of Siberia.

Cult. 1774, by Mr. James Gordon.

Fl. June and July. H. ♃.

*** *Stipulis pinnatifidis; stigmate urceolato.*

19. V. caule triquetro diffuso, foliis oblongis incisis, stipulis pinnatifidis. *Syst. veget.* 803. *Curtis lond.* *tricolor.*

α Viola bicolor arvensis. *Bauh. pin.* 200.

Three-colour'd Violet, Pansies, or Heart's-ease.

β Viola tricolor hortensis repens. *Bauh. pin.* 199.

Three-colour'd Garden Violet.

Nat. of Britain.

Fl. May——September. H. ⊙.

20. V. caule triquetro simplici, foliis oblongiusculis, stipulis pinnatifidis. *Syst. veget.* 803. *grandiflora.*

Great yellow Violet.

Nat. of Britain.

Fl. May——September. H. ♃.

calcarata. 21. V. caule abbreviato, foliis ſubovatis, ſtipulis pinnatifidis, nectariis calyce longioribus. *Syſt. veget.* 803.
Alpine Violet.
Nat. of the Alps of Switzerland.
Cult. 1739, by Mr. Philip Miller. *Mill. dict. vol.* 2. *n.* 11.
Fl. March. H. ♃.

cornuta. 22. V. caule elongato, foliis oblongo-ovatis, ſtipulis pinnatifidis, nectariis ſubulatis corolla longioribus. *Syſt. veget.* 803.
Pyrenean Violet.
Nat. of the Pyrenees.
Introd. 1776, by Caſimir Gomez Ortega, M. D.
Fl. May. H. ♃.

arboreſcens. 23. V. caule fruticoſo, foliis lanceolatis integerrimis. *Sp. pl.* 1325.
Shrubby Violet.
Nat. of Spain.
Introd. 1784, by Mr. John Fairbairn.
Fl. G. H. ♄.

IMPATIENS. *Gen. pl.* 1008.

Cal. 2-phyllus. *Cor.* 5-petala, irregularis, nectario cucullato. *Capſ.* ſupera, 5-valvis.

Balſamina. 1. I. pedunculis unifloris aggregatis, foliis lanceolatis: ſuperioribus alternis, nectariis flore brevioribus. *Sp. pl.* 1328.
Garden Balſam.
Nat. of the Eaſt Indies.

Cult.

Cult. 1596, by Mr. John Gerard. *Hort. Ger.*
Fl. July——October. S. ☉.

2. I. pedunculis multifloris solitariis, foliis ovatis, geniculis caulinis tumentibus. *Sp. pl.* 1329. *noli tangere.*
Common yellow Balsam.
Nat. of England.
Fl. June. H. ☉.

Classis XX.

GYNANDRIA

DIANDRIA.

ORCHIS. *Gen. pl.* 1009.

Nectarium corniforme pone florem.

* *Corollæ galea calcarata.*

carnea. 1. O. bulbis indivisis, corollæ galea bicalcarata, bracteis erectis.

Great-flower'd Cape Orchis.

Nat. of the Cape of Good Hope. Mr. *Fr. Masson.*

Introd. 1787.

Fl. September. G. H. ♃.

Obs. In hac *folia* subrotunda, subtus sulcata; *spica* compacta; *bracteæ* subrotundæ, acutæ, erectæ; *flores* inodori, intus albi, extus carnei; *petala* duo suprema et infimum reliquis angustiora, omnia extus carinata, carina decurrente in germen, unde hoc infra quinquecostatum; *columna* fructificationis longitudine galeæ.

In sequenti *folia* ovato-oblonga, subtus lineata; *spica* laxa; *bracteæ* ovato-lanceolatæ, reflexæ; *flores* fragrantissimi, odore caryophyllorum, e viridi flavescentes; *petala* tria inferiora angustiora, breviora; *columna* fructificationis galea brevior.

bicornis. 2. O. bulbis indivisis, corollæ galea bicalcarata, bracteis reflexis.

Orchis bicornis. *Sp. pl.* 1330.

Yellow-

Yellow-flower'd Cape Orchis.
Nat. of the Cape of Good Hope.
Introd. 1787, by Mr. Francis Masson.
Fl. September. G. H. ♃.

** *Bulbis indivisis.*

3. O. bulbis indivisis, nectarii labio lanceolato integerrimo: cornu longissimo, petalis patentibus. *Sp. pl.* 1331. *bifolia.*
Butterfly Orchis.
Nat. of Britain.
Fl. May and June. H. ♃.

4. O. bulbis indivisis, nectarii labio bicorni trifido æquali integerrimo: cornu longo, petalis sublanceolatis. *Syst. veget.* 808. *Jacqu. austr.* 3. *p.* 37. *t.* 266. *pyramidalis.*
Pyramidal Orchis.
Nat. of Britain.
Fl. June and July. H. ♃.

5. O. bulbis indivisis, nectarii labio trifido reflexo crenato: cornu brevi, petalis conniventibus. *Sp. pl.* 1332. *Jacqu. austr.* 2. *p.* 14. *t.* 122. *coriophora.*
Lizard Orchis.
Nat. of England.
Fl. June. H. ♃.

6. O. bulbis indivisis, nectarii labio quadrifido crenulato: cornu obtuso adscendente, petalis obtusis conniventibus. *Syst. veget.* 808. *Curtis lond.* *Morio.*
Female Orchis.
Nat. of Britain.
Fl. May and June. H. ♃.

mascula. 7. O. bulbis indivisis, nectarii labio quadrilobo crenulato: cornu obtuso, petalis dorsalibus reflexis. *Sp. pl.* 1333. *Curtis lond. Jacqu. ic. miscell.* 2. *p.* 375.
Male Orchis.
Nat. of Britain.
Fl. May. H. ♃.

ustulata. 8. O. bulbis indivisis, nectarii labio quadrifido punctis scabro: cornu obtuso, petalis distinctis. *Sp. pl.* 1333.
Dwarf Orchis.
Nat. of England.
Fl. May and June. H. ♃.

militaris. 9. O. bulbis indivisis, nectarii labio quinquefido punctis scabro: cornu obtuso, petalis confluentibus. *Sp. pl.* 1333.
α Orchis galea & alis fere cinereis. *Bauh. hist.* 2. *p.* 757.
Man Orchis.
β Orchis magna, latis foliis, galea fusca vel nigricante. *Bauh. hist.* 2. *p.* 759.
Orchis fusca. *Jacqu. austr.* 4. *p.* 4. *t.* 307.
Purple Man Orchis.
Nat. of England.
Fl. May. H. ♃.

*** *Bulbis palmatis.*

latifolia. 10. O. bulbis subpalmatis rectis, nectarii cornu conico: labio trilobo lateribus reflexo, bracteis flore longioribus. *Sp. pl.* 1334. *Curtis lond.*
Broad-leav'd Orchis.
Nat. of Britain.
Fl. May and June. H. ♃.

11. O.

11. O. bulbis palmatis patentibus, nectarii cornu germinibus breviore: labio plano, petalis dorsalibus erectis. *Syst. veget.* 810. *maculata.*

Spotted Orchis.

Nat. of Britain.

Fl. June. H. ♃.

12. O. bulbis palmatis, nectarii cornu setaceo germinibus longiore: labio trifido, petalis duobus patentissimis. *Sp. pl.* 1335. *conopsea.*

Red-handed Orchis.

Nat. of Britain.

Fl. June. H. ♂.

**** *Bulbis fasciculatis.*

13. O. bulbis fasciculatis, nectarii cornu germinibus longiore: labio tripartito ciliari, petalis patentibus, foliis oblongis. *fimbriata.*

Fringed Orchis.

Nat. of Canada and Newfoundland.

Introd. 1777, by William Pitcairn, M.D.

Fl. July. H. ♃.

Descr. *Caulis* erectus, glaber, ex ancipiti acute tetragonus. *Folia* caulina, nonnulla, (3-5) alterna, sessilia, oblonga, acuta, glabra, integerrima, nervosa, carinata, basi vaginantia, biuncialia. *Spica* ovato-oblonga, multiflora. *Flores* e cæruleo purpurascentes. *Bracteæ* lanceolatæ, nervosæ, germinibus paulo longiores. *Petala* quinque, plana, longitudine æqualia, trilinearia: supremum seu dorsale ovatum, obtusum, erectum; lateralia exteriora ovata, acuta, patentissima; lateralia interiora oblonga, obtusa, juxta petalum dorsale erecta, infra medium dilatata, ibique denticulata, basi attenuata. *Nectarii Labium* petalis paulo longius, tripartitum: laciniæ latæ, cunei-

cuneiformes, æquales, planæ, ad medium ſubdiviſæ in cilia ſubulata: laterales divaricatæ; intermedia patens. *Germen* ſemunciale.

SATYRIUM. *Gen. pl.* 1010.

Nectarium ſcrotiforme ſ. inflato-didymum pone florem.

hircinum. 1. S. bulbis indiviſis, foliis lanceolatis, nectarii labio trifido: intermedia lineari obliqua præmorſa. *Syſt. veget.* 811. *Jacqu. auſtr.* 4. *p.* 35. *t.* 367.
Lizard Satyrion.
Nat. of England.
Fl. June and July. H. ♃.

viride. 2. S. bulbis palmatis, foliis oblongis obtuſis, nectarii labio lineari trifido: intermedia obſoleta. *Sp. pl.* 1337.
Frog Satyrion, or Orchis.
Nat. of Britain.
Fl. June. H. ♃.

nigrum. 3. S. bulbis palmatis, foliis linearibus, nectarii labio reſupinato indiviſo. *Syſt. veget.* 811. *Jacqu. auſtr.* 4. *p.* 35. *t.* 368.
Black-flower'd Satyrion.
Nat. of the Alps of Switzerland, Auſtria, and Lapland.
Introd. 1779, by the Rev. Samuel Goodenough, LL.D.
Fl. June. H. ♃.

repens. 4. S. bulbis fibroſis, foliis ovatis radicalibus, floribus ſecundis. *Sp. pl.* 1339. *Jacqu. auſtr.* 4. *p.* 36. *t.* 369.
Creeping Satyrion.
Nat. of Scotland.
Fl. Auguſt. H. ♃.

OPHRYS.

OPHRYS. *Gen. pl.* 1011.

Nectarium ſubtus ſubcarinatum.

* *Bulbis ramoſis.*

1. O. bulbis fibroſo-faſciculatis, caule vaginato aphyllo, nectarii labio bifido. *Syſt. veget.* 812. *Nidus avis.*
Bird's-neſt Ophrys.
Nat. of Britain.
Fl. May. H. ♃.

2. O. bulbis aggregatis oblongis, caule ſubfolioſo, floribus ſpirali-ſecundis, nectarii labio indiviſo crenulato. *Syſt. veget.* 812. *Curtis lond.* *ſpiralis.*
Triple Ophrys, or Lady's Traces.
Nat. of Britain.
Fl. Auguſt. H. ♃.

3. O. bulbo fibroſo, caule bifolio, foliis ovatis, nectarii labio bifido. *Sp. pl.* 1340. *Curtis lond.* *ovata.*
Common Ophrys, or Twayblade.
Nat. of Britain.
Fl. May and June. H. ♃.

** *Bulbis rotundis.*

4. O. bulbo ſubrotundo, ſcapo nudo, foliis lanceolatis, nectarii labio integro, petalis dorſalibus linearibus. *Sp. pl.* 1341. (excluſis ſynonymis hort. cliff. et fl. ſuec.) *lilifolia.*
Lily-leav'd Ophrys.
Nat. of North America.
Cult. 1758, by Peter Collinſon, Eſq. *Philoſ. tranſact.* 1763. *p.* 81.
Fl. June. H. ♃.

5. O.

Monorchis. 5. O. bulbo globoſo, ſcapo nudo, nectarii labio trifido cruciato. *Sp. pl.* 1342.
Yellow, or Muſk Ophrys.
Nat. of England.
Fl. July. H. ♃.

anthropophora. 6. O. bulbis ſubrotundis, ſcapo folioſo, nectarii labio lineari tripartito: medio elongato bifido. *Sp. pl.* 1343.
Man Ophrys.
Nat. of England.
Fl. June. H. ♃.

insectifera. 7. O. bulbis ſubrotundis, ſcapo folioſo, nectarii labio ſubquinquelobo. *Sp. pl.* 1343.
α Ophrys muſcifera. *Hudſ. angl.* 391.
Ophrys myodes. *Jacqu. ic. miſcell.* 2. *p.* 373.
Fly Ophrys, or Orchis.
β Ophrys apifera. *Hudſ. angl.* 391. *Curtis lond.*
Bee Ophrys, or Orchis.
γ Ophrys araniſera. *Hudſ. angl.* 392.
Spider Ophrys, or Orchis.
Nat. of England.
Fl. May——Auguſt. H. ♃.

SERAPIAS. *Gen. pl.* 1012.

Nectarium ovatum, gibbum, labio ovato.

latifolia. 1. S. bulbis fibroſis, foliis ovatis amplexicaulibus, floribus pendulis. *Syſt. veget.* 814.
Broad-leav d Helleborine.
Nat. of Britain.
Fl. July and Auguſt. H. ♃.

longifolia. 2. S. bulbis fibroſis, foliis enſiformibus ſeſſilibus, floribus pendulis. *Syſt. veget.* 815.

Long-

Long-leav'd Helleborine.
Nat. of Britain.
Fl. July. H. ♃.

3. S. bulbis fibrosis, foliis ensiformibus, floribus erectis, nectarii labio obtuso petalis breviore. *Syst. nat. edit.* 12. *p.* 594. *grandiflora.*
Great-flower'd Helleborine.
Nat. of Britain.
Fl. June. H. ♃.

4. S. bulbis subrotundis, nectarii labio trifido acuminato glabro petalis longiore. *Sp. pl.* 1344. *Lingua.*
Narrow-leav'd Helleborine.
Nat. of the South of Europe.
Introd. 1786, by Sir Francis Drake, Bart.
Fl. May. G. H. ♃.

LIMODORUM. *Gen. pl.* 1013.

Nectarium monophyllum, concavum, pedicellatum, intra petalum infimum.

1. L. floribus subspicatis barbatis. *tuberosum.*
Limodorum tuberosum. *Sp. pl.* 1345. (exclusis synonymis Martyni, Milleri et Plumieri, quæ sequentis.)
Tuberous-rooted Limodorum.
Nat. of North America.
Introd. 1787, by Mr. William Curtis.
Fl. H. ♃.

2. L. floribus imberbibus, spicis subpaniculatis. *altum.*
Limodorum altum. *Syst. veget.* 816.
Tall Limodorum.
Nat. of the West Indies.

Introd.

Introd. before 1733, by William Houſtoun, M.D. *Mill. dict. edit.* 8.
Fl. Moſt part of the Year. S. ♃.

Tankervilliæ. 3. L. floribus racemoſis imberbibus. TAB. 12.
Chineſe Limodorum.
Nat. of China.
Introd. about 1778, by John Fothergill, M.D.
Fl. March and April. S. ♃.

ARETHUSA. *Gen. pl.* 1014.

Nectarium tubuloſum intra corollæ fundum: labio inferiore ſtylo adnato.

bulboſa. 1. A. radice globoſa, ſcapo vaginato, ſpatha diphylla. *Sp. pl.* 1346.
Bulbous-rooted Arethuſa.
Nat. of North America.
Introd. 1784, by Mr. William Young.
Fl. May. H. ♃.

ciliaris. 2. A. radice carnoſa, folio reniformi-orbiculato, labio ciliari. *Linn. ſuppl.* 405.
Orchis burmanniana. *Sp. pl.* 1334.
Fringed-flower'd Arethuſa.
Nat. of the Cape of Good Hope.
Introd. 1787, by Mr. Francis Maſſon.
Fl. October. G. H. ♃.

CYPRIPEDIUM. *Gen. pl.* 1015.

Nectarium ventricoſum, inflatum, cavum.

Calceolus. 1. C. radicibus fibroſis, foliis ovato-lanceolatis caulinis, petalis acuminatis.

Cypripedium

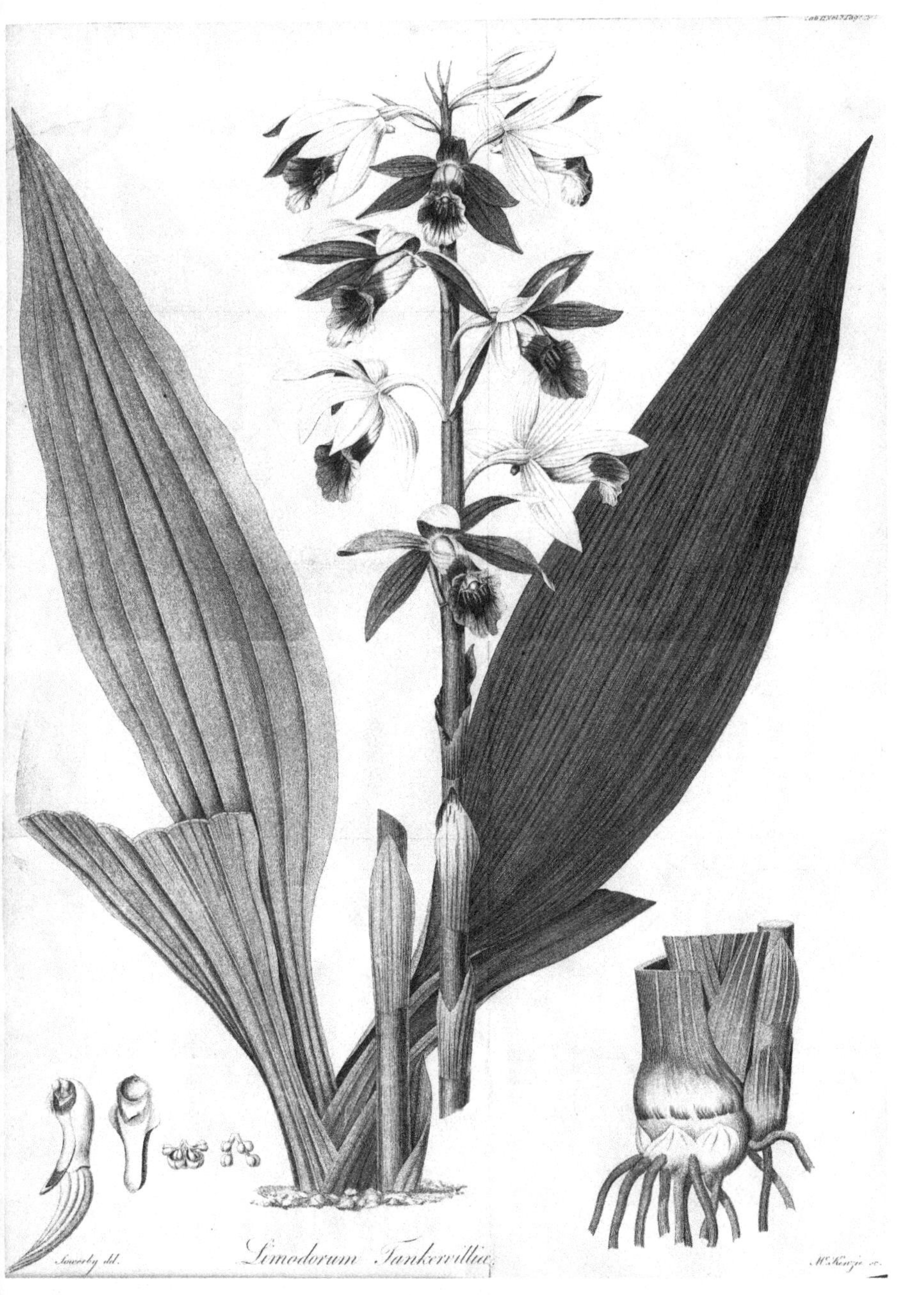

The material originally positioned here is too large for reproduction in this reissue. A PDF can be downloaded from the web address given on page iv of this book, by clicking on 'Resources Available'.

Cypripedium Calceolus. *Sp. pl.* 1346.
Common Lady's Slipper.
Nat. of England.
Fl. May——July. H. ♃.

2. C. radicibus fibrofis, foliis ovato-lanceolatis caulinis, petalis obtufis. *album.*
Helleborine Calceolus dicta, mariana, flore gemello candido, venis purpureis ftriato. *Pluk. mant.* 101. *t.* 418. *f.* 3.
White Lady's Slipper.
Nat. of North America.
Introd. about 1770, by Mr. William Young.
Fl. June and July. H. ♃.

3. C. radicibus fibrofis, foliis oblongis radicalibus. *acaule.*
Helleborine Calceolus dicta, mariana, foliis binis e radice ex adverfo prodeuntibus, flore purpureo. *Pluk. mant.* 101. *t.* 418. *f.* 1.
Two-leav'd Lady's Slipper.
Nat. of North America.
Introd. 1786, by William Hamilton, Efq.
Fl. May. H. ♃.

EPIDENDRUM. *Gen. pl.* 1016.

Nectarium turbinatum, obliquum, reflexum.

1. E. foliis oblongis geminis glabris ftriatis bulbo innatis, fcapo multifloro, nectario cordato. *Sp. pl.* 1351. *cochleatum.*
Many-flower'd Epidendrum.
Nat. of the Weft Indies.
Introd. 1786, by Mr. Alexander Anderfon.
Fl. January and February. S. ♃.

2. E.

fragrans. 2. E. folio lato-lanceolato enervi bulbo innato, ſcapo multifloro abbreviato, labio cordato. *Swartz prodr.* 123.
Sweet-ſcented Epidendrum.
Nat. of Jamaica.
Introd. 1786, by Hinton Eaſt, Eſq.
Fl. October. S. ♃.

GUNNERA. *Linn. mant.* 16.

Amentum ſquamis unifloris. *Cal.* 0. *Cor.* 0. *Germen* 2-dentatum. *Styli* 2. *Sem.* 1.

perpenſa. 1. GUNNERA. *Linn. mant.* 121.
Marſh Marygold-leav'd Gunnera.
Nat. of the Cape of Good Hope.
Introd. 1767, by Mr. William Malcolm.
Fl. G. H. ♃.

TRIANDRIA.

SISYRINCHIUM.

Monogyna. *Spatha* 2-phylla. *Pet.* 6, plana. *Capſ.* 3-locularis.

Bermudiana. 1. S. foliis enſiformibus enerviis. *Syſt. veget.* 820.
Small Siſyrinchium.
Nat. of North America.
Cult. 1693, by Mr. Jacob Bobart. *Br. Muſ. Sloan. mſſ.* 3343.
Fl. May and June. H. ♃.

latifolium. 2. S. foliis enſiformibus nervoſis plicatis. *Swartz prodr.* 17.

Bermudiana

Bermudiana Palmæ folio, radice bulbosa. *Plum. ic.* 35. *tab.* 46. *f.* 2.
Broad-leav'd Sisyrinchium.
Nat. of the West Indies.
Cult. 1739, by Mr. Philip Miller. *Rand. chel.* Bermudiana 2.
Fl. June——August. S. ♃.

FERRARIA. *Gen. pl.* 1018.

Monogyna. *Spathæ* unifloræ. *Petala* 6, undulato-crispata. *Stigmata* cucullata. *Caps.* 3-locularis, infera.

1. FERRARIA. *Sp. pl.* 1353. *undulata.*
Cape Ferraria.
Nat. of the Cape of Good Hope.
Cult. 1759, by Mr. Ph. Miller. *Mill. ic.* 187. *t.* 280.
Fl. March and April. G. H. ♃.

PENTANDRIA.

AYENIA. *Gen. pl.* 1020.

Monogyna. *Cal.* 5-phyllus. *Petala* connata in stellam unguibus longis. *Antheræ* 5, sub stella. *Caps.* 5-locularis.

1. A. foliis cordatis glabris. *Sp. pl.* 1354. *pusilla.*
Smooth Ayenia.
Nat. of Jamaica and Peru.
Cult. 1756, by Mr. Ph. Miller. *Mill. ic.* 79. *t.* 118.
Fl. July——September. S. ♂.

PASSIFLORA. *Gen. pl.* 1021.

Trigyna. *Cal.* 5-phyllus. *Petala* 5. *Nectarium* corona. *Bacca* pedicellata.

* *Foliis indivisis.*

serratifolia. 1. P. foliis indivisis ovatis serratis. *Syst. veget.* 821. *Jacqu. hort.* 1. *p.* 4. *t.* 10.
Notched-leav'd Passion-flower.
Nat. of the West Indies.
Introd. 1731, by William Houstoun, M.D. *Mart. dec.* 4. *p.* 36.
Fl. May——October. S. ♄.

maliformis. 2. P. foliis indivisis cordato-oblongis integerrimis, petiolis biglandulosis, involucris integerrimis. *Sp. pl.* 1355.
Apple-fruited Passion-flower.
Nat. of Dominica.
Cult. 1731, by Mr. Philip Miller. *Mill. dict. edit.* 1. Granadilla 9.
Fl. S. ♄.

quadrangularis. 3. P. foliis indivisis ovalibus subcordatis glabris multinerviis, petiolis glandulosis, caule membranaceo tetragono, stipulis ovali-oblongis.
Passiflora quadrangularis. *Sp. pl.* 1356.
Square-stalk'd Passion-flower.
Nat. of Jamaica.
Cult. 1768, by Mr. Philip Miller. *Mill. dict. edit.* 8.
Fl. August and September. S. ♄.

alata. 4. P. foliis indivisis ovatis subcordatis lævibus paucinerviis, petiolis glandulosis, caule membranaceo tetragono, stipulis lanceolatis serratis.

Passiflora alata. *Curtis magaz.* 66.
Wing-stalk'd Passion-flower.
Nat. of the West Indies.
Introd. 1772, by Mr. William Malcolm.
Fl. August. S. ♄.

5. P. foliis indivisis ovatis integerrimis, petiolis biglandulosis, involucris dentatis. *Sp. pl.* 1356. *Jacqu. hort.* 2. *p.* 76. *t.* 162. *laurifolia.*
Laurel-leav'd Passion-flower, or Water Lemon.
Nat. of the West Indies.
Introd. 1690, by Mr. Bentick. *Br. Mus. Sloan. mss.* 3370.
Fl. June and July. S. ♄.

** *Foliis bilobis.*

6. P. foliis bilobis cordatis acuminatis subtus subtomentosis. *Syst. veget.* 822. *Jacqu. ic. collect.* 1. *p.* 136. *rubra.*
Red-fruited Passion-flower.
Nat. of the West Indies.
Cult. 1748, by Mr. Philip Miller. *Mill. dict. edit.* 5. Granadilla 10.
Fl. April and May. S. ♄.

7. P. foliis bilobis basi rotundatis glandulosisque: lobis acutis divaricatis subtus punctatis. *Sp. pl.* 1357. *Vespertilio.*
Bat-winged Passion-flower.
Nat. of the West Indies.
Cult. 1732, by James Sherard, M. D. *Dill. elth.* 164. *t.* 137. *f.* 164.
Fl. May and June. S. ♄.

*** *Foliis trilobis.*

8. P. foliis subtrilobis obtusis subrotundis subtus punctatis. *Syst. veget.* 822. *rotundifolia.*
Round-leav'd Passion-flower.

Nat. of the West Indies.
Introd. 1786, by Baron Hake.
Fl. S. ♄.

punctata. 9. P. foliis subtrilobis oblongis subtus punctatis: intermedio minore. *Syst. veget.* 822.
Dotted Passion-flower.
Nat. of Peru.
Introd. 1784, by William Wright, M. D.
Fl. May and June. S. ♄.

lutea. 10. P. foliis trilobis cordatis glabris: lobis ovatis, petiolis eglandulosis.
Passiflora lutea. *Sp. pl.* 1358. *Jacqu. ic. vol.* 2.
Yellow Passion-flower.
Nat. of Jamaica and Virginia.
Cult. 1714. *Philosoph. transact. n.* 346. *p.* 359. *n.* 112.
Fl. May and June. S. ♃.

glauca. 11. P. foliis cordatis glabris trilobis: lobis ovatis æqualibus, petiolis glandulosis, stipulis semiovatis.
Glaucous-leav'd Passion-flower.
Nat. of Cayenne.
Introd. 1779, by Messrs. Kennedy and Lee.
Fl. August and September. S. ♄.
DESCR. Tota planta lævissima. *Folia* subtus glauca, impunctata. *Petioli* infra medium glandulis 2 vel 4 instructi. *Stipulæ* acutæ, integerrimæ, semuncia longiores. *Calycis* foliola lanceolata, subtus viridia, supra albida, uncia paulo longiora. *Petala* lanceolata, obtusa, alba, longitudine calycis. *Nectarium* petalis dimidio brevius: radii violacei, apice albi. *Flores* suaveolentes.

12. P. foliis glabris trilobis: lobis lanceolatis integerrimis: intermedio productiore, petiolis biglandulosis, caule lævi inferne ſuberoſo. *minima.*
Paſſiflora minima. *Sp. pl.* 1359.
Dwarf Paſſion-flower.
Nat. of Curaſſao.
Introd. 1690, by Mr. Bentick. *Br. Muſ. Sloan. mſſ.* 3370.
Fl. July. S. ♄.

13. P. foliis indiviſis lineari-oblongis trilobiſque glabris integerrimis, petiolis biglanduloſis. *heterophylla.*
Narrow-leav'd Paſſion-flower.
Nat. of the Weſt Indies.
Introd. about 1773.
Fl. June——September. S. ♄.

14. P. foliis trilobis ſubpeltatis, cortice ſuberoſo. *Sp. pl.* 1358. *Jacqu. hort.* 2. *p.* 77. *t.* 163. *ſuberoſa.*
Cork-bark'd Paſſion-flower.
Nat. of the Weſt Indies.
Cult. 1759, by Mr. Ph. Miller. *Mill. dict. edit.* 7. *n.* 5.
Fl. June——September. S. ♄.

15. P. foliis trilobis: baſi utrinque denticulo reflexo. *Syſt. veget.* 823. *holoſericea.*
Silky-leav'd Paſſion-flower.
Nat. of Vera Cruz.
Introd. before 1733, by William Houſtoun, M.D. *Mart. dec.* 5. *p.* 51.
Fl. Moſt part of the Summer. S. ♄.

16. P. foliis trilobis villoſis: inferioribus ſupra glabris: lobis oblongis integerrimis: intermedio productiore, petiolis biglanduloſis. *hirſuta.*

Passiflora hirsuta. *Sp. pl.* 1359.
Hairy Passion-flower.
Nat. of the West Indies.
Introd. 1778, by Mr. Gilbert Alexander.
Fl. September. S. ♄.

fœtida. 17. P. foliis trilobis cordatis pilosis, involucris multifido-capillaribus. *Sp. pl.* 1359.
Stinking Passion-flower.
Nat. of the West Indies.
Cult. 1731, by Mr. Philip Miller. *Mill. dict. edit.* 1. Granadilla 7.
Fl. July and August. S. ♂.

ciliata. 18. P. foliis trilobis glabris ciliato-serratis: intermedio longissimo, petiolis eglandulosis.
Ciliated Passion-flower.
Nat. of Jamaica.
Cult. 1783, by Mrs. Norman.
Fl. S. ♄.

incarnata. 19. P. foliis trilobis serratis æqualibus, petiolis biglandulosis.
Passiflora incarnata. *Sp. pl.* 1360. *Jacqu. ic. collect.* 1. *p.* 107.
Three-leav'd Passion-flower.
Nat. of America.
Cult. 1629. *Park. parad.* 395. *f.* 7.
Fl. July and August. G. H. ♃.

**** *Foliis multifidis.*

cærulea. 20. P. foliis palmatis integerrimis. *Sp. pl.* 1360. *Curtis magaz.* 28.
Common Passion-flower.
Nat. of Brazil.

Cult.

Cult. 1699, by the Dutchess of Beaufort. *Br. Mus. Sloan. mss.* 525 & 3349.
Fl. June——October. H. ♄.

21. P. foliis pedatis serratis. *Sp. pl.* 1360. *pedata.*
Curl'd-flower'd Passion-flower.
Nat. of the West Indies.
Introd. 1781, by Mr. Francis Masson.
Fl. S. ♄.

HEXANDRIA.

ARISTOLOCHIA. *Gen. pl.* 1022.

Hexagyna. *Cal.* 0. *Cor.* 1-petala, lingulata, integra. *Caps.* 6-locularis, infera.

1. A. foliis trilobis, caule volubili, floribus maximis. *Sp. pl.* 1361. *trilobata.*
Three-lob'd Birthwort.
Nat. of South America.
Introd. about 1775.
Fl. June and July. S. ♄.

2. A. foliis cordatis, caule volubili fruticoso, pedunculis solitariis, labio corollis majore. *Sp. pl.* 1362. *odoratissima.*
Sweet-scented Birthwort.
Nat. of Jamaica.
Cult. 1752, by Mr. Ph. Miller. *Mill. dict. edit.* 6. *n.* 8.
Fl. S. ♄.

3. A. foliis cordatis petiolatis, floribus solitariis: limbo trifido æquali, bractea ovata, caule volubili frutescente. *sipho.*

Ariſtolochia ſipho. *L'Herit. ſtirp. nov. p.* 13. *tab.* 7.
Broad-leav'd Birthwort.
Nat. of North America. Mr. *John Bartram.*
Introd. about 1763.
Fl. June and July. H. ♄.

ſempervirens. 4. A. foliis cordato-oblongis acuminatis undatis, caule infirmo, floribus ſolitariis. *Syſt. veget.* 824.
Evergreen Birthwort.
Nat. of Candia.
Cult. 1739, in Chelſea Garden. *Rand. chel. n.* 4.
Fl. May and June. H. ♄.

rotunda. 5. A. foliis cordatis ſubſeſſilibus obtuſis, caule infirmo, floribus ſolitariis. *Sp. pl.* 1364.
Round-rooted Birthwort.
Nat. of Italy, Spain, and the South of France.
Cult. 1596, by Mr. John Gerard. *Hort. Ger.*
Fl. Moſt part of the Year. G. H. ♃.

longa. 6. A. foliis cordatis petiolatis obtuſis, caule infirmo, floribus ſolitariis, fructibus ovatis.
Ariſtolochia longa. *Sp. pl.* 1364.
Long-rooted Birthwort.
Nat. of the South of Europe.
Cult. 1596, by Mr. John Gerard. *Hort. Ger.*
Fl. June——October. G. H. ♃.

clematitis. 7. A. foliis cordatis, caule erecto, floribus axillaribus confertis. *Sp. pl.* 1364.
Upright Birthwort.
Nat. of England.
Fl. May——July. H. ♃.

DECAN-

DECANDRIA.

HELICTERES. *Gen. pl.* 1025.

Pentagyna. *Cal.* 1-phyllus, obliquus. *Petala* 5. *Nectarium* foliolis 5. *Capſulæ* 5, intortæ.

1. H. decandra, foliis cordatis ſerratis, fructu contorto apicibus rectis. *Syſt. veget.* 825. *baruenſis.*
Small-fruited Helicteres.
Nat. of the Weſt Indies.
Cult. 1739, by Mr. Ph. Miller. *Rand. chel.* Iſora 2.
Fl. S. ♄.

2. H. decandra, foliis cordatis ſerratis, fructu toto contorto. *Syſt. veget.* 826. *Iſora.*
Great-fruited Helicteres, or Screw-tree.
Nat. of Jamaica.
Cult. 1739, by Mr. Ph. Miller. *Rand. chel.* Iſora 1.
Fl. June and July. S. ♄.

POLYANDRIA.

GREWIA. *Gen. pl.* 1026.

Monogyna. *Cal.* 5-phyllus. *Petala* 5: baſi ſquama nectarifera. *Bacca* 4-locularis.

1. G. foliis ſubovatis, floribus ſolitariis. *Syſt. veget.* 826. *occidentalis.*
Elm-leav'd Grewia.
Nat. of the Cape of Good Hope.

Cult.

Cult. 1692, in the Royal Garden at Hampton-court. *Pluk. phyt. t.* 237. *f.* 1.
Fl. Moſt part of the Summer. G. H. ♄.

orientalis. 2. G. foliis ſublanceolatis, floribus ſolitariis. *Syſt. veget.* 826.
Oriental Grewia.
Nat. of the Eaſt Indies.
Introd. 1767, by Mr. William Malcolm.
Fl. July and Auguſt. G. H. ♄.

ſalvifolia. 3. G. foliis oblongis integerrimis, floribus axillaribus pluribus pedicellatis, petalis recurvis linearibus, *Linn. ſuppl.* 409.
Sage-leav'd Grewia.
Nat. of the Eaſt Indies.
Introd. 1779, by John Gerard Koenig, M. D.
Fl. S. ♄.

A R U M. *Gen. pl.* 1028.

Spatha monophylla, cucullata. *Spadix* ſupra nudus, inferne femineus, medio ſtamineus.

* *Acaulia, foliis compoſitis.*

crinitum. 1. A. acaule, foliis pedatis: laciniis lateralibus involutis, ſpatha interne piloſa, ſpadice ſuperne ramentaceo.
Arum muſcivorum. *Linn. ſuppl.* 410.
Hairy-ſheath'd Arum.
Nat. of Minorca.
Introd. 1777, by Mr. William Malcolm.
Fl. March. G. H. ♃.

Dracunculus. 2. A. foliis pedatis: foliolis lanceolatis integerrimis, lamina ovata ſpadice longiore.
Arum Dracunculus. *Sp. pl.* 1367.

Long-

Long-ſheath'd Arum, or Common Dragon.
Nat. of the South of Europe.
Cult. 1596, by Mr. John Gerrard. *Hort. Ger.*
Fl. June and July. H. ♃.

3. A. foliis pedatis: foliolis lanceolatis integerrimis ſuperantibus ſpatham ſpadice breviorem. *Sp. pl.* 1368. *Dracontium.*
Short ſheath'd Arum, or Green Dragon.
Nat. of North America.
Cult. 1759, by Mr. Ph. Miller. *Mill. dict. edit.* 7. *n.* 9.
Fl. June. H. ♃.

4. A. acaule, foliis pedatis: foliolis ſubovalibus integerrimis, lamina lanceolata ſpadice longiore. *venoſum.*
Purple-flower'd Arum.
Nat.
Introd. 1774, by Mr. William Malcolm.
Fl. March. S. ♃.

5. A. acaule, foliis ternatis, lamina lanceolata acuminata longitudine ſpadicis. *triphyllum.*
Arum triphyllum. *Sp. pl.* 1368.
Three-leav'd Green-ſtalk'd Arum.
Nat. of North America.
Cult. 1664. *Evelyn's kalend. hort.* 83.
Fl. June. H. ♃.

6. A. acaule, foliis ternatis, lamina ovata ſpadice dimidio breviore. *atrorubens.*
Arum ſ. Ariſarum triphyllum minus, pene atrorubente virginianum. *Pluk. alm.* 52. *t.* 77. *f.* 5.
Three-leav'd Purple-ſtalk'd Arum.
Nat. of Virginia.
Cult. 1758, by Mr. Philip Miller.
Fl. June and July. H. ♃.

** *Acaulia,*

** *Acaulia, foliis simplicibus.*

Colocasia. 7. A. acaule, foliis peltatis ovatis repandis: basi semibifidis. *Sp. pl.* 1368.
Egyptian Arum.
Nat. of the Levant.
Cult. 1690, in the Royal Garden at Hampton-court. *Catal. mss.*
Fl. S. ♃.

bicolor. 8. A. acaule, foliis peltatis sagittatis disco coloratis, spatha medio coarctata: basi subglobosa; lamina subrotunda acuminata erecta subconvoluta.
Painted Arum.
Nat. (Cultivated in Madeira.)
Introd. 1773, by Messrs. Kennedy and Lee.
Fl. June and July. S. ♃.

esculentum. 9. A. acaule, foliis peltatis ovatis integerrimis: basi emarginatis. *Syst. veget.* 827.
Eatable Arum.
Nat. of America.
Cult. 1739, by Mr. Philip Miller. *Rand. chel. n.* 8.
Fl. S. ♃.

trilobatum. 10. A. acaule, foliis sagittato-trilobis, flore sessili. *Sp. pl.* 1369.
Three-lob'd Arum.
Nat. of Ceylon.
Introd. 1752, by Mr. Ph. Miller. *Mill. ic.* 35. *t.* 52. *f.* 2.
Fl. May and June. S. ♃.

sagittaefolium. 11. A. acaule, foliis sagittatis triangulis: angulis divaricatis acutis. *Sp. pl.* 1369. *Jacqu. hort.* 2. *p.* 73. *t.* 157.
Arrow-leav'd Arum.

Nat.

Nat. of the West Indies.
Cult. 1731, by Mr. Ph. Miller. *Mill. dict. edit.* 1. *n.* 2.
Fl. S. ♃.

12. A. acaule, foliis hastatis integerrimis, spadice clavato. *Sp. pl.* 1370. *Curtis lond.* *maculatum.*
α Arum vulgare non maculatum. *Bauh. pin.* 195.
Common Arum, or Wake Robin.
β Arum maculatum, maculis candidis, vel nigris. *Bauh. pin.* 195.
Spotted Arum, or Wake Robin.
γ Arum *italicum*, foliis hastatis acutis, petiolis longissimis, spatha maxima erecta. *Mill. dict.*
Italian Arum.
Nat. of Britain; γ. of Italy.
Fl. June and July. H. ♃.

13. A. acaule, foliis cordato-oblongis, spathæ apertura ovata. *Syst. veget.* 828. *Arisarum.*
Hooded Arum, or Frier's Cowl.
Nat. of the South of Europe.
Cult. 1596, by Mr. John Gerard. *Hort. Ger.*
Fl. April and May. G. H. ♃.

14. A. acaule, foliis lanceolatis, spadice setaceo declinato. *Sp. pl.* 1370. *tenuifolium.*
Grass-leav'd Arum.
Nat. of the South of Europe.
Cult. 1570. *Lobel. adv.* 261.
Fl. April and May. H. ♃.

*** *Caulescentia.*

15. A. caulescens suberectum, foliis lanceolato-ovatis. *Sp. pl.* 1371. *seguinum.*
Dumb Arum, or Cane.

Nat.

Nat. of America.
Cult. 1759, by Mr. Ph. Miller. *Mill. ic.* 197. *t.* 295.
Fl. May. S. ♄.

auritum. 16. A. caulescens radicans, foliis ternatis: lateralibus unilobatis. *Sp. pl.* 1371.
Ear-leav'd Arum.
Nat. of America.
Cult. 1748, by Mr. Philip Miller. *Mill. dict. edit.* 5. Dracunculus 8.
Fl. S. ♄.

DRACONTIUM. *Gen. pl.* 1029.

Spatha cymbiformis. *Spadix* tectus. *Cal.* 0. *Petala* 5. *Baccæ* polyspermæ.

pertusum. 1. D. foliis pertusis, caule scandente. *Sp. pl.* 1372.
Perforated Dragon.
Nat. of the West Indies.
Cult. 1752, by Mr. Philip Miller. *Mill. dict. edit.* 6. Appendix. Arum.
Fl. April——June. S. ♄.

CALLA. *Gen. pl.* 1030.

Spatha plana. *Spadix* tectus flosculis. *Cal.* 0. *Petala* 0. *Baccæ* polyspermæ.

æthiopica. 1. C. foliis sagittato-cordatis, spatha cucullata, spadice superne masculo. *Sp. pl.* 1373.
Ethiopian Calla.
Nat. of the Cape of Good Hope.
Cult. 1731, by Mr. Philip Miller. *Mill. dict. edit.* 1. Arum 1.
Fl. January——May. G. H. ♃.

2. C.

2. C. foliis cordatis, ſpatha plana, ſpadice undique hermaphrodito. *Sp. pl.* 1373. *paluſtris.*
Marſh Calla.
Nat. of the North of Europe.
Introd. 1770, by Daniel Charles Solander, LL.D.
Fl. July and Auguſt. H. ♃.

P O T H O S. *Gen. pl.* 1031.

Spatha. *Spadix* ſimplex, tectus. *Cal.* 0. *Petala* 4. *Stam.* 4. *Baccæ* diſpermæ.

1. P. foliis cordatis, ſpadice digitiformi. *cordata.*
Pothos cordata. *Sp. pl.* 1373.
Heart-leav'd Pothos.
Nat. of America.
Introd. about 1770, by John Hope, M.D.
Fl. April. S. ♃.

2. P. foliis cordatis, ſpadice ſubgloboſo. *fœtida.*
Dracontium fœtidum. *Sp. pl.* 1372.
Stinking Pothos, or Scunkweed.
Nat. of North America.
Cult. 1759, by Mr. Philip Miller. *Mill. dict. edit.* 7. Arum 12.
Fl. March and April. H. ♃.

Classis XXI.

MONOECIA

MONANDRIA.

CASUARINA. *Linn. suppl.* 62.

MASC. *Cal.* Amentum verticillato-imbricatum. *Cor.* biglumis.

FEM. *Cal.* et *Cor.* maris. *Pist.* 1. *Sem.* 1, alatum, corolla indurata inclusum.

equisetifolia. 1. C. monoica, ramulellis flaccidis teretibus, strobilorum squamis inermibus villosis, vaginis masculis septempartitis ciliatis.

Casuarina equisetifolia. *Linn. suppl.* 412.

Horse-tail Casuarina.

Nat. of the East Indies, and the South Sea Islands.

Introd. 1766, by Admiral Byron.

Fl. October and November. S. ♄.

stricta. 2. C. dioica, ramulellis erectis, strobilorum squamis inermibus glabriusculis, vaginis masculis multifidis glabris.

Upright Casuarina.

Nat. of New South Wales.

Introd. 1775, by Messrs. Kennedy and Lee.

Fl. November and December. G. H. ♄.

torulosa. 3. C. dioica, ramulellis flaccidis, strobilorum squamis villosis tuberculis exasperatis, vaginis masculis quadrifidis.

Cork-

Cork-bark'd Casuarina.
Nat. of New South Wales. Sir *Joseph Banks*, Bart.
Introd. 1772.
Fl. S. ♄.

ARTOCARPUS. *Linn. suppl.* 61.

MASC. *Spadix.* *Cal.* 0. *Cor.* 1-pet. 2-fida.
FEM. *Cal.* 0. *Cor.* 0. *Stylus* 1. *Baccæ* 1-spermæ, connatæ in fructum subrotundum.

1. A. foliis integris. *Linn. suppl.* 412. *integrifolia.*
Indian Jaca Tree.
Nat. of the East Indies.
Introd. 1778, by Sir Edward Hughes, K. B.
Fl. S. ♄.

ZANNICHELLIA. *Gen. pl.* 1034.

MASC. *Cal.* 0. *Cor.* 0.
FEM. *Cal.* 1-phyllus. *Cor.* 0. *Germina* circiter 4. *Sem.* totidem.

1. ZANNICHELLIA. *Sp. pl.* 1375. *palustris.*
March Zannichellia.
Nat. of Britain.
Fl. July. H. ☉.

CHARA. *Linn. mant.* 23.

MASC. *Cal.* 0. *Cor.* 0. *Anthera* germini subjecta.
FEM. *Cal.* 4-phyllus. *Cor.* 0. *Stigma* 5-fidum. *Sem.* 1.

1. C. caulibus lævibus, frondibus interne dentatis. *Sp. pl.* 1624. *vulgaris.*
Common, or stinking Chara.

Nat. of Britain.
Fl. July and August. H. ☉.

hispida. 2. C. caulinis aculeis capillaribus confertis. *Sp. pl.* 1624.
Prickly Chara.
Nat. of Britain.
Fl. June——September. H. ☉.

DIANDRIA.

LEMNA. *Gen. pl.* 1038.

MASC. *Cal.* 1-phyllus. *Cor.* nulla.
FEM. *Cal.* 1-phyllus. *Cor.* 0. *Stylus* 1. *Caps.* unilocularis.

trisulca. 1. L. foliis petiolatis lanceolatis. *Sp. pl.* 1376.
Ivy-leav'd Duck's-meat.
Nat. of Britain.
Fl. May and June. H. ♃.

minor. 2. L. foliis sessilibus utrinque planiusculis, radicibus solitariis. *Sp. pl.* 1376.
Least Duck's-meat.
Nat. of Britain.
Fl. June. H. ♃.

polyrhiza. 3. L. foliis sessilibus, radicibus confertis. *Sp. pl.* 1377.
Greater Duck's-meat.
Nat. of Britain.
Fl. May——September. H. ☉.

TRIAN-

TRIANDRIA.

TYPHA. *Gen. pl.* 1040.

MASC. Amentum cylindricum. *Cal.* obſoletus, 3-phyllus. *Cor.* 0.

FEM. Amentum cylindricum, infra maſculos. *Cal.* capillo villoſo. *Cor.* 0. *Sem.* 1, inſidens pappo capillari.

1. T. foliis ſubenſiformibus, ſpica maſcula feminеaque approximatis. *Sp. pl.* 1377. *Curtis lond.* *latifolia.*
Great Cat's-tail, or Reed-mace.
Nat. of Britain.
Fl. July. H. ♃.

2. T. foliis ſemicylindricis, ſpica maſcula femineaque remotis. *Sp. pl.* 1377. *Curtis lond.* *anguſtifolia.*
Narrow-leav'd Cat's-tail.
Nat. of England.
Fl. July. H. ♃.

SPARGANIUM. *Gen. pl.* 1041.

MASC. *Amentum* ſubrotundum. *Cal.* 3-phyllus. *Cor.* 0.
FEM. *Amentum* ſubrotundum. *Cal.* 3-phyllus. *Cor.* 0. *Stigma* 2-fidum. *Drupa* exſucca, 1-ſperma.

1. S. foliis enſiformibus triquetris, caule ramoſo. *Hudſ. angl.* 401. *Curtis lond.* *ramoſum.*
Sparganium erectum α. *Sp. pl.* 1378.
Great Bur-reed.
Nat. of Britain.
Fl. July. H. ♃.

ſimplex. 2. S. foliis enſiformibus planis, caule ſimplici. *Hudſ. angl.* 401. *Curtis lond. Fl. dan.* 932.
Sparganium erectum β. *Sp. pl.* 1378.
Small Bur-reed.
Nat. of Britain.
Fl. July. H. ♃.

ZEA. *Gen. pl.* 1042.

MASC. in diſtinctis ſpicis. *Cal.* Gluma 2-flora, mutica. *Cor.* Gluma mutica.

FEM. *Cal.* Gluma 2-valvis. *Cor.* Gluma 2-valvis. *Stylus* 1, filiformis, pendulus. *Sem.* ſolitaria, receptaculo oblongo immerſa.

Mays. 1. ZEA. *Sp. pl.* 1378.
Indian Corn.
Nat. of America.
Cult. 1562. *Turn. herb. part* 2. *fol.* 58.
Fl. June and July. H. ⊙.

TRIPSACUM. *Gen. pl.* 1044.

MASC. *Cal.* Gluma 4-flora. *Cor.* Gluma membranacea.

FEM. *Cal.* Gluma ſinubus perforatis. *Cor.* Gluma 2-valvis. *Styli* 2. *Sem.* 1.

dactyloides. 1. T. ſpicis androgynis. *Sp. pl.* 1378.
Rough-ſeeded Tripſacum.
Nat. of Virginia.
Introd. before 1640, by Mr. John Tradeſcant, Jun. *Park. theat.* 1163. *n.* 9.
Fl. Auguſt. H. ♃.

hermaphroditum. 2. T. ſpica hermaphrodita. *Sp. pl.* 1379.
Hermaphrodite Tripſacum.

Nat.

Nat. of Jamaica.
Introd. 1776, by Monſ. Thouin.
Fl. Auguſt and September. S. ☉.

COIX. *Gen. pl.* 1043.

MASC. in ſpicis remotis. *Cal.* Gluma 2-flora, mutica. *Cor.* Gluma mutica.

FEM. *Cal.* Gluma 2-flora. *Cor.* Gluma mutica. *Stylus* 2-partitus. *Sem.* calyce oſſificato tectum.

1. C. ſeminibus ovatis. *Sp. pl.* 1378. *Lacryma Jobi.*
Job's Tears.
Nat. of the Eaſt Indies.
Cult. 1596, by Mr. John Gerard. *Hort. Ger.*
Fl. July. S. ♃.

OLYRA. *Gen. pl.* 1045.

MASC. *Cal.* Gluma uniflora, ariſtata. *Cor.* Gluma mutica.

FEM. *Cal.* Gluma uniflora, patula, ovata. *Stylus* 2-fidus. *Sem.* cartilagineum.

1. OLYRA. *Sp. pl.* 1379. *latifolia.*
Broad-leav'd Olyra.
Nat. of the Weſt Indies.
Introd. 1783, by Mr. William Forſyth.
Fl. S. ♃.

CAREX. *Gen. pl.* 1046.

Masc. Amentum imbricatum. *Cal.* 1-phyllus. *Cor.* 0.
Fem. Amentum imbricatum. *Cal.* 1-phyllus *Cor.* 0. *Nectarium* inflatum, 3-dentatum. *Stigm.* 3. *Sem.* triquetrum, intra nectarium.

* *Spica unica simplici.*

dioica. 1. C. spica simplici dioica. *Sp. pl.* 1379.
Small Carex.
Nat. of Britain.
Fl. June. H. ♃.

capitata. 2. C. spica simplici androgyna ovata: superne mascula, capsulis imbricato-patulis. *Sp. pl.* 1379.
Round-headed Carex.
Nat. of England.
Fl. June. H. ♃.

pulicaris. 3. C. spica simplici androgyna: superne mascula, capsulis divaricatis retroflexis. *Sp. pl.* 1380.
Flea Carex.
Nat. of Britain.
Fl. July. H. ♃.

** *Spicis androgynis.*

disticha. 4. C. spica composita subdisticha: spiculis ovatis subimbricatis androgynis foliolo longiori instructis, culmo triquetro. *Hudf. angl.* 403.
Soft Carex.
Nat. of Britain.
Fl. May and June. H. ♃.

5. C.

5. C. ſpica compoſita: ſpiculis androgynis: inferioribus remotioribus foliolo longiori inſtructis, culmo triquetro. *Sp. pl.* 1381. *arenaria.*
Sea Carex.
Nat. of Britain.
Fl. June and July. H. ♃.

6. C. ſpica compoſita: ſpiculis ovatis ſeſſilibus approximatis alternis androgynis nudis. *Sp. pl.* 1381. *leporina.*
Naked Carex.
Nat. of Britain.
Fl. June and July. H. ♃.

7. C. ſpica ſupradecompoſita inferne laxiore: ſpiculis androgynis ovatis ſeſſilibus glomeratis: ſuperne maſculis. *Sp. pl.* 1382. *vulpina.*
Great Carex.
Nat. of Britain.
Fl. July. H. ♃.

8. C. ſpiculis ſubrotundis androgynis contiguis, capſulis ovatis acutis. *Hudſ. angl.* 405. *ſpicata.*
Small-ſpik'd Carex.
Nat. of Britain.
Fl. June. H. ♃.

9. C. ſpica compoſita: ſpiculis rotundis remotiuſculis androgynis foliolo longiori inſtructis, capſulis acutis, culmo teretiuſculo. *Hudſ. angl.* 405. *diviſa.*
Marſh Carex.
Nat. of Britain.
Fl. June and July. H. ♃.

10. C. ſpiculis ſuubovatis ſeſſilibus remotis androgynis, capſulis acutis divergentibus ſpinoſis. *Sp. pl.* 1382. *muricata.*

Prickly Carex.
Nat. of Britain.
Fl. June. H. ♃.

remota. 11. C. ſpicis ovatis ſubſeſſilibus remotis androgynis, bracteis culmum æquantibus. *Syſt. veget.* 844.
Remote Carex.
Nat. of Britain.
Fl. Auguſt. H. ♃.

caneſcens. 12. C. ſpiculis ſubrotundis remotis ſeſſilibus obtuſis androgynis, capſulis ovatis obtuſiuſculis. *Sp. pl.* 1383.
Gray Carex.
Nat. of Britain.
Fl. May——Auguſt. H. ♃.

paniculata. 13. C. ſpiculis androgynis, racemo compoſito. *Sp. pl.* 1383.
Panicled Carex.
Nat. of England.
Fl. June and July. H. ♃.

*** *Spicis ſexu diſtinctis: femineis ſeſſilibus.*

flava. 14. C. ſpicis confertis ſubſeſſilibus ſubrotundis: maſcula lineari, capſulis acutis recurvis. *Sp. pl.* 1384.
Yellow Carex.
Nat. of Britain.
Fl. June. H. ♃.

digitata. 15. C. ſpicis linearibus erectis: maſcula breviore inferioreque, bracteis aphyllis, capſulis diſtantibus. *Sp. pl.* 1384.
Digitated Carex.

Nat.

Nat. of England.
Fl. May and June. H. ♃.

16. C. ſpicis femineis ſeſſilibus ſubſolitariis ovatis maſculæ approximatis, culmo nudo, capſulis pubeſcentibus. *Sp. pl.* 1385. *montana.*
Mountain Carex.
Nat. of Britain.
Fl. June. H. ♃.

17. C. ſpicis terminalibus confertis ſubrotundis: maſcula oblonga. *Sp. pl.* 1385. *pilulifera.*
Ball-bearing Carex.
Nat. of Britain.
Fl. June and July. H. ♃.

**** *Spicis ſexu diſtinctis: femineis pedunculatis.*

18. C. ſpicis androgynis terminalibus pedunculatis: florentibus erectis; fructiferis pendulis. *Sp. pl.* 1386. *atrata.*
Black Carex.
Nat. of Britain.
Fl. June and July. H. ♃.

19. C. ſpicis pendulis: maſcula erecta; femineis ovatis imbricatis, capſulis confertis obtuſis. *Sp. pl.* 1386. *palleſcens.*
Pale Carex.
Nat. of Britain.
Fl. April. H. ♃.

20. C. ſpicis pedunculatis erectis remotis: femineis linearibus, capſulis obtuſiuſculis inflatis. *Sp. pl.* 1387. *panicea.*
Pink Carex.

Nat.

Nat. of Britain.
Fl. June and July. H. ♃.

Pſeudo-Cyperus. 21. C. ſpicis pendulis, pedunculis geminatis. *Sp. pl.* 1387.
Baſtard Carex.
Nat. of Britain.
Fl. July. H. ♃.

pendula. 22. C. ſpicis ſubſeſſilibus pendulis: maſcula erecta; femineis cylindricis longiſſimis, capſulis ſubrotundis acuminatis. *Hudſ. angl.* 411. *Curtis lond.*
Carex Agaſtachys. *Linn. ſuppl.* 414.
Pendulous Carex.
Nat. of Britain.
Fl. May and June. H. ♃.

ſylvatica. 23. C. ſpicis pedunculatis pendulis: maſcula erecta; femineis filiformibus laxis pedunculo brevioribus, capſulis ovatis ariſtatis furcatis. *Hudſ. angl.* 411.
Wood Carex.
Nat. of Britain.
Fl. May and June. H. ♃.

diſtans. 24. C. ſpicis remotiſſimis ſubſeſſilibus, bractea vaginante, capſulis angulatis mucronatis. *Sp. pl.* 1387.
Diſtant-flowering Carex.
Nat. of Britain.
Fl. June. H. ♃.

inflata. 25. C. ſpicis remotis ſubpedunculatis erectiuſculis: maſcula erecta lanceolata, capſulis ovatis acuminatis calycis duplo longioribus. *Hudſ. angl.* 412.
Bottle Carex.

Nat.

Nat. of England.
Fl. June and July. H. ♃.

26. C. ſpicis erectis cylindricis ternis ſubſeſſilibus: maſcula terminali, culmo triquetro. *Sp. pl.* 1388. *cæſpitoſa.*
Turfy Carex.
Nat. of Britain.
Fl. May and June. H. ♃.

27. C. ſpicis confertis pedunculatis cylindricis ſubpendulis: maſcula terminali, capſulis imbricatis obtuſiuſculis. *Hudſ. angl.* 413. *recurva.*
Heath Carex.
Nat. of England.
Fl. May and June. H. ♃.

***** *Spicis ſexu diſtinctis: maſculis pluribus.*

28. C. ſpicis maſculis pluribus triquetris nigricantibus acutis, ſquamis ariſtato-acuminatis, capſulis ſubinflatis bicornibus. *Curtis lond.* *riparia.*
Great, or Common Carex.
Nat. of Britain.
Fl. May and June. H. ♃.

29. C. ſpicis maſculis pluribus obtuſis, ſquamis obtuſiuſculis, caule angulato. *Curtis lond.* *acuta.*
Carex acuta. *Sp. pl.* 1388.
Acute Carex.
Nat. of Britain.
Fl. May and June. H. ♃.

30. C. ſpicis maſculis et femineis pluribus ſubfiliformibus, floribus digynis. *Curtis lond.* *gracilis.*
Slender-ſpik'd Carex.

Nat.

Nat. of England.
Fl. May and June. H. ♃.

veſicaria. 31. C. ſpicis maſculis pluribus; femineis pedunculatis, capſulis inflatis acuminatis. *Sp. pl.* 1388.
Bladder Carex.
Nat. of Britain.
Fl. June. H. ♃.

hirta. 32. C. ſpicis remotis: maſculis pluribus; femineis ſubpedunculatis erectis, capſulis hirtis. *Sp. pl.* 1389.
Hairy Carex.
Nat. of Britain.
Fl. June. H. ♃.

AXYRIS. *Gen. pl.* 1047.

MASC. *Cal.* 3-partitus. *Cor.* nulla.
FEM. *Cal.* 2-phyllus. *Cor.* o. *Styli* 2. *Sem.* 1.

Ceratoides. 1. A. foliis lanceolatis tomentoſis, floribus femineis lanatis. *Sp. pl.* 1389. *Jacqu. ic. miſcell.* 2. *p.* 355.
Shrubby Axyris.
Nat. of Siberia.
Introd. 1780, by Peter Simon Pallas, M.D.
Fl. Auguſt. H. ♄.

Amaranthoides. 2. A. foliis ovatis, caule erecto, ſpicis ſimplicibus. *Sp. pl.* 1389.
Simple-ſpik'd Axyris.
Nat. of Siberia.
Cult. 1758, by Mr. Philip Miller.
Fl. June and July. H. ⊙.

OMPHA-

OMPHALEA. *Gen. pl.* 1039.

MASC. *Cal.* 5-phyllus. *Cor.* 0. *Pelta* pedicellata, antherifera.
FEM. *Cal.* et *Cor.* maris. *Capſ.* carnoſa, 3-cocca.

1. O. foliis oblongis. *Sp. pl.* 1377. *triandra.*
Long-leav'd Omphalea.
Nat. of Jamaica.
Cult. 1763, by Mr. Philip Miller.
Fl. June and July. S. ♄.

TRAGIA. *Gen. pl.* 1048.

MASC. *Cal.* 3-partitus. *Cor.* 0.
FEM. *Cal.* 5-partitus. *Cor.* 0. *Stylus* 3-fidus. *Capſ.* 3-cocca, 3-locularis. *Sem.* ſolitaria.

1. T. foliis lanceolatis obtuſis ſubdentatis. *Syſt. veget.* 847. *urens.*
Stinging Tragia.
Nat. of Virginia.
Introd. 1778, by John Fothergill, M.D.
Fl. Auguſt. S. ⊙.

HERNANDIA. *Gen. pl.* 1049.

MASC. *Cal.* 3-partitus. *Cor.* 3-petala.
FEM. *Cal.* truncatus, integerrimus. *Cor.* 6-petala. *Drupa* cava, ore aperta: nucleo mobili.

1. H. foliis peltatis. *Sp. pl.* 1391. *ſonora.*
Jack in a Box.
Nat. of the Weſt Indies.

Cult.

Cult. 1714, by the Dutchess of Beaufort. *Br. Mus.* *H. S.* 136. *fol.* 14.
Fl. S. ♄.

PHYLLANTHUS. *Gen. pl.* 1050.

MASC. *Cal.* 6-partitus, campanulatus. *Cor.* o.
FEM. *Cal.* 6-partitus. *Cor.* o. *Styli* 3, bifidi. *Caps.* 3-locularis. *Sem.* solitaria.

Niruri. 1. P. foliis pinnatis floriferis, floribus pedunculatis, caule herbaceo erecto. *Sp. pl.* 1392.
Annual Phyllanthus.
Nat. of the East Indies.
Cult. 1692, in the Royal Garden at Hampton-court. *Pluk. phyt. t.* 183. *f.* 5.
Fl. June——September. S. ☉.

Emblica. 2. P. foliis pinnatis floriferis, caule arboreo, fructu baccato. *Sp. pl.* 1393.
Shrubby Phyllanthus.
Nat. of the East Indies.
Cult. 1768, by Mr. Philip Miller. *Mill. dict. edit.* 8.
Fl. S. ♄.

COMPTONIA. *L'Herit. stirp. nov.*

MASC. Amentum. *Cal.* 2-phyllus. *Cor.* o. *Antheræ* bipartitæ.
FEM. Amentum. *Cal.* 6-phyllus. *Cor.* o. *Styli* 2. *Nux* ovata.

asplenifolia. 1. COMPTONIA. *L'Herit. stirp. nov. tom.* 2. *tab.* 58.
Liquidambar asplenifolium. *Sp. pl.* 1418.
Liquidambar peregrinum. *Syst. veget.* 860.
Fern-leav'd Comptonia.

Nat.

a

Nat. of North America.
Cult. 1714, by the Dutchess of Beaufort. *Br. Mus.* *H. S.* 141. *fol.* 37.
Fl. March——May. H. ♄.

TETRANDRIA.

AUCUBA. *Thunb. japon.* 4.

MASC. *Cal.* 4-dentatus. *Cor.* 4-petala.
FEM. *Nect.* 0. *Nux* 1-locularis.

1. AUCUBA. *Thunb. japon.* 64. *tab.* 12, 13. *Ic. Kæmpfer.* *t.* 6. *japonica.*
Japan Aucuba.
Nat. of Japan.
Introd. 1783, by Mr. John Græfer.
Fl. S. ♄.

LITTORELLA. *Linn. mant.* 160.

MASC. *Cal.* 4-phyllus. *Cor.* 4-fida. *Stam.* longa.
FEM. *Cal.* nullus. *Cor.* 4-fida. *Stylus* longus. *Sem.* nucleus.

1. LITTORELLA. *Syst. veget.* 848. *lacustris.*
Plantago uniflora. *Sp. pl.* 167.
Small Littorella.
Nat. of Britain.
Fl. July and August. H. ♃.

BETULA.

BETULA. *Gen. pl.* 1052.

MASC. *Cal.* 1-phyllus, 3-fidus, 3-florus. *Cor.* 4-partita.

FEM. *Cal.* 1-phyllus, sub 3-fidus, 2-florus. *Sem.* membrana alata.

alba. 1. B. foliis deltoidibus acutis duplicato-ſerratis glabris, ſtrobilorum ſquamis lobis lateralibus rotundatis, petiolis glabris pedunculo longioribus.

Betula alba. *Sp. pl.* 1393.

vulgaris. α ramis adſcendentibus.

Common Birch Tree.

pendula. β Betula pendulis virgulis. *Loeſ. pruſſ.* 26.

Weeping Birch.

Nat. of Britain.

Fl. July. H. ♄.

populifolia. 2. B. foliis deltoidibus longe acuminatis inæqualiter ſerratis glaberrimis, ſtrobilorum ſquamis lobis lateralibus ſubrotundis, petiolis glabris.

Poplar-leav'd Birch Tree.

Nat. of North America.

Cult. 1750, by Archibald Duke of Argyle.

Fl. July. H. ♄.

nigra. 3. B. foliis rhombeo-ovatis duplicato-ſerratis acutis ſubtus pubeſcentibus baſi integris, ſtrobilorum ſquamis villoſis: laciniis linearibus æqualibus.

Betula nigra. *Sp. pl.* 1394.

Black Birch Tree.

Nat. of Virginia and Canada.

Introd. 1736, by Peter Collinſon, Eſq. *Collinſ. mſſ.*

Fl. July and Auguſt. H. ♄.

4. B.

4. B. foliis ovatis acuminatis duplicato-ſerratis: venis ſubtus hirſutis. *papyracea.*
Paper Birch Tree.
Nat. of North America.
Cult. 1750, by Archibald Duke of Argyle.
Fl. June. H. ♄.

5. B. foliis cordatis oblongis acuminatis ſerratis. *Sp. pl.* 1394. *lenta.*
Soft Birch Tree.
Nat. of North America.
Cult. 1759, by Mr. Ph. Miller. *Mill. dict. edit.* 7. *n.* 3.
Fl. H. ♄.

6. B. foliis ovatis acutis ſerratis, ſtrobilorum ſquamis lobis lateralibus rotundatis, petiolis pubeſcentibus pedunculo brevioribus. *excelſa.*
Tall Birch Tree.
Nat. of North America.
Introd. about 1767, by Mr. James Gordon.
Fl. May. H. ♄.

7. B. foliis orbiculatis crenatis. *Sp. pl.* 1394. *nana.*
Smooth Dwarf Birch.
Nat. of Scotland.
Fl. May. H. ♄.

8. B. foliis obovatis crenatis. *Linn. mant.* 124. *Jacqu. hort.* 2. *p.* 56. *t.* 122. *pumila.*
Hairy Dwarf Birch.
Nat. of North America.
Introd. 1762, by Mr. James Gordon.
Fl. April and May. H. ♄.

oblongata. 9. B. pedunculis ramosis, foliis ovalibus obtusiusculis glutinosis: axillis venarum subtus nudis.

Alnus foliis ovato-lanceolatis marginibus dentatis. *Mill. dict. edit.* 7. (in editione octava omissa, cum toto genere.)

Alnus folio oblongo viridi. *Bauh. pin.* 428.

α foliis oblongis.

Oblong-leav'd Turky Alder Tree.

β foliis ellipticis.

Oval-leav'd Turky Alder Tree.

Nat. of the South of Europe.

Cult. 1759, by Mr. Ph. Miller. *Mill. dict. loc. cit.*

Fl. July. H. ♃.

Alnus. 10. B. pedunculis ramosis, foliis subrotundo-cuneiformibus obtusissimis glutinosis: axillis venarum subtus villosis.

glutinosa. α foliis indivisis.

Betula Alnus *α.* glutinosa. *Sp. pl.* 1394.

Common Alder Tree.

laciniata. β foliis pinnatifidis.

Alnus foliis eleganter incisis. *Du Hamel arb.* 42. *n.* 4.

Cut-leav'd Alder Tree.

Nat. of Britain.

Fl. July. H. ♄.

serrulata. 11. B. pedunculis ramosis, foliis obovatis acutis: venis et axillis venarum subtus villosis, stipulis ovalibus obtusis.

Notch'd-leav'd Alder Tree.

Nat. of Pensylvania.

Cult. 1769, by Peter Collinson, Esq.

Fl. H. ♄.

12. B.

12. B. pedunculis ramoſis, foliis ſubrotundo-ellipticis acutis ſubtus pubeſcentibus: axillis venarum nudis, ſtipulis lanceolatis. *incana.*

Betula incana. *Linn. ſuppl.* 417.

Betula Alnus β. incana. *Sp. pl.* 1394.

α foliis ſubtus glaucis, petiolis rubris. glauca.

Glaucous-leav'd Alder Tree.

β foliis ſubtus viridibus, petiolis viridibus. angulata.

Elm-leav'd Alder Tree.

Nat. of Europe.

Introd 1780, by Mr. John Buſh.

Fl. June. H. ♄.

13. B. pedunculis ramoſis, foliis ovatis acutis ſubundulatis: venis ſubtus piloſis: axillis nudis, ſtipulis ſubrotundo-ovatis. *criſpa.*

Curl'd-leav'd Alder.

Nat. of Newfoundland and Hudſon's Bay.

Introd. 1782, by the Hudſon's Bay Company.

Fl. H. ♄.

B U X U S. *Gen. pl.* 1053.

MASC. *Cal.* 3-phyllus. *Petala* 2. *Germinis* rudimentum.

FEM. *Cal.* 4-phyllus. *Petala* 3. *Styli* 3. *Capſ.* 3-roſtris, 3-locularis. *Sem.* 2.

1. BUXUS. *Sp. pl.* 1394. *ſempervirens.*

α Buxus arboreſcens, foliis ovatis. *Mill. dict.* arboreſcens.

Common Tree Box.

β Buxus arboreſcens, foliis lanceolatis. *Mill. dict.* anguſtifolia.

Narrow-leav'd Tree Box.

γ Buxus humilis, foliis orbiculatis. *Mill. dict.* ſuffruticoſa.

 Dwarf

Dwarf Box.
Nat. of England.
Fl. April. H. ♄.

E M P L E U R U M.

MASC. *Cal.* 4-fidus. *Cor.* nulla.
FEM. *Cal.* 4-fidus, inferus. *Cor.* nulla. *Stigma* cylindraceum, denticulo laterali germinis insidens. *Caps.* latere dehiscens. *Sem.* 1, arillatum.

serrulatum. 1. EMPLEURUM.
Diosma unicapsularis. *Linn. suppl.* 155.
Cape Empleurum.
Nat. of the Cape of Good Hope. Mr. *Fr. Masson.*
Introd. 1774.
Fl. G. H. ♄.

U R T I C A. *Gen. pl.* 1054.

MASC. *Cal.* 4-phyllus. *Cor.* 0. *Nectarium* centrale, cyathiforme.
FEM. *Cal.* 2-valvis. *Cor.* 0. *Sem.* 1, nitidum.

* *Oppositifoliæ.*

pilulifera. 1. U. foliis oppositis ovatis serratis, amentis fructiferis globosis. *Sp. pl.* 1395.
Roman Nettle.
Nat. of England.
Fl. July and August. H. ⊙.

Dodartii. 2. U. foliis oppositis ovatis subintegerrimis, amentis fructiferis globosis. *Sp. pl.* 1395.
Pellitory-leav'd Nettle.
Nat. of the South of Europe.
Cult. 1713. *Philosoph. transact. n.* 337. *p.* 57. *n.* 82.
Fl July and August. H. ⊙.

3. U.

3. U. foliis oppositis ovalibus. *Sp. pl.* 1396. *urens.*
Lesser, or Stinging Nettle.
Nat. of Britain.
Fl. August. H. ☉.

4. U. foliis oppositis cordatis hispidis, racemis geminis. *dioica.*
Urtica dioica. *Sp. pl.* 1396.
Common Nettle.
Nat. of Britain.
Fl. July——September. H. ♃.

5. U. foliis oppositis tripartitis incisis. *Sp. pl.* 1396. *cannabina.*
Hemp-leav'd Nettle.
Nat. of Siberia.
Cult. 1752, by Mr. Ph. Miller. *Mill. dict. edit.* 6. *n.* 10.
Fl. July——September. H. ♃.

6. U. foliis oppositis ovato-lanceolatis nudiusculis, caule petiolisque hispidis, racemis geminis. *gracilis.*
Slender-stalk'd Nettle.
Nat. of Hudson's Bay.
Introd. 1782, by the Hudson's Bay Company.
Fl. June——August. H. ♃.

7. U. foliis oppositis ovato-oblongis, caule petiolisque nudiusculis, racemis cylindricis solitariis sessilibus. *cylindrica.*
Urtica cylindrica. *Sp. pl.* 1396.
Cylindrical Nettle.
Nat. of Virginia and the West Indies.
Cult. 1759, by Mr. Ph. Miller. *Mill. dict. edit.* 7. *n.* 6.
Fl. June——August. H. ♃.

** *Alternifoliæ.*

8. U. foliis alternis cordato-ovatis, amentis ramosis distichis erectis. *Sp. pl.* 1397. *canadensis.*

 Canada

Canada Nettle.
Nat. of Siberia and Canada.
Cult. 1713. *Philosoph. transact. n.* 337. *p.* 57. *n.* 83.
Fl. August——October. H. ♃.

nivea. 9. U. foliis alternis suborbiculatis utrinque acutis subtus tomentosis. *Sp. pl.* 1398. *Jacqu. hort.* 2. *p.* 78. *t.* 166.
Chinese, or white-leav'd Nettle.
Nat. of China and India.
Cult. 1739, by Mr. Philip Miller. *Rand. chel. n.* 7.
Fl. August and September. G. H. ♄.

MORUS. *Gen. pl.* 1055.

MASC. *Cal.* 4-partitus. *Cor.* 0.
FEM. *Cal.* 4-phyllus. *Cor.* 0. *Styli* 2. *Cal.* baccatus. *Sem.* 1.

alba. 1. M. foliis oblique cordatis lævibus. *Sp. pl.* 1398.
White Mulberry Tree.
Nat. of China.
Cult. 1596, by Mr. John Gerard. *Hort. Ger.*
Fl. June. H. ♄.

nigra. 2. M. foliis cordatis scabris. *Sp. pl.* 1398.
Common Mulberry Tree.
Nat. of Italy.
Cult. 1596, by Mr. John Gerard. *Hort. Ger.*
Fl. June. H. ♄.

papyrifera. 3. M. foliis palmatis, fructibus hispidis. *Sp. pl.* 1399.
Paper Mulberry Tree.
Nat. of Japan and the South Sea Islands.
Cult. before 1759, by Hugh Duke of Northumberland. *Mill. dict. edit.* 7. *n.* 6.
Fl. February——September. H. ♄.

4. M.

4. M. foliis cordatis ſubtus villoſis, amentis cylindricis. *rubra.*
Sp. pl. 1399.
Red Mulberry Tree.
Nat. of Virginia and Carolina.
Cult. 1629. *Park. parad.* 596. *n.* 3.
Fl. June and July. H. ♄.

5. M. foliis oblongis baſi hinc productioribus, ſpinis axillaribus. *Sp. pl.* 1399. *tinctoria.*
Dier's Mulberry, or Fuſtick-wood.
Nat. of Jamaica and Brazil.
Cult. 1739, by Mr. Philip Miller. *Mill. dict. vol.* 2. Addenda.
Fl. S. ♄.

PENTANDRIA.

XANTHIUM. *Gen. pl.* 1056.

MASC. *Cal.* communis imbricatus. *Cor.* 1-petalæ, 5-fidæ, infundibulif. *Recept.* paleaceum.
FEM. *Cal.* involucrum 2-phyllum, 2-florum. *Cor.* 0. *Drupa* ſicca, muricata, 2-fida. *Nucleus* 2-locularis.

1. X. caule inermi, foliis cordatis trinervatis. *Sp. pl.* 1400. *ſtrumarium.*
Leſſer Xanthium, or Burdock.
Nat. of England.
Fl. July and Auguſt. H. ⊙.

2. X. caule inermi, foliis cuneiformi-ovatis ſubtrilobis. *Sp. pl.* 1400. *orientale.*
Oriental Xanthium.
Nat. of China, Japan, and Ceylon.

Cult. 1713, by Bishop Compton. *Philosoph. transact.* *n.* 337. *p.* 57. *n.* 84.
Fl. July and August. S. ☉.

spinosum. 3. X. spinis ternatis, foliis trilobis. *Syst. veget.* 852.
Spiny Xanthium.
Nat. of the South of Europe.
Cult. 1713, by Bishop Compton. *Philosoph. transact.* *n.* 337. *p.* 36. *n.* 10. S. ☉.

fruticosum. 4. X. foliis pinnatifidis: laciniis incisis, caule fruticoso. *Linn. suppl.* 418. *L'Herit. stirp. nov. tom.* 2. *tab.* 91.
Ambrosia arborescens. *Mill. dict.*
Shrubby Xanthium.
Nat. of Peru.
Cult. 1759, by Mr. Philip Miller. *Mill. dict. edit.* 7. Ambrosia 5.
Fl. July. G. H. ♄.

AMBROSIA. *Gen. pl.* 1057.

MASC. *Cal.* communis, 1 phyllus. *Cor.* 1-petalæ, 3-fidæ, infundibulif. *Recept.* nudum.
FEM. *Cal.* 1-phyllus, integer, ventre 5-dentato, 1-florus. *Cor.* 0. *Nux* e calyce indurato, 1-sperma.

trifida. 1. A. foliis trilobis serratis. *Sp. pl.* 1401.
Trifid-leav'd Ambrosia.
Nat. of Virginia and Canada.
Cult. 1699, by Mr. Jacob Bobart. *Moris. hist.* 3. *p.* 4. *n.* 5. *s.* 6. *t.* 1. *f.* 4.
Fl. July——September. H. ☉.

elatior. 2. A. foliis bipinnatifidis, racemis paniculatis terminalibus glabris. *Sp. pl.* 1401.
Tall Ambrosia.

Nat.

Nat. of Virginia and Canada.
Cult. 1696, by Mr. Samuel Doody. *Pluk. alm.* 28.
Fl. July and August. H. ☉.

3. A. foliis bipinnatifidis: primoribus ramulorum indivisis integerrimis. *Sp. pl.* 1401. *artemisifolia.*
Mugwort-leav'd Ambrosia.
Nat. of North America.
Cult. 1759, by Mr. Ph. Miller. *Mill. dict. edit.* 7. *n.* 4.
Fl. July and August. H. ☉.

4. A. foliis multifidis, spicis solitariis pilosis subsessilibus. *Syst. veget.* 852. *maritima.*
Sea Ambrosia.
Nat. of Italy and the Levant.
Cult. 1570. *Lobel. adv.* 341.
Fl. July and August. H. ☉.

PARTHENIUM. *Gen. pl.* 1058.

MASC. *Cal.* communis, 5-phyllus. *Cor.* disci monopetalæ.
FEM. *Cor.* radii 5, singulæ 2, masculæ: intermedia feminea supera. *Sem.* nudum.

1. P. foliis composito-multifidis. *Sp. pl.* 1402. *Hysterophorus.*
Cut-leav'd Parthenium, or Bastard Feverfew.
Nat. of Jamaica.
Cult. 1739, by Mr. Philip Miller. *Rand. chel.* Partheniastrum.
Fl. July and August. S. ☉.

2. P. foliis ovatis crenatis. *Sp. pl.* 1402. *integrifolium.*
Entire-leav'd Parthenium.
Nat. of Virginia.

Cult.

Cult. about 1661, by Mr. Walker. *Pluk. phyt. t.* 53. *f.* 5.
Fl. June——October. H. ♃.

I V A. *Gen. pl.* 1059.

MASC. *Cal.* communis, 3-phyllus. *Cor.* disci 1-petalæ, 5-fidæ. *Recept.* pilis distinctum.
FEM. *Cal.* flosculi radii 5. *Cor.* 0. *Styli* 2, longi. *Sem.* nudum, obtusum.

frutescens. 1. I. foliis lanceolatis, caule fruticoso. *Sp. pl.* 1402.
Bastard Jesuit's Bark Tree.
Nat. of Virginia and Peru.
Cult. 1711, by the Dutchess of Beaufort. *Br. Mus. H. S.* 133. *fol.* 57.
Fl. August. H. ♄.

A M A R A N T H U S. *Gen. pl.* 1060.

MASC. *Cal.* 3-s. 5-phyllus. *Cor.* 0. *Stam.* 3 s. 5.
FEM. *Cal.* 3-s. 5-phyllus. *Cor.* 0. *Styli* 3. *Caps.* 1-locularis, circumscissa. *Sem.* 1.

* *Triandri.*

albus. 1. A. glomerulis triandris axillaribus bipartitis, bracteis subulatis, foliis ovatis emarginatis lineatis strictis. *Syst. veget.* 853.
White Amaranthus.
Nat. of North America.
Introd. 1778, by Monf. Thouin.
Fl. July and August. H. ☉

græcizans. 2. A. glomerulis triandris axillaribus, foliis lanceolatis obtusis repandis. *Sp. pl.* 1405.

Pellitory-

Pellitory-leav'd Amaranthus.
Nat. of North America.
Cult. 1723, in Chelſea Garden. *R. S. n.* 58.
Fl. July——September. H. ⊙.

3. A. glomerulis triandris axillaribus ſubrotundis ſeſſilibus, foliis lanceolatis acuminatis. *Sp. pl.* 1403. *melancholicus.*
Two-colour'd Amaranthus.
Nat. of the Eaſt Indies.
Cult. 1731, by Mr. Philip Miller. *Mill. dict. edit.* 1. *n.* 11.
Fl. June——September. S. ⊙.

4. A. glomerulis triandris axillaribus ſubrotundis amplexicaulibus, foliis lanceolato-ovatis coloratis. *Syſt. veget.* 853. *tricolor.*
Three-colour'd Amaranthus.
Nat. of the Eaſt Indies.
Cult. 1596, by Mr. John Gerard. *Hort. Ger.*
Fl. June——September. S. ⊙.

5. A. glomerulis diandris ſubſpicatis ovatis, floribus hermaphroditis femineiſque, foliis lanceolatis. *Sp. pl.* 1403. *polygamus.*
Hermaphrodite Amaranthus.
Nat. of the Eaſt Indies.
Introd. 1780, by Sir Joſeph Banks, Bart.
Fl. July and Auguſt. S. ⊙.

6. A. glomerulis triandris ſubſpicatis ovatis, foliis lanceolato-ovatis emarginatis. *Sp. pl.* 1403. *gangeticus.*
Oval-ſpik'd Amaranthus.
Nat. of India.
Introd. 1778, by Monſ. Thouin.
Fl. July——September. S. ⊙.

tristis. 7. A. glomerulis triandris rotundatis subspicatis, foliis ovato-cordatis emarginatis petiolo brevioribus. *Sp. pl.* 1404.
Round-headed Amaranthus.
Nat. of China.
Cult. 1759, by Mr. Ph. Miller. *Mill. dict. edit.* 7. *n.* 3.
Fl. June——August. H. ⊙.

lividus. 8. A. glomerulis triandris subspicatis rotundatis, foliis rotundatq-ovatis retusis. *Sp. pl.* 1404.
Livid Amaranthus.
Nat. of North America.
Cult. 1768, by Mr. Philip Miller. *Mill. dict. edit.* 8.
Fl. July——September. H. ⊙.

oleraceus. 9. A. glomerulis triandris pentandrisque, foliis ovatis obtusissimis emarginatis rugosis. *Sp. pl.* 1403.
Eatable Amaranthus.
Nat. of the East Indies.
Cult. 1768, by Mr. Philip Miller. *Mill. dict. edit.* 8.
Fl. July. H. ⊙.

Blitum. 10. A. glomerulis lateralibus: floribus trifidis, foliis ovatis retusis, caule diffuso. *Syst. veget.* 854.
Least Amaranthus, or Blite.
Nat. of England.
Fl. August. H. ⊙.

viridis. 11. A. glomerulis triandris: floribus masculis trifidis, foliis ovatis emarginatis, caule erecto. *Sp. pl.* 1405.
Green Amaranthus.
Nat. of Europe and Brazil.
Cult. 1768, by Mr. Philip Miller. *Mill. dict. edit.* 8.
Fl. August and September. H. ⊙.

12. A.

12. A. glomerulis triandris axillaribus, foliis ovatis emarginatis, floribus femineis infundibuliformibus obtusis. *Sp. pl.* 1405. *polygonoides.*
Spotted-leav'd Amaranthus.
Nat. of Jamaica.
Introd. 1778, by Monf. Thouin.
Fl. August. S. ☉.

** *Pentandri.*

13. A. racemis pentandris decompositis congestis nudis, spiculis conjugatis. *Sp. pl.* 1406. *hybridus.*
Cluster'd Amaranthus.
Nat. of Virginia.
Cult. 1656, by Mr. John Tradescant, Jun. *Mus. Trad.* 78.
Fl. June——September. H. ☉.

14. A. racemis pentandris compositis erectis: lateralibus patentissimis, foliis ovato-oblongis. *Sp. pl.* 1407. *sanguineus.*
Spreading, or Bloody Amaranthus.
Nat. of the Bahama Islands.
Cult. 1755, by Mr. Ph. Miller. *Mill. ic.* 15. *t.* 22.
Fl. July——September. S. ☉.

15. A. racemis pentandris lateralibus terminalibusque, caule flexuoso villoso, ramis retrocurvatis. *Sp. pl.* 1407. *retroflexus.*
Hairy Amaranthus.
Nat. of Pensylvania.
Cult. 1759, by Mr. Philip Miller. *Mill. dict. edit.* 7. *n.* 14.
Fl. July——September. H. ☉.

16. A. racemis pentandris compositis: summo infimisque nutantibus, foliis obovatis mucronatis. *Syst. veget.* 854. *flavus.*

Pale

Pale Amaranthus.
Nat. of India.
Cult. 1768, by Mr. Ph. Miller. *Mill. dict. edit.* 8.
Fl. July——September. S. ⊙.

hypochondriacus. 17. A. racemis pentandris compositis confertis erectis, foliis ovatis mucronatis. *Sp. pl.* 1407.
Prince's-feather, or Amaranthus.
Nat. of Virginia.
Cult. 1739, by Mr. Philip Miller. *Rand. chel. n.* 6.
Fl. July——September. H. ⊙.

cruentus. 18. A. racemis pentandris decompositis remotis patulo-nutantibus, foliis lanceolato-ovatis. *Sp. pl.* 1406.
Various-leav'd Amaranthus.
Nat. of China.
Cult. 1728. *Martyn dec.* 1. *p.* 6.
Fl. June——August. S. ⊙.

caudatus. 19. A. racemis pentandris decompositis cylindricis pendulis longissimis. *Sp. pl.* 1406.
α Amaranthus caudatus. *Mill. dict.*
Pendulous Amaranthus, or Love-lies-bleeding.
β Amaranthus maximus. *Mill. dict.*
Tree Amaranthus.
Nat. of the East Indies.
Cult. 1683, by Mr. James Sutherland. *Sutherl. hort. edin.* 20. *n.* 2.
Fl. August and September. H. ⊙.

spinosus. 20. A. racemis pentandris cylindricis erectis, axillis spinosis. *Sp. pl.* 1407.
Prickly Amaranthus.
Nat. of the West Indies.

Cult.

Cult. 1683, by Mr. James Sutherland. *Sutherl. hort. edin.* 51. *n.* 5.
Fl. July——September. S. ⊙.

POLYANDRIA.

CERATOPHYLLUM. *Gen. pl.* 1065.
MASC. *Cal.* multipartitus. *Cor.* 0. *Stam.* 16-20.
FEM. *Cal.* multipartus. *Cor.* 0. *Pist.* 1. *Styl.* 0. *Sem.* 1, nudum.

1. C. foliis dichotomo-bigeminis, fructibus trispinosis. *Sp. pl.* 1409. *demersum.*
Prickly-seeded Hornwort.
Nat. of Britain.
Fl. July. H. ♃.

2. C. foliis dichotomo-trigeminis, fructibus muticis. *Sp. pl.* 1409. *submersum.*
Smooth-seeded Hornwort.
Nat. of Britain.
Fl. July. H. ♃.

MYRIOPHYLLUM. *Gen. pl.* 1066.
MASC. *Cal.* 4-phyllus. *Cor.* 0. *Stam.* 8.
FEM. *Cal.* 4-phyllus. *Cor.* 0. *Pist.* 4. *Stylus* 0. *Sem.* 4, nuda.

1. M. floribus masculis interrupte spicatis. *Sp. pl.* 1409. *spicatum.*
Spik'd Water-millfoil.
Nat. of Britain.
Fl. June and July. H. ♃.

2. M.

verticillatum. 2. M. floribus omnibus verticillatis. *Sp. pl.* 1410.
Whorl'd Water-millfoil.
Nat. of England.
Fl. July. H. ♃.

SAGITTARIA. *Gen. pl.* 1067.

MASC. *Cal.* 3-phyllus. *Cor.* 3-petala. *Filamenta* fere 24.

FEM. *Cal.* 3-phyllus. *Cor.* 3-petala. *Pist.* multa. *Sem.* multa, nuda.

sagittifolia. 1. S. foliis sagittatis acutis. *Sp. pl.* 1410.
α Sagitta aquatica major. *Bauh. pin.* 194.
Great, or Common Arrow-head.
β Sagitta aquatica omnium minima. *Raj. syn.* 258.
Small Arrow-head.
Nat. of England.
Fl. June. H. ♃.

BEGONIA. *Gen. pl.* 1156.

MASC. *Cal.* 0. *Cor.* 4-petala: Petalis cordatis: 2 oppositis obcordatis. *Stam.* numerosa.

FEM. *Cal.* 0. *Cor.* 5-petala. *Styli* 3, bifidi: *Caps.* infera, triangularis, inæqualis, 3-locularis, polysperma.

nitida. 1. B. fruticosa erecta, foliis glaberrimis inæqualiter cordatis acutis obsolete denticulatis, stipulis carinatis.
Begonia obliqua. *L'Herit. stirp. nov. p.* 95. *tab.* 46. (exclusis synonymis plurimis.)
Begonia minor. *Jacqu. collect.* 1. *p.* 126.
Begonia purpurea. *Swartz prodr.* 86.
Shining-leav'd Begonia.
Nat. of Jamaica.

Introd.

Introd. 1777, by William Brown, M.D.
Fl. May——December. S. ♄.

2. B. caulescens erecta, foliis hispidis semicordatis acuminatis duplicato-serratis, capsulæ alis rotundatis parum inæqualibus. *humilis.*
Rough-leav'd Begonia.
Nat. of the West Indies. Mr. *Alex. Anderson.*
Introd. 1788, by Messrs. Lee and Kennedy.
Fl. October. S. ☉.
Obs. Herba similis Begoniæ hirsutæ *Aubl. guian.* 913. *tab.* 348; sed in illa capsulæ ala una maxima falcata, et flores dioici, qui in hac monoici.

THELIGONUM. *Gen. pl.* 1068.

Masc. *Cal.* 2-fidus. *Cor.* 0. *Stam.* fere 12.
Fem. *Cal.* 2-fidus. *Cor.* 0. *Pist.* 1. *Caps.* coriacea, 1-locularis, 1-sperma.

1. Theligonum. *Sp. pl.* 1411. *Cynocrambe.*
Purslain-leav'd Theligonum.
Nat. of the South of Europe.
Cult. 1759, by Mr. Philip Miller. *Mill. dict. edit.* 7.
Fl. July. H. ☉.

POTERIUM. *Gen. pl.* 1069.

Masc. *Cal.* 4-phyllus. *Cor.* 4-partita. *Stam.* 30-40.
Fem. *Cal.* 4-phyllus. *Cor.* 4-partita. *Pist.* 2. *Bacca* e tubo corollæ indurato.

1. P. inerme, caulibus subangulosis. *Sp. pl.* 1411. *Curtis lond.* *Sanguisorba.*
Common Burnet.

Nat. of England.
Fl. July. H. ♃.

hybridum. 2. P. inerme, caulibus teretibus ſtrictis. *Sp. pl.* 1412.
Sweet Burnet.
Nat. of France and Italy.
Cult. 1683, by Mr. James Sutherland. *Sutherl. hort. edin.* 271. *n.* 2.
Fl. June and July. H. ♃.

caudatum. 3. P. inerme fruteſcens, ramis teretibus villoſis, ſpicis elongatis laxis.
Smooth ſhrubby Burnet.
Nat. of the Canary Iſlands. Mr. *Francis Maſſon.*
Introd. 1779.
Fl. March and April. G. H. ♄.

ſpinoſum. 4. P. ſpinis ramoſis. *Sp. pl.* 1412.
Prickly ſhrubby Burnet.
Nat. of the Levant.
Cult. 1596, by Mr. John Gerard. *Hort. Ger.*
Fl. Moſt part of the Summer. G. H. ♄.

QUERCUS. *Gen. pl.* 1070.

MASC. *Cal.* 5-fidus fere. *Cor.* 0. *Stam.* 5-10.
FEM. *Cal.* 1-phyllus, integerrimus, ſcaber. *Cor.* 0. *Styli* 2-5. *Sem.* 1, ovatum.

Phellos. 1. Q. foliis deciduis lanceolatis integerrimis.
Quercus Phellos *α*. *Sp. pl.* 1412.
viridis. *α* foliis utrinque viridibus.
Common Willow-leav'd Oak Tree.
ſericea. *β* foliis ſubtus ſericeis.
Dwarf Willow-leav'd Oak Tree.
Nat. of North America.

Cult.

Cult. 1724. *Furber's catal.*

Fl. May and June. H. ♄.

2. Q. foliis ſempervirentibus lanceolatis oblongiſve ſubtus tomentoſis, calycibus ciliatis, antheris ovatis, cortice æquali. *Ilex.*

Quercus Ilex. *Sp. pl.* 1412.

α foliis lanceolatis integerrimis. integrifolia.

Common Evergreen Oak Tree.

β foliis lanceolatis ſerratis. ſerrata.

Ilex oblongo ſerrato folio. *Duham. arb.* 1. *p.* 314. *t.* 123.

Saw-leav'd Evergreen Oak Tree.

γ foliis ovato-oblongis ſerrato-denticulatis: denticulis pungentibus. oblonga.

Ilex folio rotundiore molli modiceque ſinuato. *Duham. arb.* 1. *p.* 314. *t.* 124.

Long-leav'd Evergreen Oak Tree.

Nat. of the South of Europe.

Cult. 1597, in Whitehall Garden. *Ger. herb.* 1161.

Fl. May and June. H. ♄.

3. Q. foliis ſubrotundo-ovatis baſi cordatis ſinuato-denticulatis pungentibus undulatis ſubtus tomentoſis, antheris ſubrotundis. *gramuntia.*

Quercus gramuntia. *Sp. pl.* 1413.

Holly-leav'd Evergreen Oak Tree.

Nat. of the South of France.

Cult. 1730. *Hort. angl.* 41. Ilex 3.

Fl. June. H. ♄.

4. Q. foliis ſempervirentibus ovato-oblongis ſubtus tomentoſis undulatis, cortice rimoſo fungoſo. *Suber*

Quercus Suber. *Sp. pl.* 1413.

Cork-bark'd Oak, or Cork Tree.

Nat. of the South of Europe.

Cult. 1699, by the Dutchess of Beaufort. *Br. Mus. Sloan. mss.* 525 & 3349.

Fl. June. H. ♄.

Coccifera. 5. Q. foliis ovatis basi cordatis dentato-spinosis utrinque glabris undulatis.

Quercus coccifera. *Sp. pl.* 1413.

Kermes Oak Tree.

Nat. of the South of Europe and the Levant.

Cult. 1683, by Mr. James Sutherland. *Sutherl. hort. edin.* 170. *n.* 2.

Fl. May. H. ♄.

virens. 6. Q. foliis sempervirentibus coriaceis lanceolato-oblongis subtus subtomentosis indivisis sinuatisque.

Quercus Phellos β. *Sp. pl.* 1412.

Quercus virginiana. *Mill. dict.*

Quercus sempervirens, foliis oblongis non sinuatis. *Catesb. car.* 1. *p.* 17. *t.* 17.

Live Oak Tree.

Nat. of North America.

Cult. 1739, by Mr. Philip Miller. *Rand. chel. n.* 9.

Fl. H. ♄.

Prinus. 7. Q. foliis deciduis ovato-ellipticis subtus pubescentibus profunde dentatis: dentibus latissimis obtusis subæqualibus.

Quercus Prinus. *Sp. pl.* 1413.

lata. α foliis ovatis.

Broad Chesnut-leav'd Oak Tree.

oblongata. β foliis oblongis.

Long Chesnut-leav'd Oak Tree.

Nat. of North America.

Cult. 1730. *Hort. angl.* 62. *n.* 6.

Fl. May and June. H. ♄.

8. Q.

8. Q. foliis annuis ſubcuneiformibus baſi attenuatis lobatis glabris. *aquatica.*

Quercus nigra *α*. *Sp. pl.* 1413.

α foliis cuneiformibus apice ſublobatis. cuneata.

Quercus folio non ſerrato in ſummitate quaſi triangulo. *Cateſb. car.* 1. *p.* 20. *t.* 20.

Common Water Oak Tree.

β foliis cuneiformi-oblongis acutis ſinuatis. hetero-phylla.

Various-leav'd Water Oak Tree.

γ foliis cuneiformi-oblongis obſolete ſinuatis ſubundatis. elongata.

Long-leav'd Water Oak Tree.

δ foliis cuneiformi-oblongis ſubintegerrimis. indiviſa.

Intire-leav'd Water Oak Tree.

ε foliis oblongo-lanceolatis ſubcuneatis acutis leviter ſinuatis. attenuata.

Narrow-leav'd Water Oak Tree.

Nat. of North America.

Cult. 1748, by Mr. Philip Miller. *Mill. dict. edit.* 5. *n.* 31.

Fl. H. ♄.

9. Q. foliis annuis cuneiformibus baſi ſubcordatis obſolete lobatis: lobis dilatatis. *nigra.*

Quercus nigra *β*. *Sp. pl.* 1413.

Black Oak Tree.

Nat. of North America.

Cult. 1739, by Mr. Philip Miller. *Rand. chel. n.* 7.

Fl. H. ♄.

10. Q. foliis annuis utrinque glabris obtuſe ſinuatis: ſinubus divaricatis, laciniis acutis ſetaceo-mucronatis. *rubra.*

Quercus rubra *α*. *Sp. pl.* 1413.

α foliis latis, calycibus abbreviatis ſubtus planiuſculis. latifolia.

Champion Oak Tree.

β foliis mediocribus, calycibus urceolatis. coccinea.

 Scarlet

Scarlet Oak Tree.

montana. γ foliis minoribus, calycibus abbreviatis ſubtus planis, glandibus ſubrotundis.
Mountain Red Oak Tree.
Nat. of North America.
Cult. 1691, by Biſhop Compton. *Pluk. phyt. t.* 54. *f.* 5.
Fl. May. H. ♄.

diſcolor. 11. Q. foliis annuis ſubtus pubeſcentibus ſinuatis: ſinubus patentibus; laciniis ſetaceo-mucronatis.
Quercus rubra β. *Sp. pl.* 1414.
Downy-leav'd Oak Tree.
Nat. of North America.
Introd. 1763, by Mr. Murdock Middleton.
Fl. H. ♄.

alba. 12. Q. foliis annuis pinnatifidis: ſinubus anguſtatis; laciniis oblongo-linearibus muticis.
Quercus alba. *Sp. pl.* 1414.
White Oak Tree.
Nat. of Virginia.
Cult. 1724. *Furber's catal.*
Fl. H. ♄.

Eſculus. 13. Q. foliis pinnatifidis ſubtus pubeſcentibus: laciniis lanceolatis acutis, ramentis axillaribus filiformibus, glandibus oblongis, calycibus muricatis.
Quercus Eſculus. *Sp. pl.* 1414.
Italian, or Small Prickly-cup'd Oak Tree.
Nat. of the South of Europe.
Cult. 1739, by Mr. Ph. Miller. *Mill. dict. vol.* 2. *n.* 1.
Fl. May. H. ♄.

Robur. 14. Q. foliis oblongis glabris ſinuatis: lobis rotundatis, glandibus oblongis.

Quercus

Quercus Robur. *Sp. pl.* 1414.

α arborea, pedunculis elongatis. peduncu-lata.

Quercus femina. *Mill. dict.*

Quercus cum longo pediculo. *Bauh. pin.* 420.

Female Oak Tree.

β arborea, fructibus ſubſeſſilibus. ſeſſilis.

Quercus Robur. *Mill. dict.*

Quercus latifolia mas, quæ brevi pediculo eſt. *Bauh. pin.* 419.

Common Oak Tree.

γ fruteſcens, ramis virgatis, fructibus ſeſſilibus. humilis.

Quercus humilis. *Mill. dict.*

Quercus humilis, gallis binis, ternis aut pluribus ſimul junctis. *Bauh. pin.* 421.

Dwarf Common Oak Tree.

Nat. α and β of Britain; and γ of the South of Europe.

Fl. April and May. H. ♄.

15. Q. foliis ovato-oblongis ſubtus tomentoſis ſinuatis repandis: laciniis acuminatis, calycibus maximis ſquamoſo-ſquarroſis. *Ægilops.*

Quercus Ægilops. *Sp. pl.* 1414.

Great Prickly-cup'd Oak Tree.

Nat. of the Levant.

Cult. 1731, by Mr. Ph. Miller. *Mill. dict. edit.* 1. *n.* 5.

Fl. H. ♄.

16. Q. foliis ſinuato-pinnatifidis ſubtus pubeſcentibus: laciniis acutiuſculis, ramentis axillaribus filiformibus, calycibus echinato-ramentaceis. *Cerris.*

Quercus Cerris. *Sp. pl.* 1415.

α foliis ovato-oblongis leviter ſinuatis planiuſculis. frondoſa.

Common Turky Oak Tree.

β foliis ovato-oblongis leviter ſinuatis ſubbullatis. bullata.

Rough-leav'd Turky Oak Tree.

sinuata. γ foliis oblongis profunde sinuatis: sinubus inæqualibus.
Narrow-leav'd Turky Oak Tree.
Nat. of the South of Europe.
Cult. 1739, by Mr. Ph. Miller. *Mill. dict. vol.* 2. *n.* 2.
Fl. May. H. ♄.

JUGLANS. *Gen. pl.* 1071.

MASC. *Cal.* 1-phyllus, squamiformis. *Cor.* 6-partita. *Filamenta* 18.

FEM. *Cal.* 4-fidus, superus. *Cor.* 4-partita. *Styli* 2. *Drupa* nucleo sulcato.

regia. 1. J. foliolis ovalibus glabris subserratis subæqualibus. *Sp. pl.* 1415.
Common Walnut Tree.
Nat. of Persia.
Fl. April and May. H. ♄.

alba. 2. J. foliolis septenis lanceolatis serratis: impari sessili. *Sp. pl.* 1415.
White Hickery, or Walnut Tree.
Nat. of North America.
Cult. 1699, by the Dutchess of Beaufort. *Br. Mus. Sloan. mss.* 525 & 3349.
Fl. H. ♄.

nigra. 3. J. foliolis quindenis lanceolatis serratis: exterioribus minoribus, gemmulis superaxillaribus. *Sp. pl.* 1415. *Jacqu. ic. miscell.* 2. *p.* 3.
Black Walnut Tree.
Nat. of North America.
Cult. 1656, by Mr. John Tradescant, Jun. *Mus. Trad.* 146.
Fl. April and May. H. ♄.

4. J. foliolis undenis lanceolatis basi altera breviore. *cinerea.*
Sp. pl. 1415. *Jacqu. ic. miscell.* 2. *p.* 7.
Shell-bark Walnut Tree.
Nat. of North America.
Cult. 1656, by Mr. John Tradescant, Jun. *Mus. Trad.* 147.
Fl. H. ♄.

5. J. foliolis tredecim lineari-lanceolatis serratis sessilibus basi æqualibus, nucibus ellipticis. *angustifolia.*
Narrow-leav'd Walnut Tree.
Nat. of North America.
Introd. 1766, by Messrs. Kennedy and Lee.
Fl. H. ♄.

F A G U S. *Gen. pl.* 1072.

MASC. *Cal.* 5-fidus, campanulatus. *Cor.* 0. *Stam.* 12.
FEM. *Cal.* 4-dentatus. *Cor.* 0. *Styli* 3. *Capsula* (Calyx antea) muricata, 4-valvis. *Sem.* 2.

1. F. foliis lanceolatis acuminato-serratis subtus nudis. *Castanea.*
Sp. pl. 1416.
Common Chesnut Tree.
Nat. of England.
Fl. July and August. H. ♄.

2. F. foliis lanceolato-ovatis acute serratis subtus tomentosis, amentis filiformibus nodosis. *Sp. pl.* 1416. *pumila.*
Chinquapine, or Dwarf Chesnut Tree.
Nat. of North America.
Cult. 1699, by the Dutchess of Beaufort. *Br. Mus. Sloan. mss.* 3358. *fol.* 17.
Fl. July. H. ♄.

ferruginea. 3. F. foliis ovato-oblongis remote acute ſerratis acuminatis ſubtus tomentoſis.
Fagus americana latifolia. *Du Roi hort. harbecc.* 1. *p.* 269.
American Beech Tree.
Nat. of North America.
Introd. 1766, by Meſſrs. Kennedy and Lee.
Fl. H. ♄.

ſylvatica. 4. F. foliis ovatis obſolete ſerratis. *Sp. pl.* 1416.
vulgaris. α foliis viridibus.
Common Beech Tree.
purpurea. β foliis atro-rubentibus. *Du Roi hort. harbecc.* 1. *p.* 268.
Purple Beech Tree.
Nat. α. of Britain; β. of Germany.
Fl. April and May. H. ♄.

CARPINUS. *Gen. pl.* 1073.

MASC. *Cal.* 1-phyllus: ſquama ciliata. *Cor.* 0. *Stam.* 20.

FEM. *Cal.* 1-phyllus: ſquama ciliata. *Cor.* 0. *Germina* 2. *Styli* ſingulis 2. *Nux* ovata.

Betulus. 1. C. ſquamis ſtrobilorum planis. *Sp. pl.* 1416.
vulgaris. α foliis ovatis ſerratis.
Common Hornbeam Tree.
incifa. β foliis oblongis inciſo-ſerratis.
Cut-leav'd Hornbeam Tree.
Nat. of Britain.
Fl. March——May. H. ♄.

Oſtrya. 2. C. ſquamis ſtrobilorum inflatis. *Sp. pl.* 1417.
Hop Hornbeam Tree.
Nat. of Italy.

Cult.

Cult. 1730. *Hort. angl.* 16. *n.* 4.
Fl. May. H. ♄.

3. C. foliis lanceolatis acuminatis, ſtrobilis longiſſimis. *Mill. dict. du Roi hort. harbecc.* 1. *p.* 130. *virginiana.*
Flowering Hornbeam Tree.
Nat. of Virginia.
Cult. 1692, by Biſhop Compton. *Pluk. phyt. t.* 156. *f.* 1.
Fl. H. ♄.

CORYLUS. *Gen. pl.* 1074.

MASC. *Cal.* 1-phyllus: 3-fidus, ſquamiformis, 1-florus. *Cor.* 0. *Stam.* 8.
FEM. *Cal.* 2-phyllus : lacerus. *Cor.* 0. *Styli* 2. *Nux* ovata.

1. C. ſtipulis lanceolatis, foliis ſubrotundis cordatis ex obtuſo acuminatis, ramulis piloſis. *Avellana.*
Corylus Avellana. *Sp. pl.* 1417.
α Corylus ſylveſtris. *Bauh. pin.* 418. ſylveſtris.
Common Hazel Nut Tree.
β Corylus ſativa, fructu albo minore ſ. vulgaris. *Bauh. pin.* 417. alba.
White Filbert Nut Tree.
γ Corylus ſativa fructu oblongo rubente. *Bauh. pin.* 418. rubra.
Red Filbert Nut Tree.
δ Corylus ſativa, fructu rotundo maximo. *Bauh. pin.* 418. grandis.
Cob Nut Tree.
ε Corylus nucibus in racemum congeſtis. *Bauh. pin.* 418. glomerata.
Cluſter'd Nut Tree.

Nat.

Nat. of Britain.
Fl. February and March. H. ♄.

rostrata. 2. C. stipulis lanceolatis, foliis oblongis cordatis acutis, ramulis glabris, calycibus fructus rostratis.
Corylus sylvestris calyce longiore, fructum etiam maturum omnino tegente. *Gron. virg.* 151.
American Cuckold Nut Tree.
Nat. of North America.
Cult. 1745, by Archibald Duke of Argyle.
Fl. March and April. H. ♄.

Colurna. 3. C. stipulis linearibus acutis. *Sp. pl.* 1417.
Constantinople Hazel Nut Tree.
Nat. of Constantinople.
Cult. 1665, by Mr. John Rea. *Rea's flora* 215.
Fl. March and April. H. ♄.

PLATANUS. *Gen. pl.* 1075.

MASC. *Cal.* Amentum globosum. *Cor.* vix manifesta. *Antheræ* filamentum circumnatæ.
FEM. *Cal.* Amentum globosum. *Cor.* polypetala. *Styli* stigmate recurvo. *Sem.* subrotunda, stylo mucronata, basi papposa.

orientalis. 1. P. foliis subpalmatis, nervis subtus glabriusculis.
Platanus orientalis. *Sp. pl.* 1417.
elongata. α foliis basi attenuatis planis.
Oriental Plane Tree.
acerifolia. β foliis transversis.
Spanish Plane Tree.
undulata. γ foliis basi attenuatis undulatis.
Wave-leav'd Plane Tree.
Nat. of the Levant.

Cult.

Cult. 1562. *Turn. herb. part* 2. *fol.* 95 *verſo.*
Fl. April and May. H. ♄.

2. P. foliis lobato-angulatis, nervis ſubtus tomentoſis. *occidentalis.*
Platanus occidentalis. *Sp. pl.* 1418.
American Plane Tree.
Nat. of North America.
Introd. before 1640, by Mr. John Tradeſcant, Jun. *Park. theat.* 1427. *f.* 2.
Fl. April and May. H. ♄.

LIQUIDAMBAR. *Gen. pl.* 1076.

MASC. *Cal.* communis 4-phyllus. *Cor.* o. *Filam.* numeroſa.

FEM. *Cal.* in globum, 4-phylli. *Cor.* o. *Styli* 2. *Capſ.* multæ in globum, bivalves, polyſpermæ.

1. L. foliis palmato-lobatis: ſinubus baſeos venarum villoſis. *Styraciflua.*
Liquidambar Styraciflua. *Sp. pl.* 1418.
Maple-leav'd Liquidamber, or Sweet-gum.
Nat. of North America.
Cult. 1688, by Biſhop Compton. *Raj. hiſt.* 2. *p.* 1681.
Fl. H. ♄.

2. L. foliis palmato-lobatis: ſinubus baſeos venarum lævibus. *imberbe.*
Liquidamber orientalis. *Mill. dict.*
Oriental Liquidamber.
Nat.
Cult. 1759, by Mr. Ph. Miller. *Mill. dict. edit.* 7. *n.* 2.
Fl. H. ♄.

MONA-

MONADELPHIA.

PINUS. *Gen. pl.* 1077.

MASC. *Cal.* 4-phyllus. *Cor.* o. *Stam.* plurima. *Antheræ* nudæ.

FEM. *Cal.* ſtrobili: ſquama 2-flora. *Cor.* o. *Piſt.* 1. *Nux* ala membranacea excepta.

* *Foliis pluribus ex eadem baſi vaginali.*

ſylveſtris. 1. P. foliis geminis rigidis, conis ovàto-conicis longitudine foliorum ſubgeminis baſi rotundatis.

Pinus ſylveſtris. *Sp. pl.* 1418.

communis. α Pinus *rubra*, foliis geminis brevioribus glaucis, conis parvis mucronatis. *Mill. dict.*

Pinus ſylveſtris foliis brevibus glaucis, conis parvis albicantibus. *Du Hamel arb.* 2. *p.* 125. *t.* 30.

Scotch Fir, or Pine Tree.

tatarica. β Pinus *tartarica*, foliis geminis brevioribus latiuſculis glaucis, conis minimis. *Mill. dict.*

Tartarian Pine Tree.

montana. γ Pinus *montana*, foliis ſæpius ternis tenuioribus viridibus, conis pyramidatis, ſquamis obtuſis. *Mill. dict.*

Pinus ſylveſtris montana. *Du Hamel arb.* 2. *p.* 125. *t.* 31.

Mountain, or Mugho Pine Tree.

divaricata. δ foliis divaricatis obliquis.

Hudſon's Bay Pine Tree.

maritima. ε Pinus maritima major. *Du Hamel arb.* 2. *p.* 125. *t.* 28.

Sea Pine Tree.

Nat. of Scotland.

Fl. May. H. ♄.

2. P.

2. P. foliis geminis margine ſubaſperis, conis oblongo-conicis folio brevioribus baſi attenuatis: ſquamis echinatis. *Pinaſter.*

Pinus ſylveſtris γ. *Sp. pl.* 1418.
Pinus ſylveſtris. *Mill. dict.*
Pinus maritima altera. *Du Ham. arb.* 2. *p.* 125. *t.* 29.
Pinaſter, or Cluſter Pine Tree.
Nat. of the South of Europe.
Cult. 1596, by Mr. John Gerard. *Hort. Ger.*
Fl. April and May. H. ♄.

3. P. foliis geminis, conis oblongo-conicis longitudine foliorum ſolitariis baſi rotundatis: ſquamis echinatis. *inops.*

Pinus virginiana. *Mill. dict.*
Jerſey Pine Tree.
Nat. of North America.
Cult. 1748, by Mr. Philip Miller. *Mill. dict. edit.* 5. *n.* 13.
Fl. May. H. ♄.

4. P. foliis geminis, conis ovato-conicis baſi rotundatis ſolitariis folio dimidio brevioribus: ſquamis inermibus. *reſinoſa.*

American Pitch Pine Tree.
Nat. of North America.
Cult. 1756, by Hugh Duke of Northumberland.
Fl. May. H. ♄.

5. P. foliis geminis, conis ovato-conicis baſi rotundatis folio ſubbrevioribus: ſquamis obtuſis. *halepenſis.*

Pinus *halepenſis*, foliis geminis tenuiſſimis, conis obtuſis, ramis patulis. *Mill. dict. ic.* 139. *t.* 208.
Aleppo Pine Tree.
Nat. of the Levant.
Introd. 1732, by Conſul Cox. *Mill. ic. loc. cit.*
Fl. May. H. ♄.

6. P.

Pinea. 6. P. foliis geminis: primordialibus ciliatis, conis ovatis obtusis subinermibus folio longioribus, nucibus duris.

Pinus Pinea. *Sp. pl.* 1419.

Stone Pine Tree.

Nat. of the South of Europe.

Cult. 1570. *Lobel. adv.* 449.

Fl. May. H. ♄.

Tæda. 7. P. foliis trinis, conis oblongo-conicis folio brevioribus aggregatis, squamis echinatis.

Pinus Tæda. *Sp. pl.* 1419.

tenuifolia. α Pinus *Tæda*, foliis longioribus tenuioribus ternis, conis maximis laxis. *Mill. dict.*

Frankincense Pine Tree.

rigida. β Pinus *rigida*, foliis ternis, conis longioribus, squamis rigidioribus. *Mill. dict.*

Three-leav'd Virginian Pine Tree.

variabilis. γ foliis binatis ternatisque.

Two and three-leav'd Pine Tree.

alopecuroidea. δ foliis ternis patulo-squarrosis.

Fox-tail Pine Tree.

Nat. of North America.

Introd. 1736, by Peter Collinson, Esq. *Collins. mss.*

Fl. May and June. H. ♄.

palustris. 8. P. foliis ternis longissimis, conis subcylindraceis echinatis, ramis stipulis ramentaceis exasperatis.

Pinus palustris. *Mill. dict.*

Pinus americana palustris trifolia, foliis longissimis. *Du Ham. arb.* 2. *p.* 126.

Swamp Pine Tree.

Nat. of Carolina and Georgia.

Cult. 1730. *Hort. angl.* 88.

Fl. H. ♄.

9. P.

9. P. foliis quinis, conis ovatis obtusis: squamis adpressis, nucibus duris. *Cembra.*

Pinus Cembra. *Sp. pl.* 1419.
Siberian Stone Pine Tree.
Nat. of Switzerland and Siberia.
Cult. 1746, by Archibald Duke of Argyle.
Fl. May. H. ♄.

10. P. foliis quinis, conis cylindraceis folio longioribus laxis. *Strobus.*

Pinus Strobus. *Sp. pl.* 1419.
Weymouth Pine Tree.
Nat. of North America.
Cult. 1705, by the Dutchess of Beaufort. *Pluk. amalth.* 171.
Fl. April. H. ♄.

11. P. foliis fasciculatis acutis. *Sp. pl.* 1420. *Cedrus.*

Cedar of Libanon.
Nat. of the Levant.
Cult. 1683. *Ray's letters, p.* 171.
Fl. May. H. ♄.

12. P. foliis fasciculatis mollibus obtusiusculis, squamis strobilorum bracteas tegentibus. *pendula.*

Black Larch Tree.
Nat. of North America.
Cult. 1739, by Peter Collinson, Esq. *Mill. dict. vol.* 2. Larix.
Fl. May. H. ♄.

13. P. foliis fasciculatis mollibus obtusiusculis, bracteis extra squamas strobilorum exstantibus. *Larix.*

Pinus Larix. *Sp. pl.* 1420.
Common White Larch Tree.
Nat. of Switzerland, Germany, and Siberia.

Cult. 1629. *Park. parad.* 608.
Fl. March and April. H. ♄.

** *Foliis solitariis, et basi distinctis.*

Picea. 14. P. foliis solitariis planis emarginatis pectinatis, squamis coni obtusissimis adpressis.
Pinus Picea. *Sp. pl.* 1420.
Silver Fir Tree.
Nat. of Switzerland and Germany.
Cult. 1739, in Chelsea Garden. *Rand. chel.* Abies 1.
Fl. May. H. ♄.

Balsamea. 15. P. foliis solitariis planis emarginatis subpectinatis supra suberectis, squamis coni florentis acuminatis reflexis.
Pinus Balsamea. *Sp. pl.* 1421.
Balm of Gilead Fir Tree.
Nat. of Virginia.
Cult. 1696, by Bishop Compton. *Pluk. alm.* 2. *pl.* 2.
Fl. May. H. ♄.

canadensis. 16. P. foliis solitariis planis submembranaceis acutiusculis pectinatis, conis ovatis folio vix brevioribus.
Pinus canadensis. *Sp. pl.* 1421.
Hemlock Spruce Fir Tree.
Nat. of North America.
Introd. 1736, by Peter Collinson, Esq. *Collins. mss.*
Fl. May. H. ♄.

nigra. 17. P. foliis solitariis tetragonis undique sparsis rectis strictis, conis oblongis.
Abies mariana. *Mill. dict.*
Abies Piceæ foliis brevioribus, conis biuncialibus laxis. *Mill. ic.* 1. *t.* 1.
Black Spruce Fir Tree.

Nat.

Nat. of North America.
Cult. before 1700, by Bishop Compton. *Mill. ic. loc. cit.*
Fl. May. H. ♄.

18. P. foliis solitariis subtetragonis acutiusculis distichis, ramis infra nudis, conis cylindraceis. *Abies.*
Pinus Abies. *Sp. pl.* 1421.
Norway Spruce Fir Tree.
Nat. of the North of Europe.
Cult. 1739, in Chelsea Garden. *Rand. chel. n.* 3.
Fl. April. H. ♄.

19. P. foliis solitariis tetragonis lateralibus incurvis, ramis subtus nudiusculis, conis subcylindraceis. *alba.*
Abies canadensis. *Mill. dict.*
White Spruce Fir Tree.
Nat. of North America.
Cult. before 1700, by Bishop Compton. *Mill. ic.* 1.
Fl. May and June. H. ♄.

THUJA. *Gen. pl.* 1078.

MASC. *Cal.* Amenti squama. *Cor.* 0. *Stam.* 4.
FEM. *Cal.* strobili: squama 2-flora. *Cor.* 0. *Pist.* 1. *Nux* 1, cincta ala emarginata.

1. T. strobilis lævibus: squamis obtusis. *Sp. pl.* 1421. *occidentalis.*
American Arbor-vitæ Tree.
Nat. of Siberia and Canada.
Cult. 1596, by Mr. John Gerard. *Hort. Ger.*
Fl. May. H. ♄.

2. T. strobilis squarrosis: squamis acutis. *Syst. veget.* 861. *orientalis.*
China Arbor-vitæ Tree.
Nat. of China.
Cult. 1752, by Mr. Ph. Miller. *Mill. dict. edit.* 6. *n.* 3.
Fl. May. H. ♄.

CUPRESSUS. *Gen. pl.* 1079.

MASC. *Cal.* Amenti ſquama. *Cor.* o. *Antheræ* 4, ſeſſiles abſque filamentis.

FEM. *Cal.* ſtrobili: ſquama 1-flora. *Cor.* o. *Styli?* puncta concava. *Nux* angulata.

ſempervirens. 1. C. foliis imbricatis, frondibus quadrangulis. *Sp. pl.* 1422.

ſtricta. α ramis erectis ſtrictis.
Upright Cypreſs Tree.

horizontalis. β Cupreſſus foliis imbricatis acutis, ramis horizontalibus. *Mill. dict.*
Male ſpreading Cypreſs Tree.
Nat. of Candia.
Cult. 1551, in Sion Garden. *Turn. herb. part.* 1. *ſign. N iiij.*
Fl. May. H. ♄.

diſticha. 2. C. foliis diſtichis patentibus. *Sp. pl.* 1422.

patens. α foliis approximatis ſtricte diſtichis.
Common Deciduous Cypreſs.

nutans. β foliis remotioribus ſubſparſis.
Long-leav'd Deciduous Cypreſs Tree.
Nat. of North America.
Introd. before 1640, by Mr. John Tradeſcant, Sen. *Park. theat.* 1477. *f.* 2.
Fl. May. H. ♄.

Thyoides. 3. C. foliis imbricatis, frondibus ancipitibus. *Sp. pl.* 1422.
White Cedar, or Arbor-vitæ-leav'd Cypreſs Tree.
Nat. of Canada and Maryland.
Introd. 1736, by Peter Collinſon, Eſq. *Collinſ. mſſ.*
Fl. April and May. H. ♄.

4. C.

4. C. foliis imbricatis glandulosis, frondibus quadrangulis glaucis, ramis dependentibus. *L'Heritier stirp. nov. p.* 15. *t.* 8. *pendula.*
Cupressus lusitanica. *Mill. dict.*
Portugal Cypress.
Nat.
Cult. 1683. *Ray's letters, p.* 171. Cedrus ex Goa.
Fl. H. ♄.

5. C. foliis oppositis decussatis subulatis patulis. *Sp. pl.* 1422. *juniperoides.*
African Cypress.
Nat. of the Cape of Good Hope.
Cult. 1756, by Mr. Ph. Miller. *Mill. dict. edit.* 7. *n.* 6.
Fl. G. H. ♄.

DALECHAMPIA. *Gen. pl.* 1081.

Involucrum commune exterius foliolis 4, interius foliis 2 trifidis.

MASC. *Umbellula* 10-flora: *involucello* 2-phyllo: paleis numerosis. *Perianth. proprium* 5-phyllum. *Cor.* nulla. *Filam.* plurima, connata.

FEM. *Flosculi* 3. *Involucello* 3-phyllo. *Perianth. proprium* foliolis 11. *Cor.* 0. *Stylus* filiformis. *Caps.* 3-cocca.

1. D. foliis trifidis serratis. *Syst. veget.* 862. *scandens.*
Climbing Dalechampia.
Nat. of the West Indies.
Introd. 1785, by Mr. Alexander Anderson.
Fl. June and July. S. ♄.

ACALYPHA. *Gen. pl.* 1082.

MASC. *Cal.* 3 f. 4-phyllus. *Cor.* o. *Stam.* 8-16.
FEM. *Cal.* 3-phyllus. *Cor.* o. *Styli* 3. *Capf.* 3-cocca, 3-locularis. *Sem.* 1.

virginica. 1. A. involucris femineis cordatis incifis, foliis ovato-lanceolatis petiolo longioribus. *Sp. pl.* 1423.
Virginian Acalypha.
Nat. of both Indies and North America.
Cult. 1759, by Mr. Ph. Miller. *Mill. dict. edit.* 7. *n.* 1.
Fl. July and Auguft. S. ☉.

indica. 2. A. involucris femineis cordatis fubcrenatis, foliis ovatis petiolo brevioribus. *Sp. pl.* 1424.
Indian Acalypha.
Nat. of the Eaft Indies.
Cult. 1759, by Mr. Ph. Miller. *Mill. dict. edit.* 7. *n.* 2.
Fl. July. S. ☉.

CROTON. *Gen. pl.* 1083.

MASC. *Cal.* cylindricus, 5-dentatus. *Cor.* 5-petala. *Stam.* 10-15.
FEM. *Cal.* polyphyllus. *Cor.* o. *Styli* 3, bifidi. *Capf.* 3-locularis. *Sem.* 1.

lineare. 1. C. foliis linearibus integerrimis obtufis fubtus tomentofis, caule fruticofo.
Croton lineare. *Jacqu. hift.* 256. *t.* 162. *f.* 4.
Croton Cafcarilla. *Sp. pl.* 1424. (exclufis fynonymis Catefbæi et Plumerii.)
Willow-leav'd Croton.
Nat. of Jamaica.
Introd. about 1772, by Mr. William Malcolm.
Fl. July. S. ♄.

2. C.

2. C. foliis ovatis obtusiusculis integerrimis lævibus, fructibus pedunculatis. *Sp. pl.* 1425. *glabellum.*
Laurel-leav'd Croton.
Nat. of Jamaica.
Introd. 1778, by Thomas Clark, M. D.
Fl. S. ♄.

3. C. foliis rhombeis repandis, capsulis pendulis, caule herbaceo. *Sp. pl.* 1425. *tinctorium.*
Officinal Croton.
Nat. of Spain and the South of France.
Cult. 1570. *Lobel. adv.* 101.
Fl. July. H. ⊙.

4. C. foliis cordato-ovatis subtus tomentosis integris serratis. *Syst. veget.* 863. *argenteum.*
Silvery-leav'd Croton.
Nat. of South America.
Introd. before 1733, by William Houstoun, M. D. *Mill. dict. edit.* 8.
Fl. S. ⊙.

5. C. foliis rhombeo-ovatis acuminatis integerrimis glabris. *Syst. veget.* 863. *sebiferum.*
Poplar-leav'd Croton, or Tallow-tree.
Nat. of China.
Introd. 1755, by Hugh Duke of Northumberland.
Fl. September. H. ♄.

6. C. foliis ovalibus subcordatis integerrimis utrinque stellato-tomentosis, ramulis densius tomentosis. *Astroites.*
Woolly Croton.
Nat. of the West Indies. Mr. *Francis Masson.*
Introd. 1782.
Fl. July and August. S. ♄.

lobatum. 7. C. foliis inermi-ſerratis: inferioribus quinquelobis; ſuperioribus trilobis. *Sp. pl.* 1427.
Various-leav'd Croton.
Nat. of Vera Cruz.
Introd. 1730, by William Houſtoun, M.D. *Martyn dec.* 5. *p.* 46.
Fl. July and Auguſt. S. ☉.

JATROPHA. *Gen. pl.* 1084.

MASC. *Cal.* 0. *Cor.* 1-petala, infundibuliformis. *Stam.* 10: alterna breviora.
FEM. *Cal.* 0. *Cor.* 5-petala, patens. *Styli* 3, bifidi. *Capſ.* 3-locularis. *Sem.* 1.

Curcas. 1. J. foliis cordatis angulatis. *Sp. pl.* 1429. *Jacqu. hort.* 3. *p.* 36. *t.* 63.
Angular-leav'd Phyſic-nut.
Nat. of South America.
Cult. 1731, by Mr. Philip Miller. *Mill. dict. edit.* 1. Ricinoides 1.
Fl. S. ♄.

multifida. 2. J. foliis multipartitis lævibus, ſtipulis ſetaceis multifidis. *Sp. pl.* 1429.
French Phyſic-nut.
Nat. of South America.
Cult. 1696, in the Royal Garden at Hampton-court. *Catal. mſſ.*
Fl. June——Auguſt. S. ♄.

Manihot. 3. J. foliis palmatis: lobis lanceolatis integerrimis lævibus. *Sp. pl.* 1429.
Eatable-rooted Phyſic-nut, or Caſſava.
Nat. of South America.

Cult.

Cult. 1739, by Mr. Philip Miller. *Rand. chel.* Manihot 1.
Fl. July and August. S. ♄.

4. J. foliis palmatis dentatis aculeatis. *Sp. pl.* 1429. *Jacqu. hort.* 1. *p.* 8. *t.* 21. *urens.*
Stinging Physic-nut.
Nat. of Brazil.
Introd. 1690, by Mr. Bentick. *Br. Mus. Sloan. mss.* 3370.
Fl. May——July. S. ♄.

RICINUS. *Gen. pl.* 1085.

MASC. *Cal.* 5-partitus. *Cor.* 0. *Stam.* numerosa.
FEM. *Cal.* 3-partitus. *Cor.* 0. *Styli* 3, bifidi. *Caps.* 3-locularis. *Sem.* 1.

1. R. foliis peltatis subpalmatis serratis. *Sp. pl.* 1430. *communis.*
Common Palma Christi.
Nat. of both Indies.
Cult. 1562. *Turn. herb. part* 2. *fol.* 116.
Fl. July and August. S. ♂.

STERCULIA. *Gen. pl.* 1086.

MASC. *Cal.* 5-partitus. *Cor.* 0. *Filamenta* 15.
FEM. *Cal.* 5-partitus. *Cor.* 0. *Germen* columnæ insidens. *Caps.* 5-locularis, polysperma.

1. S. foliis digitatis. *Sp. pl.* 1431. *fœtida.*
Fœtid Sterculia.
Nat. of the East Indies.
Cult. 1690, in the Royal Garden at Hampton-court. *Catal. mss.*
Fl. S. ♄.

2. S.

platanifolia. 2. S. hermaphrodita, foliis cordatis lobatis, floribus paniculatis. *Linn. suppl.* 423. *L'Herit. stirp. nov. tom.* 2. *tab.* 93.

Firmiana. *Marsili in act. acad. patav.* 1. *p.* 106. *tab.* 1. 2.

Outom-chu. *Le Comte mem. de la Chine*, 1. *p.* 241. *fig.*

Ou tong chu. *Du Halde chin.* 2. *p.* 149. *fig. in tab. ad pag.* 154.

Maple-leav'd Sterculia.

Nat. of Japan and China.

Cult. 1757, by Hugh Duke of Northumberland.

Fl. G. H. ♄

HIPPOMANE. *Gen. pl.* 1088.

MASC. *Amentum. Perianth.* 2-fidum. *Cor.* 0.

FEM. *Perianth.* 3-fidum. *Cor.* 0. *Stigma* 3-partitum. *Drupa* s. *Caps.* 3-cocca.

Mancinella. 1. H. foliis ovatis serratis basi biglandulosis. *Syst. veget.* 866.

Manchineel Tree.

Nat. of the West Indies.

Cult. 1739, by Mr. Philip Miller. *Rand. chel.* Mançanilla 1.

Fl. S. ♄.

STILLINGIA. *Lin. mant.* 19.

MASC. *Cal.* hemisphæricus, multiflorus. *Cor.* tubulosa, erosa.

FEM. *Cal.* 1-florus, inferus. *Cor.* supera. *Styl.* 3-fidus. *Caps.* 3-cocca.

sylvatica. 1. STILLINGIA. *Linn. mant.* 126.

Wood Stillingia.

Nat.

Nat. of Carolina.
Introd. 1787, by Thomas Walter, Esq.
Fl. G. H. ♄.

HURA. *Gen. pl.* 1087.

MASC. *Amentum* imbricatum. *Perianth.* truncatum. *Cor.* o. *Filam.* cylindrica, apice peltata, cincta *Anth.* plurimis geminatis.
FEM. *Cal.* o. *Cor.* o. *Stylus* infundibuliformis. *Stigma* 12-fidum. *Capf.* 12-locularis. *Sem.* 1.

1. HURA. *Sp. pl.* 1431. *crepitans.*
Sand-box Tree.
Nat. of Mexico and the West Indies.
Introd. before 1733, by William Houstoun, M. D. *Mill. dict. vol.* 2.
Fl. S. ♄.

SYNGENESIA.

TRICHOSANTHES. *Gen. pl.* 1089.

MASC. *Cal.* 5-dentatus. *Cor.* 5-partita, ciliata. *Filam.* 3.
FEM. *Cal.* 5-dentatus. *Cor.* 5-partita, ciliata. *Stylus* 3-fidus. *Pomum* oblongum.

1. T. pomis teretibus oblongis incurvis. *Sp. pl.* 1432. *Anguina.*
Snake Gourd.
Nat. of China.
Cult. 1755, by Mr. Ph. Miller. *Mill. ic.* 21. *t.* 32.
Fl. May and June. S. ⊙.

MOMOR-

MOMORDICA. *Gen. pl.* 1090.

MASC. *Cal.* 5-fidus. *Cor.* 5-partita. *Filamenta* 3.
FEM. *Cal.* 5-fidus. *Cor.* 5-partita. *Stylus* 3-fidus.
Pomum elaſtice diſſiliens.

Balſamina. 1. M. pomis angulatis tuberculatis, foliis glabris patenti-palmatis. *Sp. pl.* 1433.
Common Momordica, or Male Balſam Apple.
Nat. of India.
Cult. 1568. *Turn. herb. part* 3. *p.* 16.
Fl. June and July. S. ☉.

Charantia. 2. M. pomis angulatis tuberculatis, foliis villoſis longitudinaliter palmatis. *Sp. pl.* 1433.
Hairy Momordica.
Nat. of the Eaſt Indies.
Cult. 1731, by Mr. Ph. Miller. *Mill. dict. edit.* 1. *n.* 3.
Fl. June and July. S. ☉.

Luffa. 3. M. pomis oblongis: ſulcis catenulatis, foliis inciſis. *Sp. pl.* 1433.
Egyptian Momordica.
Nat. of the Eaſt Indies.
Cult. 1739, by Mr. Philip Miller. *Mill. dict. vol.* 2. Luffa.
Fl. July and Auguſt. S. ☉.

Elaterium. 4. M. pomis hiſpidis, cirrhis nullis. *Sp. pl.* 1434.
Squirting Momordica, or Cucumber.
Nat. of the South of Europe.
Cult. 1596, by Mr. John Gerard. *Hort. Ger.*
Fl. June and July. H. ♃.

CUCUR-

CUCURBITA. *Gen. pl.* 1091.

MASC. *Cal.* 5-dentatus. *Cor.* 5-fida. *Filam.* 3.
FEM. *Cal.* 5-dentatus. *Cor.* 5-fida. *Pist.* 5-fidum.
Pomi ſemina margine tumido.

1. C. foliis ſubangulatis tomentoſis baſi ſubtus biglandu-loſis, pomis lignoſis. *Syſt. veget.* 868. *lagenaria.*
Bottle Gourd.
Nat. of both Indies.
Cult. 1597. *Ger. herb.* 777.
Fl. July——September. S. ⊙.

2. C. foliis lobatis, pomis lævibus. *Sp. pl.* 1435. *Pepo.*
Pompion, or Pumpkin Gourd.
Nat.
Cult. before 1570. *Lobel. adv.* 286.
Fl. June——Auguſt. H. ⊙.

3. C. foliis lobatis, pomis nodoſo-verrucoſis. *Sp. pl.* 1435. *verrucoſa.*
Warted Gourd.
Nat.
Cult. 1731, by Mr. Philip Miller. *Mill. dict. edit.* 1. Melopepo 4.
Fl. June and July. H. ⊙.

4. C. foliis lobatis, caule erecto, pomis depreſſo-nodoſis. *Sp. pl.* 1435. *Melopepo.*
Squaſh Gourd.
Nat.
Cult. 1597. *Ger. herb.* 774. *f.* 4.
Fl. Moſt part of the Summer. H. ⊙.

Citrullus. 5. C. foliis multipartitis. *Sp. pl.* 1435.
Water Melon.
Nat. of the South of Europe.
Cult. before 1597, by Mr. John Gerard. *Ger. herb.* 767.
Fl. Most part of the Summer. S. ☉.

CUCUMIS. *Gen. pl.* 1092.

MASC. *Cal.* 5-dentatus. *Cor.* 5-partita. *Filamenta* 3.
FEM. *Cal.* 5-dentatus. *Cor.* 5-partita. *Pist.* 3-fidum. *Pomi* semina arguta.

Colocynthis. 1. C. foliis multifidis, pomis globosis glabris. *Sp. pl.* 1435.
Bitter Cucumber, or Gourd.
Nat.
Cult. 1551. *Turn. herb. part* 1. *sign. N ij.*
Fl. May——August. S. ☉.

prophetarum. 2. C. foliis cordatis quinquelobis denticulatis obtusis, pomis globosis spinoso-muricatis. *Sp. pl.* 1436. *Jacqu. hort.* 1. *p.* 3. *t.* 9.
Globe Cucumber.
Nat. of the Levant.
Introd. 1777, by Joseph Nicholas de Jacquin, M. D.
Fl. June——September. S. ☉.

Anguria. 3. C. foliis palmato-sinuatis, pomis globosis echinatis. *Sp. pl.* 1436.
Round prickly fruited Cucumber.
Nat. of Jamaica.
Cult. 1755, by Mr. Ph. Miller. *Mill. ic.* 22. *t.* 33.
Fl. July and August. S. ☉.

4. C.

4. C. foliis rotundato-angulatis, pomis angulis decem acutis. *Syst. veget.* 869. *Jacqu. hort.* 3. *p.* 40. *t.* 73, 74. *acutangulus.*
Acute-angled Cucumber.
Nat. of India.
Cult. 1692, by Bishop Compton. *Pluk. phyt.* 172. *f.* 1.
Fl. June——September. H. ⊙.

5. C. foliorum angulis rotundatis, pomis torulosis. *Sp. pl.* 1436. *Melo.*
Common Melon.
Nat.
Cult. before 1570. *Lobel. adv.* 285.
Fl. Most part of the Summer. S. ⊙.

6. C. foliorum angulis rotundatis, pomis sphæricis umbilico retuso. *Sp. pl.* 1437. *Dudaim.*
Apple shaped Cucumber.
Nat. of the Levant.
Cult. 1732, by James Sherard, M.D. *Dill. elth.* 223. *t.* 177. *f.* 218.
Fl. July and August. S. ⊙.

7. C. hirsutus, foliorum angulis integris rotundatis, pomis fusiformibus utrinque attenuatis hirtis. *Syst. veget.* 869. *Chate.*
Hairy Cucumber.
Nat. of the Levant.
Cult. 1768, by Mr. Ph. Miller. *Mill. dict. edit.* 8.
Fl. June. S. ⊙.

8. C. foliorum angulis rectis, pomis oblongis scabris. *Syst. veget.* 869. *sativus.*
Common Cucumber.
Nat.

Cult.

Cult. 1597. *Ger. herb.* 762. *f.* 1.
Fl. April——September. H. ☉.

flexuosus. 9. C. foliis angulato-ſublobatis, pomis cylindricis ſulcatis curvatis. *Sp. pl.* 1437.
Serpent Cucumber, or Melon.
Nat.
Cult. 1597. *Ger. herb.* 763. *fig.* 3.
Fl. Moſt part of the Summer. S. ☉.

BRYONIA. *Gen. pl.* 1093.

MASC. *Cal.* 5-dentatus. *Cor.* 5-partita. *Filamenta* 3.
FEM. *Cal.* dentatus. *Cor.* 5-partita. *Stylus* 3-fidus. *Bacca* ſubgloboſa, polyſperma.

latebroſa. 1. B. foliis ſubtrilobis piloſis baſi attenuatis.
Hairy Briony.
Nat. of the Canary Iſlands. Mr. *Francis Maſſon.*
Introd. 1779.
Fl. June. G. H. ♃.
OBS. Facillime diſtinguitur a congeneribus, foliis baſi minime cordatis, ſed ſecundum petiolos ſubdecurrentibus.

alba. 2. B. foliis palmatis utrinque calloſo-ſcabris. *Sp. pl.* 1438.
White Briony.
Nat. of Britain.
Fl. May. H. ♃.

palmata. 3. B. foliis palmatis lævibus quinquepartitis: laciniis lanceolatis repando-ſerratis. *Sp. pl.* 1438.
Palmated Briony.
Nat. of Ceylon.

Introd.

Introd. 1778, by Meſſrs. Gordon and Græfer.
Fl. S. ♃.

4. B. foliis cordatis angulatis glabriuſculis baſi ſubtus glandulofis, cirrhis ſimplicibus. *grandis.*
Bryonia grandis. *Linn. mant.* 126.
Great-flower'd Briony.
Nat. of India.
Introd. 1783, by Mr. John Græfer.
Fl. S. ♃.

5. B. foliis cordatis angulatis lobatiſque calloſo-hiſpidis, cirrhis ſimplicibus, baccis globoſis, ſeminibus muricatis. *ſcabrella.*
Bryonia ſcabrella. *Linn. ſuppl.* 424.
Briſtly Briony.
Nat. of the Eaſt Indies. *John Gerard Koenig*, M.D.
Introd. 1781, by Sir Joſeph Banks, Bart.
Fl. May——July. S. ☉.

6. B. foliis cordatis angulatis ſubtus villōſis; ſupra calloſo-ſcabris, cirrhis ſimplicibus, baccis globoſis, ſeminibus lævibus. *ſcabra.*
Bryonia ſcabra. *Linn. ſuppl.* 423.
Globe-fruited Briony.
Nat. of the Cape of Good Hope. Mr. *Fr. Maſſon.*
Introd. 1774.
Fl. September and October. G. H. ♃.

7. B. foliis cordatis angulatis ſupra veniſque ſubtus calloſo-ſcabris: callis remotis, cirrhis ſimplicibus, baccis globoſis. *verrucoſa.*
Rough Briony.
Nat. of the Canary Iſlands. Mr. *Francis Maſſon.*
Introd. 1779.
Fl. G. H. ♃.

africana. 8. B. foliis palmatis quinquepartitis utrinque lævibus: laciniis pinnatifidis. *Sp. pl.* 1438.
African, or Smooth-leav'd Briony.
Nat. of the Cape of Good Hope.
Cult. 1759, by Mr. Ph. Miller. *Mill. dict. edit.* 7. *n.* 2.
Fl. July and August. G. H. ♃.

SICYOS. *Gen. pl.* 1094.

MASC. *Cal.* 5-dentatus. *Cor.* 5-partita. *Filamenta* 3.
FEM. *Cal.* 5-dentatus. *Cor.* 5-partita. *Stylus* 3-fidus. *Drupa* 1-sperma.

angulata. 1. S. foliis angulatis. *Sp. pl.* 1439.
Angular-leav'd Sicyos, or Single-seeded Cucumber.
Nat. of North America.
Cult. 1732, by James Sherard, M. D. *Dill. elth.* 58. *t.* 51. *f.* 59.
Fl. July——September. H. ☉.

GYNANDRIA.

ANDRACHNE. *Gen. pl.* 1095.

MASC. *Cal.* 5-phyllus. *Cor.* 5-petala. *Stam.* 5, *Styli* rudimento inserta.
FEM. *Cal.* 5-phyllus. *Cor.* 0. *Styli* 3. *Caps.* 3-locularis. *Sem.* 2.

Telephioides. 1. A. procumbens herbacea. *Sp. pl.* 1439.
Bastard Orpine.
Nat. of Italy and the Levant.
Cult. 1732, by James Sherard, M. D. *Dill. elth.* 377. *t.* 282. *f.* 364.
Fl. July and August. H. ☉.

Classis

Classis XXII.

D I OE C I A

M O N A N D R I A.

P A N D A N U S. *Linn. ſuppl.* 64.

MASC. *Cal.* o. *Cor.* o. *Anthera* ſeſſilis, ramificationibus thyrſi inſerta.

FEM. *Cal.* o. *Cor.* o. *Stigmata* 2. *Fruct.* compoſitus.

1. PANDANUS. *Linn. ſuppl.* 424. *odoratiſſimus.*
Sweet-ſcented Pandanus, or Screw Pine.
Nat. of the Eaſt Indies and the South Sea Iſlands.
Introd. 1771, by Monſ. Richard.
Fl. S. ♄.

B R O S I M U M. *Swartz prodr.*

MASC. *Ament.* globoſum undique tectum ſquamis orbiculatis, peltatis. *Cor.* o. *Filam.* ſolitaria, inter ſquamas erumpentia.

FEM. *Ament.* maris. *Cor.* o. *Stylus* 2-fidus. *Bacca* 1-ſperma.

1. B. foliis ovato-lanceolatis perennantibus, amentis globoſis pedicellatis ſolitariis axillaribus, fructu corticoſo. *Swartz prodr.* 12. *Alicaſtrum.*

 Alicaſ-

Alicastrum arboreum, foliis ovatis alternis, fructibus solitariis. *Brown jam.* 372.
Jamaica Bread Nut Tree.
Nat. of Jamaica.
Introd. 1776, by Messrs. Kennedy and Lee.
Fl. S. ♄.

DIANDRIA.

CECROPIA. *Gen. pl.* 1099.

Masc. *Spatha* caduca. *Amenta* imbricata squamis turbinatis, subtetragonis. *Cor.* 0.
Fem. ut in mare. *Germina* imbricata. *Styl.* 1. *Stigma* lacerum. *Bacca* monosperma.

peltata. 1. Cecropia. *Sp. pl.* 1449.
Peltated Cecropia.
Nat. of Jamaica.
Introd. 1778, by Thomas Clark, M.D.
Fl. S. ♄.

SALIX.

SALIX. *Gen. pl.* 1098.

MASC. *Amenti* ſquamæ. *Cor.* 0. *Glandula* baſeos nectarifera.

FEM. *Amenti* ſquamæ. *Cor.* 0. *Stylus* 2-fidus. *Capſ.* 1-locularis, 2-valvis. *Sem.* papposa.

* *Foliis glabris ſerratis.*

1. S. foliis ſerratis glabris, floribus hermaphroditis diandris. *Sp. pl.* 1442. *hermaphroditica.*
Shining Willow.
Nat. of England.
Fl. March. H. ♄.

2. S. foliis ſerratis glabris, floribus triandris. *Sp. pl.* 1442. *triandra.*
Smooth Willow.
Nat. of Britain.
Fl. April and May. H. ♄.

3. S. foliis ſerratis glabris, floribus pentandris. *Sp. pl.* 1442. *pentandra.*
Sweet Willow.
Nat. of Britain.
Fl. March. H. ♄.

4. S. foliis ſerratis ovatis acutis glabris: ſerraturis cartilagineis, petiolis calloſo-punctatis. *Sp. pl.* 1442. *vitellina.*
Yellow Willow.
Nat. of England.
Fl. March and April. H. ♄.

5. S. foliis ſerratis glabris lanceolatis petiolatis, ſtipulis trapeziformibus. *Sp. pl.* 1443. *amygdalina.*
Almond-leav'd Willow.

Nat. of Britain.
Fl. May. H. ♄.

hastata. 6. S. foliis serratis glabris subovatis acutis sessilibus, stipulis subcordatis. *Sp. pl.* 1443.
Halberd-leav'd Willow.
Nat. of Switzerland and Lapland.
Introd. about 1780, by Messrs. Kennedy and Lee.
Fl. May. H. ♄.

fragilis. 7. S. foliis serratis glabris ovato-lanceolatis, petiolis dentato-glandulosis. *Sp. pl.* 1443.
Crack Willow.
Nat. of Britain.
Fl. May. H. ♄.

babylonica. 8. S. foliis serratis glabris lineari-lanceolatis, ramis pendulis. *Sp. pl.* 1443.
Weeping Willow.
Nat. of the Levant.
Cult. 1730. *Hort. angl.* 71. *n.* 11.
Fl. May. H. ♄.

Helix. 9. S. foliis serratis glabris lanceolato-linearibus: superioribus oppositis obliquis. *Sp. pl.* 1444.
Rose Willow.
Nat. of Britain.
Fl. May. H. ♄.

myrsinites. 10. S. foliis serratis glabris ovatis venosis. *Sp. pl.* 1445.
Whortle-leav'd Willow.
Nat. of Scotland.
Fl. June. H. ♄.

herbacea. 11. S. foliis serratis glabris orbiculatis. *Sp. pl.* 1445.
Herbaceous

Herbaceous Willow.
Nat. of Britain.
Fl. June. H. ♄.

12. S. foliis subserratis glabris obovatis obtusissimis. *Sp. pl.* 1445. *retusa.*
Blunt-leav'd Willow.
Nat. of Switzerland and Italy.
Introd. 1763, by the Earl of Coventry.
Fl. May. H. ♄.

** *Foliis glabris integerrimis.*

13. S. foliis integerrimis glabris ovatis obtusis. *Sp. pl.* 1446. *reticulata.*
Wrinkl'd Willow.
Nat. of Britain.
Fl. June. H. ♄.

14. S. foliis integris glabris ovatis acutis. *Sp. pl.* 1446. *myrtilloides.*
Myrtle-leav'd Willow.
Nat. of Sweden and Iceland.
Introd. 1772, by Sir Joseph Banks, Bart.
Fl. May. H. ♄.

*** *Foliis villosis integerrimis.*

15. S. foliis integerrimis utrinque villosis obovatis appendiculatis. *Sp. pl.* 1446. *aurita.*
Small round-eared Willow.
Nat. of Britain.
Fl. May and June. H. ♄.

16. S. foliis utrinque lanatis subrotundis acutis. *Sp. pl.* 1446. *lanata.*
Downy Willow.
Nat. of Scotland.
Fl. May. H. ♄.

lapponum. 17. S. foliis integerrimis hirſutis lanceolatis. *Sp. pl.* 1447.
Lapland Willow.
Nat. of Scotland.
Fl. June. H. ♄.

arenaria. 18. S. foliis integris ovatis acutis: ſupra ſubvilloſis; ſubtus tomentoſis. *Sp. pl.* 1447.
Sand Willow.
Nat. of Britain.
Fl. May. H. ♄.

incubacea. 19. S. foliis integerrimis lanceolatis: ſubtus villoſis nitidis, ſtipulis ovatis acutis. *Sp. pl.* 1447.
Trailing Willow.
Nat. of Europe.
Introd. 1775, by the Doctors Pitcairn and Fothergill.
Fl. May. H. ♄

repens. 20. S. foliis integerrimis lanceolatis utrinque nudiuſculis, caule repente. *Syſt. veget.* 880.
Creeping Willow.
Nat. of Britain.
Fl. May. H. ♄.

fuſca. 21. S. foliis integerrimis ovatis ſubtus pubeſcentibus. *Syſt. veget.* 880. *Jacqu. auſtr.* 5. *p.* 4. *t.* 409.
Brown Willow.
Nat. of England.
Fl. May. H. ♄.

roſmarinifolia. 22. S. foliis integerrimis lanceolato-linearibus ſtrictis ſeſſilibus ſubtus tomentoſis. *Sp. pl.* 1448.
Roſemary-leav'd Willow.
Nat. of England.
Fl. May. H. ♄.

**** *Foliis*

**** *Foliis villosis subserratis.*

23. S. foliis ovatis rugosis subtus tomentosis undatis superne denticulatis. *Sp. pl.* 1448. *caprea.*
Black Sallow, or Common Willow.
Nat. of Britain.
Fl. May. H. ♄.

24. S. exstipulacea, foliis lineari-lanceolatis petiolatis rugosis subtus tomentosis. *tristis.*
Narrow-leav'd American Willow.
Nat. of Pensylvania. Mr. *William Young.*
Introd. 1765.
Fl. April. H. ♄.

25. S. exstipulacea, foliis lanceolato-linearibus longissimis subintegerrimis planis subtus sericeis. *viminalis.*
Salix viminalis. *Sp. pl.* 1448.
Osier Willow.
Nat. of Britain.
Fl. April and May. H. ♄.

26. S. foliis lanceolatis acuminatis serratis utrinque pubescentibus: serraturis infimis glandulosis. *Sp. pl.* 1449. *alba.*
White Willow.
Nat. of Britain.
Fl. May. H. ♄.

TRIAN-

TRIANDRIA.

EMPETRUM. *Gen. pl.* 1100.

MASC. *Cal.* 3-partitus. *Cor.* 3-petala. *Stam.* longa.
FEM. *Cal.* 3-partitus. *Cor.* 3-petala. *Styli* 9. *Bacca* 9-ſperma.

nigrum. 1. E. procumbens. *Sp. pl.* 1450.
Black-berried Heath, Crow, or Crake-berry.
Nat. of Britain.
Fl. April and May. H. ♄.

OSYRIS. *Gen. pl.* 1101.

MASC. *Cal.* 3-fidus. *Cor.* 0.
FEM. *Cal.* 3-fidus. *Cor.* 0. *Stylus* 0. *Stigma* ſubrotundum. *Drupa* 1-locularis.

alba. 1. OSYRIS. *Sp. pl.* 1450.
Poet's Caſia.
Nat. of France, Spain, and Italy.
Cult. 1739, by Mr. Philip Miller. *Mill. dict. vol.* 2. Caſia 1.
Fl. G. H. ♄.

STILAGO. *Linn. mant.* 16.

MASC. *Cal.* 1-phyllus, urceolatus. *Cor.* 0.
FEM. *Cal.* 1-phyllus, urceolatus. *Cor.* 0. *Stigmata* ſeſſilia. *Drupa* nuce 2-loculari.

Bunius. 1. STILAGO. *Linn. mant.* 122.
Noeli-tali. *Rheed. mal.* 4. *p.* 115. *tab.* 56.

Chineſe

Chinese Laurel.
Nat. of the East Indies.
Introd. about 1757, by Hugh Duke of Northumberland.
Fl. August. S. ♄.

TETRANDRIA.

VISCUM. *Gen. pl.* 1105.

MASC. *Cal.* 4-partitus. *Cor.* 0. *Filamenta* 0. *Antheræ* calyci adnatæ.
FEM. *Cal.* 4-phyllus, superus. *Cor.* 0. *Stylus* 0. *Bacca* 1-sperma. *Sem.* cordatum.

1. V. foliis lanceolatis obtusis, caule dichotomo, spicis axillaribus. *Sp. pl.* 1451. *album.*
White, or Common Misseltoe.
Nat. of England.
Fl. May. H. ♄.

MONTINIA. *Linn. suppl.* 65.

MASC. *Cal.* 4-dentatus. *Petala* 4.
FEM. *Cal.* et *Cor.* ut in mare. *Caps.* 2-locularis. *Sem.* plura, plana.

1. MONTINIA. *Thunberg in act. lund.* 1. *p.* 109. *nov. gen.* 1. *p.* 28. *caryophyllacea.*
Montinia acris. *Linn. suppl.* 427.
Caryophyllæa fruticosa, foliis alternis oblongis. *Burm. afr.* 245. *t.* 90.
Sea Purslain-leav'd Montinia.
Nat. of the Cape of Good Hope.

Introd.

Introd. 1774, by Mr. Francis Maſſon.
Fl. July. G. H. ♄.

HIPPOPHAE. *Gen. pl.* 1106.

MASC. *Cal.* 2-partitus. *Cor.* 0.
FEM. *Cal.* 2-fidus. *Cor.* 0. *Styl.* 1. *Bacca* monoſperma.

Rhamnoides. 1. H. foliis lanceolatis. *Sp. pl.* 1452.
Common Sea Buck-thorn.
Nat. of England,
Fl. April. H. ♄.

MYRICA. *Gen. pl.* 1107.

MASC. *Amenti* ſquama lunata. *Cor.* 0.
FEM. *Amenti* ſquama lunata. *Cor.* 0. *Styli* 2. *Bacca* 1-ſperma.

Gale. 1. M. foliis lanceolatis ſubſerratis, caule ſuffruticoſo. *Sp. pl.* 1453.
Sweet Gale, Willow, or Candleberry Myrtle.
Nat. of Britain.
Fl. May. H. ♄.

cerifera. 2. M. foliis lanceolatis ſubſerratis, caule arboreſcente. *Sp. pl.* 1453.
anguſtifolia. α Myrtus brabanticæ ſimilis carolinenſis baccata, fructu racemoſo ſeſſili monopyreno. *Cateſb. car.* 1. *p.* 69. *t.* 69.
Common American Candleberry Myrtle.
latifolia. β Myrtus brabanticæ ſimilis caroliniensis humilior, foliis latioribus et magis ſerratis. *Cateſb. car.* 1. *p.* 13. *t.* 13.
Broad-leav'd American Candleberry Myrtle.

Nat.

Nat. of North America.
Cult. 1699, by the Dutchess of Beaufort. *Br. Muf. Sloan. mff.* 525 & 3349.
Fl. May and June. H. ♄.

3. M. foliis elliptico-lanceolatis fubferratis, amentis mafculis compofitis, drupa nucleo quadriloculari. *Faya.*
Azorian Candleberry Myrtle.
Nat. of Madeira and the Azores. Mr. *Fr. Maffon.*
Introd. 1777.
Fl. June and July. H. ♄.

4. M. foliis oblongis oppofite finuatis. *Sp. pl.* 1453. *quercifolia.*
α Myrica *quercifolia*, foliis oblongis oppofite finuatis glabris. *Mill. dict.*
Smooth Oak-leav'd Candleberry Myrtle.
β Myrica *hirfuta*, foliis oblongis oppofite finuatis hirfutis. *Mill. dict.*
Hairy Oak-leav'd Candleberry Myrtle.
Nat. of the Cape of Good Hope.
Cult. 1752, by Mr. Ph. Miller. *Mill. dict. edit.* 6. *n.* 5.
Fl. June and July. G. H. ♄.

5. M. foliis fubcordatis ferratis feffilibus. *Sp. pl.* 1454. *cordifolia.*
Heart-leav'd Candleberry Myrtle.
Nat. of the Cape of Good Hope.
Cult. 1759, by Mr. Ph. Miller. *Mill. dict. edit.* 7. *n.* 7.
Fl. May and June. G. H. ♄.

BRUCEA. *L'Herit. ftirp. nov.*

MASC. *Cal.* 4-phyllus. *Petala* 4.
FEM. *Cal.* et *Cor.* ut in mare. *Peric.* 4, monofperma.

1. BRUCEA. *L'Herit. ftirp. nov. p.* 19. *tab.* 10. *ferruginea.*
Brucea antidyfenterica. *J. F. Miller. ic. tab.* 25.
African

African Brucea.
Nat. of Africa.
Introd. 1775.
Fl. April and May. S. ♄.

PENTANDRIA.

PISTACIA. *Gen. pl.* 1108.

MASC. Amenti. *Cal.* 5-fidus. *Cor.* 0.
FEM. diſtincta. *Cal.* 3-fidus. *Cor.* 0. *Styli* 2. *Drupa* monoſperma.

officinarum. 1. P. foliis ſimplicibus ternatis pinnatiſque: foliolis ovalibus.
Piſtacia trifolia. *Sp. pl.* 1454.
Piſtachia Tree.
Nat. of the Levant.
Cult. 1570, by Mr. Gray. *Lobel. adv.* 413.
Fl. April and May. H. ♄.

Terebinthus. 2. P. foliis impari-pinnatis: foliolis ovato-lanceolatis. *Sp. pl.* 1455.
Common Turpentine Tree.
Nat. of Barbary and the South of Europe.
Cult. 1730. *Hort. angl.* 78. Terebinthus 1.
Fl. June and July. H. ♄.

Lentiſcus. 3. P. foliis abrupte pinnatis: foliolis lanceolatis. *Sp. pl.* 1455.
α Lentiſcus vulgaris. *Bauh. pin.* 399.
Common Maſtick Tree.

β Piſtacia

β Pistacia *massiliensis*, foliis abrupte pinnatis: foliolis lineari-lanceolatis. *Mill. dict.*
Narrow-leav'd Mastick Tree.
Nat. of the South of Europe and the Levant.
Cult. 1664. *Evelyn's kalend. hort.* 63.
Fl. May. G. H. ♄.

ZANTHOXYLUM. *Gen. pl.* 1109.

MASC. *Cal.* 5-partitus. *Cor.* 0.
FEM. *Cal.* 5-partitus. *Cor.* 0. *Pist.* 5. *Caps.* 5, monospermæ.

1. Z. foliis pinnatis. *Sp. pl.* 1455. *Clava Herculis.*
Common Toothach Tree.
Nat. of North America.
Cult. 1739, by Mr. Philip Miller. *Mill. dict. edit.* 8.
Fl. March and April. H. ♄.

SPINACIA. *Gen. pl.* 1112.

MASC. *Cal.* 5-partitus. *Cor.* 0.
FEM. *Cal.* 4-fidus. *Cor.* 0. *Styli* 4. *Sem.* 1, intra calycem induratum.

1. S. fructibus sessilibus. *Sp. pl.* 1456. *oleracea.*
α Lapathum hortense seu Spinacia semine spinoso. *Bauh. pin.* 114.
Prickly Spinage.
β Lapathum hortense seu Spinacia semine non spinoso. *Bauh. pin.* 115.
Round Spinage.
Nat.
Cult. 1568. *Turn. herb. part* 3. *p.* 71.
Fl. Most part of the Summer. H. ☉.

CANNA-

CANNABIS. *Gen. pl.* 1115.

MASC. *Cal.* 5-partitus. *Cor.* o.
FEM. *Cal.* 1-phyllus, integer, latere hians. *Cor.* o. *Styli* 2. *Nux* 2-valvis, intra calycem clausum.

sativa. 1. CANNABIS. *Sp. pl.* 1457.
Common Hemp.
Nat. of India.
Fl. June and July. H. ☉.

HUMULUS. *Gen. pl.* 1116.

MASC. *Cal.* 5-phyllus. *Cor.* o.
FEM. *Cal.* 1-phyllus, oblique patens, integer. *Cor.* o. *Styli* 2. *Sem.* 1, intra calycem foliatum.

Lupulus. 1. HUMULUS. *Sp. pl.* 1457.
Hops.
Nat. of Britain.
Fl. June——August. H. ♃.

HEXANDRIA.

TAMUS. *Gen. pl.* 1119.

MASC. *Cal.* 6-partitus. *Cor.* o.
FEM. *Cal.* 6-partitus. *Cor.* o. *Stylus* 3-fidus. *Bacca* 3-locularis, infera. *Sem.* 2.

communis. 1. T. foliis cordatis indivisis. *Sp. pl.* 1458.
Common Tamus, or Black Briony.
Nat. of England.
Fl. May——August. H. ♃.

2. T.

2. T. foliis reniformibus integris. *L'Herit. sert. angl. tab.* 40. *elephantipes.*

Tuberous Tamus.

Nat. of the Cape of Good Hope. Mr. *Fr. Masson.*

Introd. 1774.

Fl. July. G. H. ♃.

SMILAX. *Gen. pl.* 1120.

MASC. *Cal.* 6-phyllus. *Cor.* 0.

FEM. *Cal.* 6-phyllus. *Cor.* 0. *Styli* 3. *Bacca* 3-locularis. *Sem.* 2.

* *Caule aculeato, angulato.*

1. S. caule aculeato angulato, foliis dentato-aculeatis cordatis novemnerviis. *Sp. pl.* 1458. *aspera.*

α foliis basi rotundatis. simplicifolia.

Common Smilax, or Rough Bindweed.

β Smilax viticulis aspera, foliis longis angustis mucronatis lævibus, auriculis ad basim rotundioribus. *Pluk. alm.* 348. *t.* 110. *f.* 3. auriculata.

Ear-leav'd Smilax, or Rough Bindweed.

Nat. of Spain and Italy.

Cult. 1656, by Mr. John Tradescant, Jun. *Mus. Trad.* 168.

Fl. September. H. ♄.

2. S. caule aculeato angulato, foliis inermibus: caulinis cordatis; rameis ovato-oblongis. *Sp. pl.* 1459. *zeylanica.*

Ceylon Smilax.

Nat. of the East Indies.

Introd. 1778, by Patrick Russell, M. D.

Fl. S. ♄.

3. S. caule aculeato angulato, foliis inermibus ovatis retuso-mucronatis trinerviis. *Sp. pl.* 1459. *Sarsaparilla.*

Medicinal Smilax, or Sarsaparilla.
Nat. of America.
Cult. 1691, by Mr. Foster. *Pluk. phyt. t.* 111. *f.* 2.
Fl. July and August. H. ♄.

** *Caule aculeato, tereti.*

China. 4. S. caule aculeato teretiusculo, foliis inermibus ovato-cordatis quinquenerviis. *Sp. pl.* 1459.
Chinese Smilax.
Nat. of China and Japan.
Cult. 1759, by Mr. Ph. Miller. *Mill. dict. edit.* 7. *n.* 5.
Fl. G. H. ♄.

rotundifolia. 5. S. caule aculeato tereti, foliis inermibus cordatis acuminatis subseptemnerviis. *Sp. pl.* 1460.
Round-leav'd Smilax.
Nat. of Canada.
Cult. before 1760, by Archibald Duke of Argyle.
Fl. July and August. H. ♄.

laurifolia. 6. S. caule aculeato tereti, foliis inermibus ovato-lanceolatis trinerviis. *Sp. pl.* 1460.
Laurel-leav'd Smilax.
Nat. of Virginia and Carolina.
Cult. 1739, by Mr. Ph. Miller. *Mill. dict. vol.* 2. *n.* 2.
Fl. July. H. ♄.

tamnoides. 7. S. caule aculeato tereti, foliis inermibus cordatis oblongis septemnerviis. *Sp. pl.* 1460.
Black Briony-leav'd Smilax.
Nat. of North America.
Cult. 1739, by Mr. Ph. Miller. *Mill. dict. vol.* 2. *n.* 4.
Fl. June and July. H. ♄.

caduca. 8. S. caule aculeato tereti, foliis inermibus ovatis trinerviis. *Sp. pl.* 1460.

Deciduous

Deciduous Smilax.
Nat. of Canada.
Introd. 1775, by Mr. William Young.
Fl. H. ♄.

*** *Caule inermi.*

9. S. caule inermi angulato, foliis ciliato-aculeatis. *Sp. pl.* 1460. *bona nox.*
Ciliated Smilax.
Nat. of North America.
Introd. about 1778, by Mr. William Young.
Fl. June and July. H. ♄.

10. S. caule inermi angulato, foliis inermibus ovatis ſeptemnerviis. *Sp. pl.* 1460. *herbacea.*
Herbaceous Smilax.
Nat. of North America.
Cult. 1699, by the Dutcheſs of Beaufort. *Br. Muſ. Sloan. mſſ.* 525 & 3349.
Fl. July. H. ♃.

11. S. caule inermi tereti, foliis inermibus lanceolatis. *Sp. pl.* 1461. *lanceolata.*
Spear-leav'd Smilax.
Nat. of North America.
Introd. 1785, by Mr. William Young.
Fl. H. ♄

RAJANIA. *Gen. pl.* 1121.

MASC. *Cal.* 6-partitus. *Cor.* o.
FEM. *Cal.* 6-partitus. *Cor.* o. *Styli* 3. *Fructus* ſubrotundus, ala obliqua, inferus.

1. R. foliis cordatis ſeptemnerviis. *Sp. pl.* 1461. *cordata.*
Black Briony-leav'd Rajania.

Nat. of the Weſt Indies.
Introd. 1786, by Mr. Alexander Anderſon.
Fl. S. ♃.

DIOSCOREA. *Gen. pl.* 1122.

MASC. *Cal.* 6-partitus. *Cor.* 0.
FEM. *Cal.* 6-partitus. *Cor.* 0. *Styli* 3. *Capſ.* 3-locularis, compreſſa. *Sem.* 2, membranacea.

alata. 1. D. foliis cordatis, caule alato bulbifero. *Sp. pl.* 1462.
Wing-ſtalk'd Dioſcorea.
Nat. of both Indies.
Cult. 1739, by Mr. Philip Miller. *Mill. dict. vol.* 2. Addenda. Ricophora 1.
Fl. S. ♃.

ſativa. 2. D. foliis cordatis alternis, caule lævi tereti. *Sp. pl.* 1463.
Cultivated Dioſcorea, or Yam.
Nat. of the Weſt Indies.
Introd. before 1733, by William Houſtoun, M. D. *Mill. dict. edit.* 8.
Fl. Auguſt. S. ♃.

villoſa. 3. D. foliis cordatis alternis oppoſitiſque, caule lævi. *Sp. pl.* 1463.
Hairy Dioſcorea.
Nat. of Florida and Maryland.
Cult. 1759, by Mr. Philip Miller. *Mill. dict. edit.* 7. *n.* 5.
Fl. Auguſt. H. ♃.

OCTAN-

OCTANDRIA.

POPULUS. *Gen. pl.* 1123.

MASC. Amenti *Cal.* lamina lacera. *Cor.* turbinata, obliqua, integra.

FEM. Amenti *Cal.* et *Cor.* maris. *Stigma* 4-fidum. *Caps.* 2-locularis. *Sem.* multa, papposa.

1. P. foliis subrotundis dentato-angulatis subtus tomentosis. *Sp. pl.* 1463. *alba.*

α foliis angulato-dentatis subtus tomentosis cinerascentibus. canescens.

Populas alba. *Mill. dict.*

Common white Poplar.

β foliis sublobatis dentatis subtus tomentosis niveis: lobis acutis patulis. nivea.

Populus major. *Mill. dict.*

Great white Abele Tree, or Poplar.

Nat. of England.

Fl. April. H. ♄.

2. P. foliis subrotundis dentato-angulatis utrinque glabris. *Sp. pl.* 1464. *tremula.*

Asp, or Trembling Poplar Tree.

Nat. of Britain.

Fl. April. H. ♄.

3. P. foliis utrinque glabris acuminatis serratis deltoidibus: diametro longitudinali longiori. *nigra.*

Populus nigra. *Sp. pl.* 1464.

Black Poplar Tree.

Nat. of Britain.

Fl. April. H. ♄.

dilatata. 4. P. foliis utrinque glabris acuminatis ſerratis deltoidibus: diametro tranſverſali longiore.
Lombardy, or Po Poplar Tree.
Nat. of Italy.
Introd. about 1758, by the Earl of Rochford.
Fl. March and April. H. ♄.

balſamifera. 5. P. foliis ovatis ſerratis ſubtus albidis, ſtipulis reſinoſis. *Syſt. veget.* 889. *Pallas roſſ.* 67. *tab.* 41.
Common Tacamahac Poplar Tree.
Nat. of North America and Siberia.
Introd. 1731. *Mill. ic.* 174. *t.* 261.
Fl. April. H. ♄.

candicans. 6. P. foliis cordatis acuminatis ſubtus albidis ſubtriplinerviis, ſtipulis reſinoſis, ramis teretibus.
Heart-leav'd Tacamahac Poplar Tree.
Nat. of Canada.
Introd. about 1772, by John Hope, M. D.
Fl. March. H. ♄.

lævigata. 7. P. foliis cordatis trinerviis glabris baſi glanduloſis inæqualiter ſerratis, petiolis compreſſis, ramis teretibus.
Smooth Poplar Tree.
Nat. of North America.
Cult. 1769, by Hugh Duke of Northumberland.
Fl. March and April. H. ♄.

monilifera. 8. P. foliis ſubcordatis glabris baſi glandulοſis: ſerraturis cartilagineis hamatis piloſiuſculis; nervis patulis, petiolis compreſſis, ramis teretibus.
Canadian Poplar Tree.
Nat. of Canada.
Introd. about 1772, by John Hope, M. D.
Fl. May. H. ♄.

9. P.

9. P. foliis cordatis glabris basi glandulosis remote crenatis, petiolis compressis, ramis teretibus. *græca.*
Athenian Poplar Tree.
Nat. of the Islands of Archipelago.
Cult. 1779, by Hugh Duke of Northumberland.
Fl. March and April. H. ♄.

10. P. foliis cordatis primoribus pubescentibus basi eglandulosis, petiolis subteretibus, ramis teretibus. *heterophylla.*
Populus heterophylla. *Sp. pl.* 1464.
Various-leav'd Poplar Tree.
Nat. of Virginia and New York.
Cult. 1765, by John Fothergill, M.D.
Fl. April and May. H. ♄.

11. P. foliis cordatis glabris, ramis alato-angulosis. *angulata.*
Populus balsamifera. *Mill. dict.*
Populus magna virginiana, foliis amplissimis, ramis nervosis quasi quadrangulis. *Duham. arb.* 2. *p.* 178. *t.* 39. *f.* 9.
Carolina Poplar Tree.
Nat. of Carolina.
Cult. 1759, by Mr. Ph. Miller. *Mill. dict. edit.* 7. *n.* 6.
Fl. March. H. ♄.

RHODIOLA. *Gen. pl.* 1124.

MASC. *Cal.* 4-partitus. *Cor.* 4-petala.
FEM. *Cal.* 4-partitus. *Cor.* 0. *Nectaria* 4. *Pist.* 4. *Caps.* 4, polyspermæ.

1. RHODIOLA. *Sp. pl.* 1465. *rosea.*
Rose-root.
Nat. of Britain.
Fl. May——July. H. ♃.

ENNEANDRIA.

MERCURIALIS. *Gen. pl.* 1125.

MASC. *Cal.* 3-partitus. *Cor.* 0. *Stam.* 9 f. 12. *Antheræ* globofæ, didymæ.

FEM. *Cal.* 3-partitus. *Cor.* 0. *Styli* 2. *Capf.* dicocca, 2-locularis, 1-fperma.

perennis. 1. M. caule fimpliciffimo, foliis fcabris. *Sp. pl.* 1465. *Curtis lond.*
Dog's Mercury.
Nat. of Britain.
Fl. May——October. H. ♃.

annua. 2. M. caule brachiato, foliis glabris, floribus fpicatis. *Sp. pl.* 1465. *Curtis lond.*
Annual Mercury.
Nat. of Britain.
Fl. Auguft. H. ⊙.

tomentofa. 3. M. caule fuffruticofo, foliis tomentofis. *Sp. pl.* 1465.
Woolly Mercury.
Nat. of Spain.
Cult. 1731, by Mr. Ph. Miller. *Mill. dict. edit.* 1. *n.* 6.
Fl. July——September. H. ♄.

HYDROCHARIS. *Gen. pl.* 1126.

MASC. *Spatha* 2-phylla. *Cal.* 3-fidus. *Cor.* 3-petala. *Filamenta* 3 interiora ſtyliſera.

FEM. *Cal.* 3-fidus. *Cor.* 3-petala. *Styli* 6. *Capſ.* 6-locularis, polyſperma, infera.

1. HYDROCHARIS. *Sp. pl.* 1466. *Curtis lond.* *Morſus ranæ.*
Frog-bit.
Nat. of Britain.
Fl. June. H. ♃.

DECANDRIA.

CARICA. *Gen. pl.* 1127.

MASC. *Cal.* ſubnullus. *Cor.* 5-fida, infundibuliformis. *Filamenta* in tubo corollæ: alterna breviora.

FEM. *Cal.* 5-dentatus. *Cor.* 5-petala. *Stigmata* 5. *Bacca* 1-locularis, polyſperma.

1. C. foliorum lobis ſinuatis. *Sp. pl.* 1466. *Papaya.*
Papaw Tree.
Nat. of both Indies.
Cult. 1690, in the Royal Garden at Hampton-court. *Catal. mſſ.*
Fl. July. S. ♄.

KIGGELARIA. *Gen. pl.* 1128.

MASC. *Cal.* 5-partitus. *Cor.* 5-petala: glandulæ 5, trilobæ. *Antheræ* apicibus perforatæ.

FEM. *Cal.* et *Cor.* maris. *Styli* 5. *Capſ.* 1-locularis, 5-valvis, polyſperma.

1. KIGGELARIA. *Sp. pl.* 1466. *africana.*
African Kiggelaria.

Nat.

Nat. of the Cape of Good Hope.
Cult. 1690, in the Royal Garden at Hampton-court. *Catal. mss.*
Fl. May and June. G. H. ♄.

SCHINUS. *Gen. pl.* 1130.

MASC. *Cal.* 5-fidus. *Petala* 5.
FEM. *Flos* maris. *Bacca* 3-cocca.

Molle. 1. S. foliis pinnatis: foliolis serratis: impari longissimo. *Sp. pl.* 1467.
Peruvian Mastick Tree.
Nat. of Peru.
Cult. before 1597, by Mr. John Gerard. *Ger. herb.* 1346.
Fl. July and August. G. H. ♄.

CORIARIA. *Gen. pl.* 1129.

MASC. *Cal.* 5-phyllus. *Cor.* 5-petala, calyci simillima, connexa. *Antheræ* bipartitæ.
FEM. *Cal.* 5-phyllus. *Cor.* maris. *Styli* 5. *Sem.* 5, petalis succulento-baccatis tecta.

myrtifolia. 1. C. foliis ovato-oblongis. *Sp. pl.* 1467.
Myrtle-leav'd Sumach.
Nat. of Spain, and the South of France.
Cult. 1629, by Mr. John Parkinson. *Park. parad.* 609. *f.* 5.
Fl. May—August. H. ♄.

DODECAN-

DODECANDRIA.

EUCLEA. *Linn. ſuppl.* 67.

MASC. *Cal.* 4- vel 5- dentatus. *Cor.* 4- vel 5-partita. *Stam.* 12-15.

FEM. *Cal.* et *Cor.* maris. *Germ.* ſuperum. *Styli* 2. *Bacca* 2-locularis.

1. EUCLEA. *Syſt. veget.* 892. *racemoſa.*
Round-leav'd Euclea.
Nat. of the Cape of Good Hope.
Introd. 1722, by Mr. Thomas Knowlton.
Fl. November and December. G. H. ♄.

DATISCA. *Gen. pl.* 1132.

MASC. *Cal.* 5-phyllus. *Cor.* 0. *Antheræ* ſeſſiles, longæ, 15.

FEM. *Cal.* 2-dentatus. *Cor.* 0. *Styli* 3. *Capſ.* 3-angularis, 3-cornis, 1-locularis, pervia, polyſperma, infera.

1. D. caule lævi. *Sp. pl.* 1469. *cannabina.*
Baſtard Hemp.
Nat. of Candia.
Cult. 1739, by Mr. Philip Miller. *Rand. chel.* Cannabina 1 & 2.
Fl. July——September. H. ♃.

MENISPERMUM. *Gen. pl.* 1131.

MASC. *Petala* 4 exteriora; 8 interiora. *Stam.* 16.

FEM. *Cor.* maris. *Stam.* 8 ſterilia. *Baccæ* binæ, monoſpermæ.

1. M. foliis peltatis cordatis ſubrotundo-angulatis. *Sp. pl.* 1468. *canadenſe.*

Canadian

Canadian Moon-feed.
Nat. of Virginia and Canada.
Cult. 1713, by Bifhop Compton. *Philofoph. tranf.* *n.* 337. *p.* 213. *n.* 133.
Fl. June and July. H. ♄.

caroli-num. 2. M. foliis cordatis fubtus villofis. *Sp. pl.* 1468.
Carolina Moon-feed.
Nat. of Carolina.
Introd. 1765, by Mr. John Cree.
Fl. H. ♄.

POLYANDRIA.

CLIFFORTIA. *Gen. pl.* 1133.

MASC. *Cal.* 3-phyllus. *Cor.* 0. *Stam.* fere 30.
FEM. *Cal.* 3-phyllus, fuperus. *Cor.* 0. *Styli* 2. *Capf.* 2-locularis. *Sem.* 1.

ilicifolia. 1. C. foliis fubcordatis dentatis. *Sp. pl.* 1469.
Ilex-leav'd Cliffortia.
Nat. of the Cape of Good Hope.
Cult. 1714, in Chelfea Garden. *Br. Muf. H. S.* 134. *fol.* 14.
Fl. Moft part of the Year. G. H. ♄.

rufcifolia. 2. C. foliis lanceolatis integerrimis. *Sp. pl.* 1469.
Butchers-broom-leav'd Cliffortia.
Nat. of the Cape of Good Hope.
Introd. 1786, by Mr. Francis Maffon.
Fl. G. H. ♄.

3. C.

3. C. foliis cuneiformibus apice ſerratis. *cuneata.*
Wedge-leav'd Cliffortia.
Nat. of the Cape of Good Hope. Mr. *Fr. Maſſon.*
Introd. 1787.
Fl. April. G. H. ♄.

FLACOURTIA. *L'Herit. ſtirp. nov.*

MASC. *Cal.* 5-partitus. *Cor.* 0. *Stam.* numeroſiſſima, calyci impoſita.
FEM. *Cal.* polyphyllus. *Cor.* 0. *Germ.* ſuperum. *Styli* 5-9. *Bacca* multilocularis.

1. FLACOURTIA. *L'Herit. ſtirp. nov. p.* 59. *tab.* 30. *Ramontchi.*
Shining-leav'd Flacourtia.
Nat. of Madagaſcar.
Introd. 1775, by Monſ. Richard.
Fl. June and July. S. ♄.

MONADELPHIA.

JUNIPERUS. *Gen. pl.* 1134.

MASC. *Amenti* Calyx ſquamæ. *Cor.* 0. *Stam.* 3.
FEM. *Cal.* 3-partitus. *Petala* 3. *Styli* 3. *Bacca* 3-ſperma, tribus tuberculis calycis inæqualis.

1. J. foliis quadrifariam imbricatis acutis. *Sp. pl.* 1471. *thurifera.*
Spaniſh Juniper.
Nat. of the South of Europe.
Cult. 1759, by Mr. Ph. Miller. *Mill. dict. ed.* 7. *n.* 13.
Fl. May and June. H. ♄.

2. J. foliis inferioribus ternis: ſuperioribus binis decurrentibus ſubulatis patulis acutis. *Sp. pl.* 1471. *bermudiana.*
Bermudas

Bermudas Cedar, or Juniper.
Nat. of Bermudas.
Cult. 1700, by the Earl of Clarendon. *Br. Muſ. Sloan. mſſ.* 3343.
Fl. May and June. H. ♄.

Sabina. 3. J. foliis oppoſitis erectis decurrentibus: oppoſitionibus pyxidatis. *Sp. pl.* 1472.
cupreſſifolia. α Sabina folio cupreſſi. *Bauh. pin.* 487.
Common Savin.
tamariſcifolia. β Sabina folio tamariſci dioſcoridis. *Bauh. pin.* 487.
Tamariſk-leav'd Savin.
Nat. of the South of Europe and the Levant.
Cult. 1562. *Turn. herb. part* 2. *fol.* 124.
Fl. May and June. H. ♄.

virginiana. 4. J. foliis ternis baſi adnatis: junioribus imbricatis; ſenioribus patulis. *Sp. pl.* 1471.
Virginian Red Cedar, or Juniper.
Nat. of North America.
Cult. 1664. *Evelyn's ſylva* 61.
Fl. May and June. H. ♄.

communis. 5. J. foliis ternis patentibus mucronatis bacca longioribus. *Sp. pl.* 1470.
vulgaris. α Juniperus vulgaris fruticoſa. *Bauh. pin.* 488.
Common Juniper.
ſuecica. β Juniperus foliis ternis patentibus, acutioribus, ramis erectioribus, bacca longioribus. *Mill. dict.*
Swediſh Juniper.
montana. γ Juniperus minor montana, folio latiore, fructuque longiore. *Bauh. pin.* 489.
Procumbent Juniper.
Nat. of Britain.
Fl. May. H. ♄.

6. J.

6. J. foliis ternis patentibus mucronatis bacca brevioribus. *Sp. pl.* 1470. *Oxycedrus.*
Brown-berried Juniper.
Nat. of Spain and Portugal.
Cult. 1739, by Mr. Philip Miller. *Rand. chel. n.* 6.
Fl. May and June. H. ♄.

7. J. foliis ternis obliteratis imbricatis obtusis. *Sp. pl.* 1471. *phœnicea.*
Phœnician Cedar, or Juniper.
Nat. of the South of Europe, and the Levant.
Cult. 1683, by Mr. James Sutherland. *Sutherl. hort. edin.* 74. *n.* 4.
Fl. May and June. H. ♄.

T A X U S. *Gen. pl.* 1135.

MASC. *Cal.* 3-phyllus gemmæ. *Cor.* 0. *Stamina* multa. *Antheræ* peltatæ, 8-fidæ.
FEM. *Cal.* 3-phyllus gemmæ. *Cor.* 0. *Stylus* 0. *Sem.* 1, calyculo baccato, integerrimo.

1. T. foliis linearibus, receptaculis masculis subglobosis. *baccata.*
Taxus baccata. *Sp. pl.* 1472.
Common Yew Tree.
Nat. of Britain.
Fl. February——April. H. ♄.

2. T. foliis lineari-lanceolatis, receptaculis masculis filiformi-cylindraceis amentiformibus, antheris numerosissimis spiraliter collocatis. *elongata.*
African Yew Tree.
Nat. of the Cape of Good Hope.
Introd. 1774, by Thomas Lucas, Esq.
Fl. July. G. H. ♄.

EPHEDRA.

EPHEDRA. *Gen. pl.* 1136.

MASC. *Amenti* Calyx 2-fidus. *Cor.* 0. *Stamina* 7. *Antheræ* 4 inferiores; 3 superiores.

FEM. *Cal.* 2-partitus, 5-tuplex. *Cor.* 0. *Pist.* 2. *Sem.* 2, calyce baccato tecta.

distachya. 1. E. pedunculis oppositis: amentis geminis. *Sp. pl.* 1472.
Great Shrubby Horse-tail.
Nat. of France and Spain.
Cult. before 1570, by Matthias de L'Obel. *Lobel. adv.* 355.
Fl. June and July. H. ♄.

monostachya. 2. E. pedunculis pluribus: amentis solitariis. *Sp. pl.* 1472.
Small Shrubby Horse-tail.
Nat. of Siberia.
Introd. about 1772, by Messrs. Kennedy and Lee.
Fl. September——November. H. ♄.

CISSAMPELOS. *Gen. pl.* 1138.

MASC. *Cal.* 4-phyllus. *Cor.* 0. *Nectarium* rotatum. *Stam.* 4: filamentis connatis.

FEM. *Cal.* monophyllus, ligulato-subrotundus, *Cor.* 0. *Styli* 3. *Bacca* 1-sperma.

smilacina. 1. C. foliis cordatis acutis angulatis. *Sp. pl.* 1473.
Smilax-leav'd Cissampelos.
Nat. of Carolina.
Introd. about 1776, by John Hope, M.D.
Fl. H. ♄.

2. C.

2. C. caule volubili, foliis ovatis obtuſis petiolatis integris. *Linn. ſuppl.* 432. *capenſis.*
Cape Ciſſampelos.
Nat. of the Cape of Good Hope.
Introd. 1775, by Mr. Francis Maſſon.
Fl. G. H. ♄.

NAPÆA. *Linn. mant.* 168.

MASC. *Cal.* 5-fidus. *Petala* 5. *Stam.* monadelpha, plurima, fertilia. *Styli* plures, ſteriles.
FEM. *Cal.* 5-fidus. *Petala* 5. *Stam.* monadelpha, plurima, ſterilia. *Styli* plures, ſtamine longiores. *Capſ.* orbicularis, depreſſa, 10-locularis. *Sem.* ſolitaria.

1. N. pedunculis nudis lævibus, foliis lobatis glabris. *Syſt. veget.* 896. *lævis.*
Napæa hermaphrodita. *Sp. pl.* 965.
Smooth Napæa.
Nat. of Virginia.
Cult. 1759, by Mr. Ph. Miller. *Mill. dict. edit.* 7. *n.* 2.
Fl. Auguſt and September. H. ♃.

2. N. pedunculis involucratis angulatis, foliis palmatis ſcabris. *Syſt. veget.* 896. *ſcabra.*
Napæa dioica. *Sp. pl.* 965.
Rough Napæa.
Nat. of Virginia.
Cult. 1759, by Mr. Ph. Miller. *Mill. dict. edit.* 7. *n.* 1.
Fl. Auguſt and September. H. ♃.

ADELIA. *Gen. pl.* 1137.

MASC. *Cal.* 3-partitus. *Cor.* 0. *Stam.* plurima, basi coalita.
FEM. *Cal.* 5-partitus. *Cor.* 0. *Styli* 3, laceri. *Caps.* 3-cocca.

Acidoton. 1. A. ramis flexuosis: spinis gemmaceis. *Sp. pl.* 1473.
Box-leav'd Adelia.
Nat. of Jamaica.
Introd. 1778, by John Fothergill, M. D.
Fl. June. S. ♄.

SYNGENESIA.

RUSCUS. *Gen. pl.* 1139.

MASC. *Cal.* 6-phyllus. *Cor.* 0. *Nectarium* centrale, ovatum, apice perforatum.
FEM. *Calyx*, *Corolla* et *Nectarium* maris. *Stylus* 1. *Bacca* 3-locularis. *Sem.* 2.

aculeatus. 1. R. foliis supra floriferis nudis. *Sp. pl.* 1474.
Prickly Butcher's-broom.
Nat. of England.
Fl. December——June. H. ♄.

Hypophyllum. 2. R. foliis subtus floriferis nudis. *Sp. pl.* 1474.
Broad-leav'd Butcher's-broom.
Nat. of Italy.
Cult. 1683, by Mr. James Sutherland. *Sutherl. hort. edin.* 190. *n.* 3.
Fl. May and June. I. ♄.

3. R.

3. R. foliis ſupra floriferis ſub foliolo. *Sp. pl.* 1474. *Hypoglossum.*
Double-leav'd Butcher's-broom.
Nat. of Hungary and Italy.
Cult. 1597, by Mr. John Gerard. *Ger. herb.* 761. *f.* 1.
Fl. April and May. H. ♄.

4. R. foliis margine floriferis. *Sp. pl.* 1474. *androgynus.*
Climbing Butcher's-broom.
Nat. of the Canary Iſlands.
Cult. 1713, in the Royal Garden at Hampton-court. *Philoſoph. tranſact. n.* 337. *p.* 199. *n.* 77.
Fl. Moſt part of the Summer. G. H. ♄.

5. R. racemo terminali hermaphroditico. *Sp. pl.* 1474. *racemoſus.*
Alexandrian Laurel, or Butcher's-broom.
Nat. of Portugal.
Cult. 1739, in Chelſea Garden. *Rand. chel. n.* 4.
Fl. June. H. ♄.

GYNANDRIA.

CLUYTIA. *Gen. pl.* 1140.

MASC. *Cal.* 5-phyllus. *Cor.* 5-petala.
FEM. *Cal.* 5-phyllus. *Cor.* 5-petala. *Styli* 3. *Capſ.* 3-locularis. *Sem.* 1.

1. C. foliis ſeſſilibus lineari-lanceolatis, floribus ſolitariis erectis. *Sp. pl.* 1475. *alaternoides.*
Narrow-leav'd Cluytia.
Nat. of the Cape of Good Hope.

Cult. 1692, in the Royal Garden at Hampton-court. *Pluk. phyt. t.* 230. *f.* 1.
Fl. December——March. G. H. ♄.

pulchella. 2. C. ſoliis ovatis integerrimis, floribus lateralibus. *Sp. pl.* 1475.
Broad-leav'd Cluytia.
Nat. of the Cape of Good Hope.
Cult. 1739, by Mr. Philip Miller. *Rand. chel. p.* 213.
Fl. Moſt part of the Year. G. H. ♄.

Classis XXIII.

POLYGAMIA

MONOECIA.

MUSA. *Gen. pl.* 1141.

HERMAPHRODITUS. *Cal.* spatha. *Cor.* 2-petala: altero erecto, 5-dentato; altero nectarifero, concavo, breviore. *Filamenta* 6: horum 5 perfecta. *Stylus* 1. *Germen* inferum, abortiens.

HERMAPHRODITA. *Calyx*, *Corolla*, *Filamenta*, *Pistillum* hermaphroditi, filamento unico perfecto. *Bacca* oblonga, 3-quetra, infera.

1. M. spadice nutante, floribus masculis persistentibus. *Sp. pl.* 1477. *paradisiaca.*
Plantain Tree.
Nat. of both Indies.
Cult. 1690, in the Royal Garden at Hampton-court. *Catal. mss.*
Fl. October——December. S. ♄.

2. M. spadice nutante, floribus masculis deciduis. *Sp. pl.* 1477. *sapientum.*
Banana Tree.
Nat. of the West Indies.
Cult. 1731, by Mr. Ph. Miller. *Mill. dict. edit.* 1. *n.* 2.
Fl. October——March. S. ♄.

VERATRUM. *Gen. pl.* 1144.

HERMAPHROD. *Cal.* 0. *Cor.* 6-petala. *Stam.* 6. *Pist.* 3. *Caps.* 3, polyspermæ.

MASC. *Cal.* 0. *Cor.* 6-petala. *Stam.* 6. *Pist.* rudimentum.

album. 1. V. racemo supradecomposito, corollis erectis. *Sp. pl.* 1479. *Jacqu. austr.* 4. *p.* 18. *t.* 335.
White Veratrum, or Hellebore.
Nat. of Italy, Switzerland, Austria, and Russia.
Cult. 1596, by Mr. John Gerard. *Hort. Ger.*
Fl. June——August. H. ♃.

viride. 2. V. racemo supradecomposito, corollis campanulatis: unguibus latere intus incrassatis.
Green-flower'd Veratrum.
Nat. of North America.
Cult. 1763, by Peter Collinson, Esq.
Fl. July and August. H. ♃.

nigrum. 3. V. racemo composito, corollis patentissimis. *Sp. pl.* 1479. *Jacqu. austr.* 4. *p.* 18. *t.* 336.
Dark-flower'd Veratrum.
Nat. of Austria and Siberia.
Cult. 1596, by Mr. John Gerard. *Hort. Ger.*
Fl. June and July. H. ♃.

luteum. 4. V. racemo simplicissimo, foliis sessilibus. *Sp. pl.* 1479.
Yellow-flower'd Veratrum.
Nat. of North America.
Cult. 1759, by Mr. Ph. Miller. *Mill. dict. edit.* 7. *n.* 3.
Fl. July and August. H. ♃.

ANDRO-

ANDROPOGON. *Gen. pl.* 1145.

HERMAPHROD. *Cal.* Gluma uniflora. *Cor.* Gluma basi aristata. *Stam.* 3. *Styli* 2. *Sem.* 1.
MASC. *Cal.* et *Cor.* prioris. *Stam.* 3.

1. A. spica solitaria, flosculis hinc masculis muticis, inde femineis longissimis unito-contortoque aristatis. *Syst. veget.* 903. *contortum.*
Twisted Andropogon.
Nat. of the East Indies.
Introd. 1779, by Anthony Chamier, Esq.
Fl. July——September. S. ♃.

2. A. spicis digitatis, calycibus persistentibus, corollis ciliatis. *Linn. mant.* 302. *barbatum.*
Bearded Andropogon.
Nat. of the East Indies.
Introd. 1777, by Daniel Charles Solander, LL. D.
Fl. June and July. S. ⊙.

3. A. spicis digitatis, calycibus subtrifloris, petalis exterioribus aristatis; hermaphroditi carina margineque ciliatis. *pubescens.*
Chloris ciliata. *Swartz prodr.* 25.
Ciliated Andropogon.
Nat. of Jamaica. *Mr. Gilbert Alexander.*
Introd. 1779.
Fl. July——September. S. ♃.

4. A. spicis digitatis plurimis, flosculis sessilibus: aristato muticoque, pedicellis lanatis. *Sp. pl.* 1483. *Jacqu. austr.* 4. *p.* 43. *t.* 384. *Ischaemum.*
Woolly Andropogon.

Nat. of the South of Europe.
Introd. 1778, by Mr. Thomas Blackie.
Fl. August. H. ♂.

fasciculatum. 5. A. spicis plurimis fasciculatis glabris, calycibus bifloris: valvulis acutis lævibus; petaloideis exterioribus aristatis, flosculo interiori sterili.
Andropogon fasciculatum. *Sp. pl.* 1483.
Many-spiked Andropogon.
Nat. of the West Indies.
Introd. 1780, by Mr. Gilbert Alexander.
Fl. August and September. S. ⊙.

HOLCUS. *Gen. pl.* 1146.

HERMAPHROD. *Cal.* Gluma 1- s. 2-flora. *Cor.* Gluma aristata. *Stam.* 3. *Styli* 2. *Sem.* 1.
MASC. *Cal.* Gluma 2-valvis. *Cor.* 0. *Stam.* 3.

Sorghum. 1. H. glumis villosis, seminibus compressis aristatis. *Syst. veget.* 905.
Indian Holcus, or Millet.
Nat. of India.
Cult. 1596, by Mr. John Gerard. *Hort. Ger.*
Fl. July. S. ⊙.

halepensis. 2. H. glumis glabris, floribus hermaphroditis muticis; femineo aristato. *Sp. pl.* 1485.
Panicl'd Holcus.
Nat. of Syria.
Cult. 1699, by Mr. Jacob Bobart. *Moris. hist.* 3. *p.* 201. *n.* 26.
Fl. July. H. ♃.

3. H.

3. H. glumis villosis, seminibus omnibus aristatis. *Syst. veget.* 905. *saccharatus.*
Yellow-seeded Holcus.
Nat. of India.
Cult. 1759, by Mr. Ph. Miller. *Mill. dict. edit.* 7. *n.* 2.
Fl. July and August. S. ♂.

4. H. glumis bifloris nudiusculis: flosculo hermaphrodito mutico; masculo arista geniculata. *Sp. pl.* 1485. *Curtis lond.* *mollis.*
Soft Holcus, or creeping Soft-grass.
Nat. of Britain.
Fl. July. H. ♃.

5. H. glumis bifloris villosis: flosculo hermaphrodito mutico; masculo arista recurva. *Sp. pl.* 1485. *Curtis lond.* *lanatus.*
Woolly Holcus, or Meadow Soft-grass.
Nat. of Britain.
Fl. June. H. ♃.

6. H. glumis trifloris muticis acuminatis: flosculo hermaphrodito diandro. *Syst. veget.* 905. *odoratus.*
Sweet-scented Holcus.
Nat. of Canada and the North of Europe.
Introd. 1777, by Brook Watson, Esq.
Fl. June and July. H. ♃.

CENCHRUS. *Gen. pl.* 1149.

Involucrum laciniatum, echinatum, 2-florum. *Cal.* Gluma biflora: altero masculo, altero hermaphrodito.

HERMA-

HERMAPHROD. *Cor.* Gluma mutica. *Stam.* 3. *Sem.* 1.

MASC. *Cor.* Gluma mutica. *Stam.* 3.

racemosus. 1. C. panicula spicata, glumis muricatis setis ciliaribus. *Sp. pl.* 1487.
Branching Cenchrus.
Nat. of the South of Europe.
Introd. 1771, by Monf. Richard.
Fl. July and August. H. ☉.

lappaceus. 2. C. paniculæ ramis simplicissimis, corollis retrorsum hispidis, calycibus trivalvibus bifloris. *Sp. pl.* 1488.
Bur Cenchrus.
Nat. of India.
Introd. 1773, by John Earl of Bute.
Fl. July. S. ☉.

capitatus. 3. C. spica ovata simplici. *Sp. pl.* 1488.
Oval-spiked Cenchrus.
Nat. of Italy and the South of France.
Introd. 1771, by Monf. Richard.
Fl. July and August. H. ☉.

echinatus. 4. C. spica oblonga conglomerata. *Sp. pl.* 1488.
Rough-spik'd Cenchrus.
Nat. of the West Indies.
Cult. 1691, by Mr. Doody. *Pluk. phyt. t.* 92. *f.* 3.
Fl. August——December. S. ☉.

ciliaris. 5. C. spica involucellis setaceis ciliatis quadrifloris. *Linn. mant.* 302.
Ciliated Cenchrus.
Nat. of the Cape of Good Hope.
Introd. 1777, by Monf. Thouin.
Fl. July and August. S. ☉.

ÆGILOPS.

ÆGILOPS. *Gen. pl.* 1150.

HERMAPHROD. *Cal.* Gluma ſubtriflora, cartilaginea: *Cor.* Gluma terminata triplici ariſta. *Stamina* 3. *Styli* 2. *Sem.* 1.

MASC. *Cal.* et *Cor.* Gluma, ut in priore. *Stam.* 3.

1. Æ. ſpica ariſtata: calycibus omnibus triariſtatis. *Sp. pl.* 1489. *ovata.*

Oval-ſpiked Hard-graſs.

Nat. of the South of Europe.

Cult. 1683, by Mr. James Sutherland. *Sutherl. hort. edin.* 10. *n.* 5.

Fl. June and July. H. ☉.

2. Æ. ſpica ariſtata: calycibus inferioribus biariſtatis. *Sp. pl.* 1489. *triuncialis.*

Long-ſpiked Hard-graſs.

Nat. of the South of Europe and the Levant.

Introd. 1776, by Monſ. Thouin.

Fl. July and Auguſt. H. ☉.

VALANTIA. *Gen. pl.* 1151.

HERMAPHROD. *Cal.* 0. *Cor.* 4-partita. *Stam.* 4. *Stylus* 2-fidus. *Sem.* 1.

MASC. *Cal.* 0. *Cor.* 3- ſ. 4-partita. *Stam.* 4 ſ. 3. *Piſtillum* obſoletum.

1. V. capſulis ſeſſilibus glabris echinatis, foliis cuneiformibus lævibus. *muralis.*

Valantia muralis. *Sp. pl.* 1490.

Wall Croſs-wort.

Nat. of the South of France and Italy.

Cult. 1768, by Mr. Philip Miller. *Mill. dict. edit.* 8.

Fl. May——July. H. ☉.

2. V.

hispida. 2. V. capsulis sessilibus hispidis echinatis, foliis cuneiformibus, floribus masculis trifidis.
Valantia hispida. *Sp. pl.* 1490.
Bristly Cross-wort.
Nat. of the South of Europe.
Cult. 1768, by Mr. Philip Miller. *Mill. dict. edit.* 8.
Fl. May and June. H. ☉.

filiformis. 3. V. capsulis pedicello longioribus cylindraceis pilosis inermibus, foliis lanceolatis glabris subciliatis.
Least Cross-wort.
Nat. of the Canary Islands. Mr. *Francis Masson.*
Introd. 1780.
Fl. July. S. ☉.

Cucullaria. 4. V. fructificationibus singulis bractea ovata deflexa obtectis. *Sp. pl.* 1491.
Hooded Cross-wort.
Nat. of the Levant.
Introd. 1780, by Monf. Thouin.
Fl. May and June. H. ☉.

Aparine. 5. V. floribus masculis trifidis pedicellatis hermaphroditici pedunculo insidentibus. *Sp. pl.* 1491.
Smooth-seeded Cross-wort.
Nat. of Germany, France, and Sicily.
Fl. July. H. ☉.

Cruciata. 6. V. floribus masculis quadrifidis, pedunculis diphyllis. *Sp. pl.* 1491.
Common Cross-wort.
Nat. of England.
Fl. May——November. H. ♃.

glabra. 7. V. floribus masculis quadrifidis, pedunculis dichotomis aphyllis, foliis ovalibus ciliatis. *Sp. pl.* 1491.

Smooth

Smooth Crofs-wort.
Nat. of the South of Europe.
Introd. 1783, by William Pitcairn, M.D.
Fl. July. H. ♃.

PARIETARIA. *Gen. pl.* 1152.

HERMAPHROD. *Cal.* 4-fidus. *Cor.* 0. *Stamina* 4. *Styl.* 1. *Sem.* 1, fuperum, elongatum.
FEM. *Cal.* 4-fidus. *Cor.* 0. *Stam.* 0. *Stylus* 1. *Sem.* 1, fuperum, elongatum.

1. P. foliis lanceolato-ovatis, pedunculis dichotomis, calycibus diphyllis. *Syft. veget.* 908. *Curtis lond.* *officinalis.*
Wall Pellitory.
Nat. of Britain.
Fl. June—September. H. ♃.

2. P. foliis ovatis obtufis, caulibus ftriatis lævibus filiformibus procumbentibus. *Sp. pl.* 1492. *lufitanica.*
Chick-weed-leav'd Pellitory.
Nat. of Spain and Portugal.
Introd. 1771, by Monf. Richard.
Fl. July. H. ⊙.

3. P. foliis ellipticis acuminatis fubtriplinerviis, caule arboreo. *L'Herit. in Journ. de Rozier* 33. *p.* 55. *arborea.*
Urtica arborea. *Linn. fuppl.* 417. *L'Herit. ftirp. nov. p.* 39. *tab.* 20.
Tree Pellitory.
Nat. of the Canary Iflands. Mr. *Francis Maffon.*
Introd. 1779.
Fl. February—May. G. H. ♄.

ATRI-

ATRIPLEX. *Gen. pl.* 1153.

HERMAPHROD. *Cal.* 5-phyllus. *Cor.* 0. *Stam.* 5. *Stylus* 2-partitus. *Sem.* 1, depreſſum.

FEM. *Cal.* 2-phyllus. *Cor.* 0. *Stam.* 0. *Stylus* 2-partitus. *Sem.* 1, compreſſum.

Halimus. 1. A. caule fruticoſo, foliis deltoidibus integris. *Sp. pl.* 1492.
Tall ſhrubby Orache, or Spaniſh Sea Purſlane.
Nat. of Spain and Portugal.
Cult. 1640. *Park. theat.* 724. *f.* 2.
Fl. H. ♄.

Portulacoides. 2. A. caule fruticoſo, foliis obovatis. *Sp. pl.* 1493.
Dwarf ſhrubby Orache, or Common Sea Purſlane.
Nat. of Britain.
Fl. Auguſt and September. H. ♄.

albicans. 3. A. caule fruticoſo erecto, foliis haſtatis integerrimis acutis, ſpicis terminalibus.
White Orache.
Nat. of the Cape of Good Hope. Mr. *Fr. Maſſon.*
Introd. 1774.
Fl. June and July. G. H. ♄.

hortenſis. 4. A. caule erecto herbaceo, foliis triangularibus. *Sp. pl.* 1493.
α Atriplex hortenſis alba ſ. pallide virens. *Bauh. pin.* 119.
White Garden Orache.
β Atriplex hortenſis rubra. *Bauh. pin.* 119.
Red Garden Orache.
Nat. of Tartary.
Cult. 1596, by Mr. John Gerard. *Hort. Ger.*
Fl. July and Auguſt. H. ⊙.

5. A.

5. A. caule herbaceo, foliis deltoidibus dentatis subtus argenteis. *Sp. pl.* 1494. *laciniata.*
Jagged Sea Orache.
Nat. of Britain.
Fl. August. H. ⊙.

6. A. caule herbaceo, valvulis femineis magnis deltoidibus sinuatis. *Sp. pl.* 1494. *Curtis lond.* *hastata.*
Wild Orache.
Nat. of Britain.
Fl. August and September. H. ⊙.

7. A. caule herbaceo patulo, foliis subdeltoideo-lanceolatis, calycibus semiuum disco dentatis. *Sp. pl.* 1494. *patula.*
Spreading Orache.
Nat. of Britain.
Fl. August and September. H. ⊙.

8. A. caule herbaceo erecto, foliis linearibus serratis. *Linn. mant.* 300. *marina.*
Atriplex serrata. *Hudſ. angl.* 444.
Serrated Sea Orache.
Nat. of Britain.
Fl. August and September. H. ⊙.

9. A. caule herbaceo erecto, foliis omnibus linearibus integerrimus. *Sp. pl.* 1494. *littoralis.*
Grass-leav'd Sea Orache.
Nat. of Britain.
Fl. August and September. H. ⊙.

10. A. caule herbaceo divaricato, foliis lanceolatis obtusis integris, calycibus femineis pedunculatis. *Sp. pl.* 1675. *pedunculata.*

Peduncu-

Pedunculated Orache.
Nat. of England.
Fl. July——September. H. ☉.

TERMINALIA. *Linn. mant.* 21.

MASC. *Cal.* 5-partitus. *Cor.* 0. *Stam.* 10.
HERMAPHROD. *Flos* masculi. *Stylus* 1. *Drupa* infera, cymbiformis.

Catappa. 1. T. foliis obovatis subtus tomentosis. *Syst. veget.* 910. *Jacqu. ic. collect.* 1. *p.* 130.
Broad-leav'd Terminalia.
Nat. of the East Indies.
Introd. 1778, by Messrs. Kennedy and Lee.
Fl. S. ♄.

angustifolia. 2. T. foliis lanceolatis pubescentibus.
Terminalia angustifolia. *Jacqu. hort.* 3. *p.* 51. *tab.* 100.
Terminalia Benzoin. *Linn. suppl.* 434.
Croton Bentzoë. *Linn. mant.* 297.
Narrow-leav'd Terminalia.
Nat. of the East Indies.
Cult. 1757, by Mr. Philip Miller.
Fl. S. ♄

C L U S I A. *Gen. pl.* 1154.

MASC. *Cal.* 4-s. 6-phyllus: foliolis oppositis, imbricatis. *Cor.* 4-s. 6-petala. *Stam.* numerosa.
FEM. *Cal.* et *Cor.* ut in masculis. *Nectarium* ex antheris coalitis germen includens. *Caps.* 5-locularis, 5-valvis, pulpa farcta.

flava. 1. C. foliis aveniis, corollis tetrapetalis. *Sp. pl.* 1495.
Succulent-leav'd Balsam Tree.

Nat.

Nat. of Jamaica.
Cult. 1768, by Mr. Ph. Miller. *Mill. dict. edit.* 8.
Fl. S. ♄.

OPHIOXYLUM. *Gen. pl.* 1142.

HERMAPHROD. *Cal.* 5-fidus. *Cor.* 5-fida, infundibul. *Stam.* 5. *Pist.* 1.

MASC. *Cal.* 2-fidus. *Cor.* 5-fida, infundibul. ore *Nectario* cylindrico. *Stam.* 2.

1. OPHIOXYLUM. *Sp. pl.* 1478. *serpentinum.*
Scarlet-flower'd Ophioxylum.
Nat. of the East Indies.
Cult. 1690, in the Royal Garden at Hampton-court. *Catal. mss.*
Fl. May and June. S. ♄.

FUSANUS. (Colpoon. *Berg. cap.* 38.)

HERMAPHROD. *Cal.* 5-fidus. *Cor.* nulla. *Stam.* 4. *Germ.* inferum. *Stigmata* 4. *Drupa.*

HERM. MASC. *Cal. Cor. Stam. Pist.* prioris. *Fructus* abortiens.

1. FUSANUS. *Syst. nat. edit.* 13. *p.* 765. *compressus.*
Thesium Colpoon. *Linn. suppl.* 161.
Flat-stalk'd Fusanus.
Nat. of the Cape of Good Hope.
Introd. 1776, by Mr. Francis Masson.
Fl. G. H. ♄.

ACER. *Gen. pl.* 1155.

HERMAPHROD. *Cal.* 5-fidus. *Cor.* 5-petala. *Stam.* 8. *Pist.* 1. *Capf.* 2 f. 3, monofpermæ, ala terminatæ.
MASC. *Cal.* 5-fidus. *Cor.* 5-petala. *Stamina* 8.

tataricum. 1. A. foliis cordatis indivifis ferratis: lobis obfoletis, floribus racemofis. *Sp. pl.* 1495.
Tartarian Maple.
Nat. of Tartary.
Cult. 1759, by Mr. Ph. Miller. *Mill. dict. edit.* 7. *n.* 12.
Fl. May and June. H. ♄.

Pfeudo-Platanus. 2. A. foliis quinquelobis inæqualiter ferratis, floribus racemofis. *Sp. pl.* 1495.
Great Maple, or Sycamore.
Nat. of Britain.
Fl. April and May. H. ♄.

rubrum. 3. A. foliis quinquelobis fubdentatis fubtus glaucis, pedunculis fimpliciffimis aggregatis. *Sp. pl.* 1496.
coccineum. α floribus coccineis.
Scarlet-flower'd Maple.
pallidum. β floribus carneis.
Pale-flower'd Maple.
Nat. of Virginia and Penfylvania.
Cult. 1656, by Mr. John Tradefcant, Jun. *Muf. Trad.* 74.
Fl. April and May. H. ♄.

faccharinum. 4. A. foliis quinquepartito-palmatis acuminato-dentatis fubtus pubefcentibus. *Syft. veget.* 911.
Sugar Maple.

Nat.

Nat. of Penſylvania.
Introd. 1735, by Peter Collinſon, Eſq.
Fl. May. H. ♄.

5. A. quinquelobis acuminatis acute dentatis glabris, floribus corymboſis. *Sp. pl.* 1496. *Platanoides.*
α foliis lobatis. lobatum.
Norway Maple.
β foliis laciniatis. laciniatum.
Cut-leav'd Maple.
Nat. of Europe.
Cult. 1724. *Furber's catal.*
Fl. May——July. H. ♄.

6. A. foliis ſubquinquelobis acutis ſerratis, racemis compoſitis, calycibus piloſis. *montanum.*
Acer penſylvanicum. *Du Roi hort. harbecc.* 1. *p.* 22. *tab.* 2. *Lauth Acer, p.* 33.
Mountain Maple.
Nat. of North America.
Cult. 1750, by Archibald Duke of Argyle.
Fl. April. H. ♄.

7. A. foliis trilobis acuminatis argute duplicato-ſerratis, racemis ſimplicibus, calycibus glabris. *penſylvanicum.*
Acer penſylvanicum. *Sp. pl.* 1496.
Acer ſtriatum. *Du Roi hort. harbecc.* 1. *p.* 8. *tab.* 1. *Lauth Acer, p.* 35.
Penſylvanian Maple.
Nat. of North America.
Introd. 1755, by Meſſrs. Kennedy and Lee.
Fl. May and June. H. ♄.

8. A. foliis lobatis obtuſis emarginatis. *Sp. pl.* 1497. *campeſtre.*
Common Maple.

Nat. of Britain.
Fl. May and June. H. ♄.

Opalus. 9. A. foliis subrotundis quinquelobis laxe serratis, capsulis ovatis glabris suberectis. *L'Herit. stirp. nov. tom.* 2. *tab.* 98.
Acer Opalus. *Mill. dict.*
Acer italum. *Lauth Acer, p.* 32.
Erable printanier. *Reynier in act. lausann.* 1. *p.* 71.
Italian Maple.
Nat. of Italy and Switzerland.
Cult. 1752, by Mr. Ph. Miller. *Mill. dict. edit.* 6. *n.* 9.

monspessulanum. 10. A. foliis trilobis integerrimis glabris annuis. *Syst. veget.* 912.
Montpellier Maple.
Nat. of the South of France.
Cult. 1739, by Mr. Philip Miller. *Rand. chel. n.* 6.
Fl. May. H. ♄.

creticum. 11. A. foliis trilobis integerrimis pubescentibus perennantibus. *Syst. veget.* 912.
Cretan Maple.
Nat. of the Levant.
Cult. 1752, by Mr. Ph. Miller. *Mill. dict. edit.* 6. *n.* 11.
Fl. May and June. H. ♄.

Negundo. 12. A. foliis compositis, floribus racemosis. *Sp. pl.* 1497.
Ash-leav'd Maple.
Nat. of North America.
Cult. 1688, by Bishop Compton. *Raj. hist.* 2. *p.* 1798. *n.* 7. conf. *Pluk. alm.* 7.
Fl. April. H. ♄.

CELTIS.

CELTIS. *Gen. pl.* 1143.

HERMAPHROD. *Cal.* 5-partitus. *Cor.* 0. *Stam.* 5. *Styli* 2. *Drupa* 1-ſperma.

MASC. *Cal.* 6-partitus. *Cor.* 0. *Stam.* 6.

1. C. foliis ovato-lanceolatis. *Sp. pl.* 1478. *auſtralis.*
European Nettle-Tree.
Nat. of the South of Europe.
Cult. 1596, by Mr. John Gerard. *Hort. Ger.*
Fl. May. H. ♄.

2. C. foliis oblique cordatis ſerratis ſubtus villoſis. *Sp. pl.* 1478. *orientalis.*
Oriental Nettle-Tree.
Nat. of the Levant.
Cult. 1748, by Mr. Ph. Miller. *Mill. dict. edit.* 5. *n.* 5.
Fl. H. ♄.

3. C. foliis oblique ovatis ſerratis acuminatis. *Sp. pl.* 1478. *occidentalis.*
American Nettle-Tree.
Nat. of Virginia.
Cult. 1656, by Mr. John Tradeſcant, Jun. *Muſ. Trad.* 135.
Fl. April and May. H. ♄.

4. C. foliis oblique cordatis ovato-lanceolatis ſerrulatis ſuperne ſcabriuſculis. *Swartz prodr.* 53. *micrantha.*
Rhamnus micranthus. *Sp. pl.* 280.
Jamaica Nettle-Tree.
Nat. of Jamaica.
Introd. 1778, by Mr. Gilbert Alexander.
Fl. Auguſt and September. S. ♄.

GOUANIA. *Gen. pl.* 1157.

HERMAPHROD. *Cal.* 5-fidus. *Cor.* 0. *Antheræ* 5, ſub calyptra tectæ. *Stylus* 3-fidus. *Fructus* inferus, tripartibilis.

MASC. ſimilis, abſque germine ſtigmateque.

domingenſis. 1. G. foliis glabris. *Sp. pl.* 1663.
Chaw-Stick.
Nat. of the Weſt Indies.
Cult. 1739, by Mr. Philip Miller. *Mill. dict. vol.* 2. Addenda. Lupulus.
Fl. S. ♄.

MIMOSA. *Gen. pl.* 1158.

HERMAPHROD. *Cal.* 5-dentatus. *Cor.* 5-fida. *Stam.* 5 ſ. plura. *Piſt.* 1. Legumen.

MASC. *Cal.* 5-dentatus. *Cor.* 5-fida. *Stam.* 5, 10, plura.

* *Foliis ſimplicibus.*

verticillata. 1. M. inermis, foliis verticillatis linearibus pungentibus. *L'Herit. ſert. angl. tab.* 41.
Whorl'd-leav'd Mimoſa.
Nat. of New South Wales. Mr. *David Nelſon.*
Introd. 1780, by Sir Joſeph Banks, Bart.
Fl. March——May. G. H. ♄.

ſimplicifolia. 2. M. inermis arborea, foliis ovatis integerrimis nervoſis obtuſis, ſpicis globoſis pedunculatis. *Linn. ſuppl.* 436.
Simple-leav'd Mimoſa.
Nat. of the Iſland of Tanna.
Introd. 1775, by John Reinhold Forſter, LL. D.
Fl. S. ♄.

** *Foliis*

** *Foliis bigeminis.*

3. M. ſpinoſa, foliis bigeminis obtuſis. *Sp. pl.* 1499. *Unguis cati.*
Four-leav'd Mimoſa.
Nat. of the Weſt Indies.
Introd. 1690, by Mr. Bentick. *Br. Muſ. Sloan. mſſ.* 3370. *part* 1. *n.* 9.
Fl. S. ♄.

*** *Foliis conjugatis (ſimulque pinnatis.)*

4. M. inermis, foliis conjugatis pinnatis: pinnis intimis minoribus. *Syſt. veget.* 914. *purpurea.*
Purple Mimoſa, or Soldier-wood.
Nat. of the Weſt Indies.
Cult. 1768, by Mr. Philip Miller. *Mill. dict. edit.* 8.
Fl. S. ♄.

5. M. aculeata, foliis conjugatis pinnatis: pinnis æqualibus, ſtipulis ſpinoſis. *Sp. pl.* 1499. *circinalis.*
Spiral Mimoſa.
Nat. of the Weſt Indies.
Introd. 1726, by Mr. Mark Cateſby. *Mill. dict. edit.* 8.
Fl. S. ♄.

6. M. aculeata, foliis conjugatis pinnatis: partialibus bijugis: intimis minimis. *Sp. pl.* 1501. *ſenſitiva.*
Senſitive Plant.
Nat. of Brazil.
Introd. before 1733, by William Houſtoun, M. D. *Mill. dict. edit.* 7. *n.* 8.
Fl. Moſt part of the Summer. S. ♄.

7. M. aculeata, foliis ſubdigitatis pinnatis, caule hiſpido. *Sp. pl.* 1501. *pudica.*
Humble Plant.

Nat. of Brazil.
Cult. 1638, by Sir John Davers. *Park. theat.* 1617. *n.* 2.
Fl. Moſt part of the Summer. S. ♄.

**** *Foliis duplicato-pinnatis.*

ſcandens. 8. M. inermis, foliis conjugatis cirrho terminatis: foliolis bijugis. *Sp. pl.* 1501.
Climbing Mimoſa.
Nat. of both Indies.
Introd. about 1780.
Fl. S. ♄.

virgata. 9. M. inermis erecta angulata, foliis bipinnatis, ſpicis decandris: inferioribus caſtratis maſculis. *Syſt. veget.* 915. *Jacqu. hort.* 1. *p.* 34. *t.* 80.
Long-twigg'd Mimoſa.
Nat. of the Weſt Indies.
Introd. 1774, by Joſeph Nicholas de Jacquin, M. D.
Fl. July and Auguſt. S. ♄.

Julibriſſin. 10. M. arboreſcens, foliis bipinnatis: pinnulis cultriformibus acuminatis, floribus omnibus perfectis. *Scopoli inſubr.* 1. *p.* 18. *tab.* 8.
Mimoſa arborea. *Forſk. deſcr.* 177. *n.* 89.
Smooth Tree Mimoſa.
Nat. of the Levant.
Introd. about 1745, by Richard Bateman, Eſq.
Fl. Auguſt. G. H. ♄.

ſpecioſa. 11. M. inermis, foliis bipinnatis ſubquadrijugis: pinnis ſubnovemjugis: foliolis oblongis glabris, glandula ſupra coſtæ baſin. *Jacqu. ic. collect.* 1. *p.* 47.
Bladder-Senna-leav'd Mimoſa.

Nat.

Nat. of the East Indies.
Cult. before 1742, by Robert James Lord Petre.
Fl. S. ♄.

12. M. inermis, foliis bipinnatis: partialibus quinquejugis, ramis flexuosis, gemmis globosis. *Sp. pl.* 1504. *latisiliqua.*
Broad-podded Mimosa.
Nat. of the West Indies.
Introd. 1777, by Daniel Charles Solander, LL. D.
Fl. Most part of the Summer. S. ♄.

13. M. inermis, foliis bipinnatis: partialibus sejugis: pinnis plurimis, glandula inter infima. *Sp. pl.* 1504. *glauca.*
Glaucous Mimosa.
Nat. of America.
Cult. 1690, in the Royal Garden at Hampton-court. *Catal. mss.*
Fl. August. S. ♄.

14. M. inermis, foliis abrupte bipinnatis multijugis: pinnulis multijugis: foliolis distinctissimis, racemo composito terminali. *L'Herit. sert. angl. tab.* 42. *grandiflora.*
Great-flower'd Mimosa.
Nat. of the East Indies.
Introd. about 1769, by Mrs. Norman.
Fl. June——September. S. ♄.

15. M. spinis stipularibus geminis connatis, foliis bipinnatis. *Sp. pl.* 1505. *cornigera.*
Horn'd Mimosa, or Cuckold-tree.
Nat. of South America.
Cult. 1691, in the Royal Garden at Hampton-court, *Pluk. phyt. t.* 122. *f.* 1.
Fl. S. ♄.

16. M.

farnesiana. 16. M. spinis stipularibus distinctis, foliis bipinnatis: partialibus octojugis, spicis globosis sessilibus. *Sp. pl.* 1506.
Sweet-scented Mimosa, or Sponge-Tree.
Nat. of St. Domingo.
Cult. 1731, by Mr. Philip Miller. *Mill. dict. edit.* 1. Acacia 4.
Fl. June——August. S. ♄.

nilotica. 17. M. spinis stipularibus patentibus, foliis bipinnatis: partialibus extimis glandula interstinctis, spicis globosis pedunculatis. *Sp. pl.* 1506.
Egyptian Mimosa.
Nat. of Arabia and Egypt.
Cult. 1664. *Evel. kalend. hort. p.* 75.
Fl. July. S. ♄.

asperata. 18. M. aculeata hirta, foliis bipinnatis opposite aculeatis: spina erecta inter singula partialia. *Sp. pl.* 1507.
Hairy-podded Mimosa.
Nat. of Vera Cruz.
Introd. before 1733, by William Houstoun, M. D. *Mill. ic.* 122. *t.* 182. *f.* 3.
Fl. S. ♄.

cæsia. 19. M. aculeata, foliis bipinnatis: pinnis ovali-oblongis oblique acuminatis. *Sp. pl.* 1507.
Gray Mimosa.
Nat. of the East Indies.
Introd. 1773, by Sir Joseph Banks, Bart.
Fl. S. ♄.

pennata. 20. M. aculeata, foliis bipinnatis numerosissimis lineari-acerosis, panicula aculeata: capitulis globosis. *Sp. pl.* 1507.

Small-

Small-leav'd Mimoſa.
Nat. of the Eaſt Indies.
Introd. 1773, by Sir Joſeph Banks, Bart.
Fl. S. ♄.

21. M. aculeata, foliis bipinnatis: pinnis incurvis, caule angulato, ſtipulis aculeo longioribus. *Syſt. veg.* 917. *Intſia.*
Angular-ſtalk'd Mimoſa.
Nat. of the Eaſt Indies.
Introd. 1778, by Patrick Ruſſell, M. D.
Fl. S. ♄.

AILANTHUS. *Desfontaines in act. pariſ.*

MASC. *Cal.* 5-partitus. *Cor.* 5-petala. *Stam.* 10.
FEM. *Cal.* et *Cor.* maris. *Germina* 3-5. *Styli* laterales. *Peric.* membranacea, 1-ſperma.
HERMAPHR. *Cal.* et *Cor.* maris. *Stam.* 2-3.

1. AILANTHUS. *Desfontaines in act. pariſ.* 1786. *p.* 265. *tab.* 8. *glanduloſa.*
Rhus Cacodendrum. *Ehrhart in Hannov. magaz.* 1783. *p.* 227.
Rhus ſinenſe foliis alatis: foliolis oblongis acuminatis ad baſin ſubrotundis et dentatis. *Ellis in philoſoph. tranſact. vol.* 49. *p.* 870. *t.* 25. *f.* 5. and *vol.* 50. *p.* 446. *t.* 17.
Tall Ailanthus.
Nat. of China.
Introd. about 1751, by Father D'Incarville. *Philoſoph. tranſact. loc. cit.*
Fl. Auguſt. H. ♄.

DIOECIA.

DIOECIA.

GLEDITSIA. *Gen. pl.* 1159.

HERMAPHROD. *Cal.* 4-fidus. *Cor.* 4-petala. *Stam.* 6. *Pistill.* 1. *Legumen.*
MASC. *Cal.* 3-phyllus. *Cor.* 3-petala. *Stam.* 6.
FEM. *Cal.* 5-phyllus. *Cor.* 5-petala. *Pist.* 1. *Legumen.*

triacanthos. 1. G. spinis triplicibus axillaribus. *Sp. pl.* 1509.
poly-sperma. α leguminibus polyspermis, foliolis lineari-oblongis.
Three-thorn'd Acacia.
mono-sperma. β leguminibus monospermis, foliolis ovato-oblongis, spinis paucis.
Acacia Abruæ foliis, triacanthos, capsula ovali unicum semen claudente. *Catesb. car.* 1. *p.* 43. *t.* 43.
Single-seeded, or Water Acacia.
horrida. γ foliolis ovato-oblongis, spinis creberrimis.
Strong-spined Acacia.
Nat. of North America.
Cult. 1700, by Bishop Compton. *Pluk. mant.* 1.
Fl. June and July. H. ♃.

FRAXINUS. *Gen. pl.* 1160.

HERMAPHROD. *Cal.* 0, s. 4-partitus. *Cor.* 0, s. 4-petala. *Stam.* 2. *Pist.* 1. *Sem.* 1, lanceolatum.
FEM. *Pist.* 1, lanceolatum.

excelsior. 1. F. foliolis lanceolatis serratis sessilibus, floribus apetalis.
Fraxinus excelsior. *Sp. pl.* 1509.

α foliis

α foliis pinnatis, ramis adſcendentibus. communis.
Common Aſh.

β foliis pinnatis, ramis pendulis. pendula.
Weeping Aſh.

γ foliis integris, trilobis, ternatiſque. diverſifolia.
Various-leav'd Aſh.

Nat. of Britain.
Fl. April and May. H. ♄.

2. F. foliolis ſubrotundis acutiuſculis duplicato-ſerratis ſubſeſſilibus, floribus corollatis. *rotundifolia.*
Fraxinus rotundifolia. *Mill. dict.*
Fraxinus alepenſis. *Pluk. alm.* 158. *t.* 182. *f.* 4.
Manna Aſh.
Nat. of Italy.
Cult. 1697, by the Dutcheſs of Beaufort. *Br. Muſ. Sloan. mſſ.* 3357. *fol.* 39.
Fl. April. H. ♄.

3. F. foliolis ovato-oblongis ſerratis petiolatis, floribus corollatis. *Ornus.*
Fraxinus Ornus. *Sp. pl.* 1510.
Flowering Aſh.
Nat. of Italy.
Introd. before 1730, by Dr. Uvedale. *Hort. angl.* 33. *n.* 4.
Fl. May and June. H. ♄.

4. F. foliolis integerrimis, petiolis teretibus. *Sp. pl.* 1510. *americana.*
American Aſh.
Nat. of North America.
Introd. 1724, by Mr. Mark Cateſby. *Mill. dict. edit.* 8.
Fl. H. ♄.

DIOSPY-

DIOSPYROS. *Gen. pl.* 1161.

HERMAPHROD. *Cal.* 4-fidus. *Cor.* urceolata, 4-fida. *Stam.* 8. *Stylus* 4-fidus. *Bacca* 8-ſperma.

MASC. *Cal. Cor. Stam.* prioris.

Lotus. 1. D. foliorum paginis diſcoloribus. *Sp. pl.* 1510.
European Date-plum.
Nat. of Italy and Barbary.
Cult. 1597, by the Earl of Eſſex. *Ger. herb.* 1310. *f.* 1.
Fl. June and July. H. ♄.

virginiana. 2. D. foliorum paginis concoloribus. *Sp. pl.* 1510.
American Date-plum.
Nat. of North America.
Cult. 1629. *Park. parad.* 569. *f.* 6.
Fl. June and July. H. ♄.

NYSSA. *Gen. pl.* 1163.

HERM. *Cal.* 5-partitus. *Cor.* 0. *Stam.* 5. *Piſt.* 1. *Drupa* infera.

MASC. *Cal.* 5-partitus. *Cor.* 0. *Stam.* 10.

integrifolia. 1. N. foliis integerrimis, nucibus ſubrotundis ſtriatis.
Nyſſa aquatica. *Sp. pl.* 1511. (excluſo ſynonymo Cateſbæi priori.)
Mountain Tupelo.
Nat. of North America.
Cult. 1750, by Archibald Duke of Argyle.
Fl. H. ♄.

denticulata. 2. N. foliis remote dentatis, nucibus oblongis ſulcatis ſubrugoſis.

Arbor

Arbor in aqua nascens, foliis latis acuminatis et dentatis, fructu Elæagni majore. *Catesb. car.* 1. *p.* 60. *t.* 60.

Water Tupelo.

Nat. of North America.

Introd. 1735, by Peter Collinson, Esq.

Fl. H. ♄.

ANTHOSPERMUM. *Gen. pl.* 1164.

HERMAPHROD. *Cal.* 4-partitus. *Cor.* o. *Stam.* 4. *Pist.* 2. *Germen* inferum.

MASC. et FEM. in eadem vel distincta planta.

1. A. foliis lævibus. *Sp. pl.* 1511. *æthiopicum.*
Amber Tree.
Nat. of the Cape of Good Hope.
Cult. 1692, by Bp. Compton. *Pluk. phyt. t.* 183. *f.* 1.
Fl. June and July. G. H. ♄.

ARCTOPUS. *Gen. pl.* 1165.

MASC. *Umbella* composita. *Involucra* 5-phylla. *Cor.* 5-petala. *Stam.* 5. *Pist.* 2, abortientia.

ANDROG. *Umbella* simplex. *Involucrum* 4-partitum, spinosum, maximum, continens flosculos masculos in disco plurimos, femineos 4 in radio.

MASC. *Petala* 5. *Stam.* 5.

FEM. *Petala* 5. *Styli* 2. *Sem.* 1, 2-loculare, inferum.

1. ARCTOPUS. *Sp. pl.* 1512. *echinatus.*
Rough Arctopus.
Nat. of the Cape of Good Hope.
Introd. 1774, by Mr. Francis Masson.
Fl. G. H. ♃.

PISO-

PISONIA. *Gen. pl.* 1162.

HERMAPHROD. *Cal.* vix ullus. *Cor.* campanulata, 5-fida. *Stam.* 5 f. 6. *Pist.* 1. *Capf.* 1-locularis, 1-fperma, 5-valvis.

MASC. et FEM. nunc in eadem, nunc in diverfa planta.

aculeata. 1. P. fpinis axillaribus patentiffimis. *Sp. pl.* 1511.
Prickly Pifonia.
Nat. of Jamaica.
Cult. 1739, by Mr. Philip Miller. *Rand. chel.*
Fl. March and April. S. ♄.

PANAX. *Gen. pl.* 1166.

HERMAPHROD. Umbella. *Cal.* 5-dentatus, fuperus. *Cor.* 5-petala. *Stam.* 5. *Styli* 2. *Bacca* difperma.

MASC. Umbella. *Cal.* integer. *Cor.* 5-petala. *Stam.* 5.

quinquefolium. 1. P. foliis ternis quinatis. *Sp. pl.* 1512.
Ginfeng.
Nat. of China and North America.
Introd. 1740, by Peter Collinfon, Efq.
Fl. June. H. ♃.

aculeatum. 2. P. foliis ternatis; fummis juxta flores confertis fimplicibus, petiolis ramulifque aculeatis, caule fruticofo. *L'Herit. ftirp. nov. tom.* 2. *tab.* 99.
Zanthoxylum trifoliatum. *Sp. pl.* 1455.
Prickly Panax.
Nat. of China.
Cult. 1773, by John Fothergill, M. D.
Fl. November. S. ♄.

TRIOECIA.

TRIOECIA.

CERATONIA. *Gen. pl.* 1167.

HERMAPHROD. *Cal.* 5-partitus. *Cor.* nulla. *Stamina* quinque. *Stylus* filiformis. *Legumen* coriaceum, polyſpermum.

DIOICA. *Maſc.* et *Fem.* diſtincta.

1. CERATONIA. *Sp. pl.* 1513. *Siliqua.*
Carob-Tree, or St. John's-bread.
Nat. of Sicily and the Levant.
Cult. 1570, *Lobel. adv.* 414.
Fl. G. H. ♄.

FICUS. *Gen. pl.* 1168.

Receptaculum commune turbinatum, carnoſum, connivens, occultans floſculos vel in eodem vel diſtincto.

MASC. *Cal.* 3-partitus. *Cor.* o. *Stam.* 3.

FEM. *Cal.* 5-partitus. *Cor.* o. *Piſt.* 1. *Sem.* 1.

1. F. foliis palmatis. *Sp. pl.* 1513. *Carica.*
Common Fig-Tree.
Nat. of the South of Europe.
Cult. 1562. *Turn. herb. part* 2. *fol.* 2.
Fl. June and July. H. ♄.

2. F. foliis cordatis ſubrotundis mucronatis integerrimis glabris ſubtus glaucis. *Linn. mant.* 305. *nymphæifolia.*
Water-lily-leav'd Fig-Tree.
Nat. of India.
Cult. 1759, by Mr. Ph. Miller. *Mill. dict. edit.* 7. *n.* 9.
Fl. S. ♄.

religiosa. 3. F. foliis cordatis oblongis integerrimis acuminatis. *Sp. pl.* 1514.
Poplar-leav'd Fig-tree.
Nat. of the East Indies.
Cult. 1731, by Mr. Ph. Miller. *Mill. dict. edit.* 1. *n.* 18.
Fl. S. ♄.

benjamina. 4. F. foliis ovatis acuminatis transverse striatis: margine lævi. *Linn. mant.* 129.
Oval-leav'd Fig-Tree.
Nat. of the East Indies.
Cult. 1757, by Mr. Philip Miller.
Fl. S. ♄.

bengalensis. 5. F. foliis ovatis cordatis integerrimis glabris obtusiusculis basi quinquenerviis, ramis deflexis.
Ficus bengalensis. *Sp. pl.* 1514.
Bengal Fig-Tree.
Nat. of India.
Cult. 1692, in the Royal Garden at Hampton-court. *Pluk. phyt. t.* 178. *f.* 1.
Fl. April. S. ♄.

pedunculata. 6. F. foliis ovato-oblongis cordatis integerrimis acutis glabris, fructibus globosis, pedunculis geminis elongatis.
Ficus arbor americana, arbuti foliis non serrata, fructu pisi magnitudine, funiculis e ramis ad terram demissis prolifera. *Pluk. alm.* 144. *t.* 178. *f.* 4.
Willow-leav'd Fig-Tree.
Nat. of South America.
Introd. 1776, by Hugh Duke of Northumberland.
Fl. S. ♄.

7. F.

7. F. foliis ovatis cordatis integerrimis glabris obtusis basi trinerviis, ramis erectis. *lucida.*
Shining-leav'd Fig-Tree.
Nat. of the East Indies.
Introd. 1772, by Mr. William Malcolm.
Fl. S. ♄.

8. F. foliis oblongis basi rotundatis lævibus integerrimis subtus subglaucis supra impresso-punctatis, fructibus subglobosis. *indica.*
Ficus indica *α.* *Sp. pl.* 1514.
Indian Fig-Tree.
Nat. of the East Indies.
Cult. 1759, by Mr. Ph. Miller. *Mill. dict. edit.* 7. *n.* 5.
Fl. S. ♄.

9. F. foliis oblongis acuminatis integerrimis lævibus basi angustato-rotundatis. *virens.*
Ficus indica maxima, folio oblongo, funiculis e summis ramis demissis radices agentibus se propagans, fructu minore sphærico sanguineo. *Sloan. jam.* 2. *p.* 140. *t.* 223.
Round-fruited Fig-Tree.
Nat. of the West Indies.
Introd. about 1762, by Mr. James Gordon.
Fl. G. H. ♄.

10. F. foliis ovatis subcordatis acutis integerrimis lævibus supra punctato-impressis. *venosa.*
Tsjakela. *Rheed. mal.* 3. *p.* 87. *t.* 64.
Waved-leav'd Fig-Tree.
Nat. of the East Indies.
Introd. 1763, by Mr. John Bush.
Fl. S. ♄.

costata. 11. F. foliis ovatis cordatis: sinu profundo angusto; integerrimis glabris acutis utrinque viridibus.
Upright heart-leav'd Fig-Tree.
Nat. of the East-Indies.
Introd. 1763, by Mr. John Bush.
Fl. S. ♄.

racemosa. 12. F. foliis ovatis integerrimis acutis impresso-punctatis, caule arboreo. *Syst. veget.* 922.
Red-wooded Fig-Tree.
Nat. of the East Indies.
Cult. 1759, by Mr. Philip Miller. *Mill. dict. edit.* 7. *n.* 7.
Fl. S. ♄.

pertusa. 13. F. foliis ovatis glabris, calycibus bifidis, baccis globosis foramine umbilicatis. *Linn. suppl.* 442.
Laurel-leav'd Fig-Tree.
Nat. of South America.
Introd. about 1780.
Fl. S. ♄.

stipulata. 14. F. foliis oblique cordatis obtusis glabris, caule decumbente squamoso. *Thunb. Ficus. n.* 7.
Trailing Fig-Tree.
Nat. of China and Japan.
Introd. about 1771.
Fl. S. ♄.

heterophylla. 15. F. foliis oblongis indivisis trilobis sinuatisque scabris, caule hispido, fructu pedunculato glabro. *Linn. suppl.* 442.
Rough-leav'd Fig-Tree.
Nat. of the East Indies.

Cult.

Cult. 1758, by Mr. Philip Miller.
Fl. S. ♄.

16. F. foliis oblongis lævibus basi attenuatis cordatis coriaceis: venis immersis. *coriacea.*
Leathery-leav'd dwarf Fig-Tree.
Nat. of the East-Indies.
Introd. 1772, by Mr. William Malcolm.
Fl. S. ♄.

Classis XXIV.

CRYPTOGAMIA

F I L I C E S.

EQUISETUM. *Gen. pl.* 1169.

Spica fructificationibus peltatis, basi dehiscentibus multivalvi.

sylvaticum. 1. E. caule spicato, frondibus compositis. *Sp. pl.* 1516.
Wood Horse-tail.
Nat. of Britain.
Fl. April and May. H. ♃.

arvense. 2. E. scapo fructificante nudo; sterili frondoso. *Sp. pl.* 1516. *Curtis lond.*
Corn Horse-tail.
Nat. of Britain.
Fl. March. H. ♃.

palustre. 3. E. caule angulato, frondibus simplicibus. *Sp. pl.* 1516.
Marsh Horse-tail.
Nat. of Britain.
Fl. April. H. ♃.

fluviatile. 4. E. caule striato, frondibus subsimplicibus. *Sp. pl.* 1517.
River Horse-tail.
Nat. of Britain.
Fl. May. H. ♃.

5. E. caule ſubnudo lævi. *Sp. pl.* 1517. *limoſum.*
Smooth Horſe-tail.
Nat. of Britain.
Fl. June. H. ♃.

6. E. caule nudo ſcabro baſi ſubramoſo. *Sp. pl.* 1517. *hyemale.*
Rough Horſe-tail, or Shave Graſs.
Nat. of Britain.
Fl. July and Auguſt. H. ♃.

ONOCLEA. *Gen. pl.* 1170.

Spica diſticha: fructificat. quinquevalvibus.

1. O. frondibus pinnatis apice ſubracemoſis. *Syſt. veget.* 926. *ſenſibilis.*
Senſitive Onoclea, or Fern.
Nat. of Virginia.
Cult. 1758, by Mr. Philip Miller.
Fl. Auguſt. H. ♃.

OPHIOGLOSSUM. *Gen. pl.* 1171.

Spica articulata, diſticha: articulis tranſverſim dehiſcentibus.

1. O. fronde ovata. *Sp. pl.* 1518. *vulgatum.*
Adder's Tongue.
Nat. of Britain.
Fl. May. H. ♃.

OSMUNDA. *Gen. pl.* 1172.

Spica ramoſa: fructific. globoſis.

* *Scapis inſidentibus cauli ad baſin frondis.*

1. O. ſcapo caulino ſolitario, fronde pinnata ſolitaria. *Sp. pl.* 1519. *Lunaria.*

 Moon-

Moon-wort.
Nat. of Britain.
Fl. May. H. ♃.

** *Fronde ipsa fructificationes ferente.*

regalis. 2. O. frondibus bipinnatis apice racemiferis. *Sp. pl.* 1521.
Osmunda-royal, or Flowering-fern.
Nat. of Britain.
Fl. July and August. H. ♃.

claytoniana. 3. O. frondibus pinnatis: pinnis pinnatifidis apice coarctato-fructificantibus. *Sp. pl.* 1521.
Virginian Osmunda.
Nat. of North America.
Introd. 1772, by Samuel Martin, M. D.
Fl. August. H. ♃.

*** *Frondibus aliis foliaceis, aliis fructificantibus.*

cinnamomea. 4. O. frondibus pinnatis: pinnis pinnatifidis, scapis hirsutis, racemis oppositis compositis. *Sp. pl.* 1522.
Woolly Osmunda.
Nat. of North America.
Introd. 1772, by Samuel Martin, M. D.
Fl. June. H. ♃.

Struthiopteris. 5. O. frondibus pinnatis: pinnis pinnatifidis, scapo fructificante disticho. *Sp. pl.* 1522.
Bird's-nest, or Russian Osmunda.
Nat. of Europe.
Cult. 1760, by Peter Collinson, Esq.
Fl. July and August. H. ♃.

Spicant. 6. O. frondibus lanceolatis pinnatifidis: laciniis confluentibus integerrimis parallelis. *Sp. pl.* 1522. *Curtis lond.*

Rough

Rough Spleen-wort.
Nat. of Britain.
Fl. July. H. ♃.

7. O. frondibus supradecompositis: pinnis alternis subrotundis incisis. *Sp. pl.* 1522. *crispa.*
Curl'd Osmunda, or Stone-fern.
Nat. of Britain.
Fl. August. H. ♃.

ACROSTICHUM. *Gen. pl.* 1173.

Fructific. discum totum frondis tegentes.

1. A. frondibus nudis linearibus laciniatis. *Sp. pl.* 1524. *septentrionale.*
Fork'd Acrostichum, or Fern.
Nat. of Britain.
Fl. August. H. ♃.

2. A. frondibus bipinnatis: pinnis omnibus ovatis cordatis latere incisis subtus hirsutissimis. *velleum.*
Woolly Acrostichum.
Nat. of Madeira. Mr. *Francis Masson.*
Introd. 1778.
Fl. August and September. G. H. ♃.

3. A. frondibus subbipinnatis: pinnis opposito-coadunatis obtusis subtus hirsutis basi integerrimis. *Sp. pl.* 1528. *ilvense.*
Hairy Acrostichum.
Nat. of Wales and Scotland.
Fl. July——September. H. ♃.

PTERIS.

PTERIS. *Gen. pl.* 1174.

Fructificationes in lineis marginalibus.

longifolia. 1. P. frondibus pinnatis: pinnis linearibus repandis basi cordatis. *Sp. pl.* 1531.
Long-leav'd Brake.
Nat. of the West Indies.
Introd. about 1770, by Mr. James Gordon.
Fl. July——September. S. ♃.

aquilina. 2. P. frondibus supradecompositis: foliolis pinnatis: pinnis lanceolatis: infimis pinnatifidis; superioribus minoribus. *Sp. pl.* 1533.
Common Brake, or Female-fern.
Nat. of Britain.
Fl. August. H. ♃.

caudata. 3. P. frondibus supradecompositis: pinnis linearibus: infimis basi pinnato-dentatis; terminalibus longissimis. *Sp. pl.* 1533. *Jacqu. ic.*
Great American Brake.
Nat. of North America.
Introd. about 1777, by Mr. William Young.
Fl. September——December. H. ♃.

atropurpurea. 4. P. frondibus decompositis pinnatis: pinnis lanceolatis: terminalibus longioribus. *Sp. pl.* 1534.
Purple Brake.
Nat. of North America.
Introd. 1770, by Mr. William Young.
Fl. August——November. H. ♃.

arguta. 5. P. fronde bipinnatifida: ramis infimis deorsum ramosis, pinnis lanceolatis serratis.
Pteris serrulata. *Forsk. descr. p.* 187.
Sharp-notch'd Brake.

Nat.

Nat. of Arabia, Madeira, and the Cape of Good Hope.
Introd. 1778, by Mr. Francis Maſſon.
Fl. Auguſt and September. G. H. ♃.

6. P. fronde ſubbipinnatifida: laciniis linearibus; ſterilibus ſerratis. *ſerrulata.*
Pteris ſerrulata. *Linn. ſuppl.* 445. (excluſis ſynonymis et loco natali.)
Filix fronde pinnata: pinnis enſiformibus ſerratis; inferioribus oppoſitis: inferne folio longo auctis. *Linn. zeyl.* 424.
Polypodium caule ſimplici, foliis ſimplicibus variis longis ſerratis. *Burm. zeyl.* 196. *t.* 87.
Filicula cheuſanica ſ. Hemionitis multifido folio tenuiſſime ſerrato, ad margines ſeminifera. *Pluk. amalth.* 94. *t.* 407. *f.* 2.
Keiſon Kuſa. *Kæmpf. amœn.* 912.
Various-leav'd Pteris.
Nat. of Japan, China, and Ceylon.
Introd. about 1770, by Mr. James Gordon.
Fl. Auguſt and September. S. ♃.

BLECHNUM. *Gen. pl.* 1175.

Fructific. in lineis duabus, coſtæ frondis approximatis, parallelis.

1. B. frondibus pinnatis: pinnis lanceolatis oppoſitis baſi emarginatis. *Sp. pl.* 1534. *occidentale.*
South American Blechnum.
Nat. of South America.
Introd. about 1777.
Fl. Moſt part of the Year. S. ♃.

2. B. frondibus pinnatis: pinnis ſubſeſſilibus cordato-lanceolatis integerrimis: infimis oppoſitis. *Syſt. veget.* 932. *auſtrale.*

Cape

Cape Blechnum.
Nat. of the Cape of Good Hope.
Introd. 1774, by Mr. Francis Maſſon.
Fl. Moſt part of the Year. G. H. ♃.

virginicum. 3. B. frondibus pinnatis: pinnis multifidis. *Linn. mant.* 307.
Virginian Blechnum.
Nat. of Carolina and Virginia.
Cult. 1774, by John Fothergill, M. D.
Fl. Auguſt and September. H. ♃.

radicans. 4. B. frondibus bipinnatis: pinnis lanceolatis crenulatis, lineolis fructificantibus interruptis. *Linn. mant.* 307.
Rooted-leav'd Blechnum.
Nat. of Madeira.
Introd. 1779, by Mr. Francis Maſſon.
Fl. September. G. H. ♃.

ASPLENIUM. *Gen. pl.* 1178.

Fructific. in lineolis diſci frondis ſparſis.

* *Fronde ſimplici.*

rhizophyllum. 1. A. frondibus cordato-enſiformibus indiviſis: apice filiformi radicante. *Sp. pl.* 1536.
Rooted-leav'd Spleen-wort.
Nat. of North America.
Introd. about 1764, by Mr. John Bartram.
Fl. H. ♃.

Hemionitis. 2. A. frondibus ſimplicibus cordato-haſtatis quinquelobis integerrimis, ſtipitibus lævibus. *Sp. pl.* 1536.
Mules Fern, or Spleen-wort.
Nat. of the South of Europe and Madeira.

Introd.

Introd. 1779, by Mr. Francis Maſſon.

Fl. G. H. ♃.

3. A. frondibus ſimplicibus cordato-lingulatis integerrimis, ſtipitibus hirſutis. *Sp. pl.* 1537. *Curtis lond.* *Scolopendrium.*

α Lingua Cervina officinarum. *Bauh. pin.* 353.
Hart's Tongue Spleen-wort.

β Phyllitis criſpa. *Bauh. hiſt.* 3. *p.* 757. criſpum.
Curl'd Hart's Tongue Spleen-wort.

γ Phyllitis ſ. Lingua Cervina maxima, undulato folio auriculato per baſim. *Pluk. phyt.* 248. *f.* 1. undulatum.
Wave-leav'd Hart's Tongue Spleen-wort.

δ Lingua Cervina multifido folio. *Bauh. pin.* 353. multifidum.
Cluſter'd Hart's Tongue Spleen-wort.

ε Phyllitis ſ. Lingua Cervina minor criſpa, folio multifido, ramoſa. *Pluk. phyt.* 248. *f.* 2. ramoſum.
Branching-cluſter'd Hart's Tongue Spleen-wort.

Nat. of Britain.

Fl. Auguſt. H. ♃.

** *Fronde pinnatifida.*

4. A. frondibus pinnatifidis: lobis alternis confluentibus obtuſis. *Syſt. veget.* 933. *Ceterach.*

Common Spleen-wort.

Nat. of Britain.

Fl. May——October. H. ♃.

*** *Fronde pinnata.*

5. A. frondibus pinnatis: pinnis ſubrotundis crenatis. *Sp. pl.* 1540. *Trichomanes.*

Common Maidenhair, or Spleen-wort.

Nat. of Britain.

Fl. May——October. H. ♃.

6. A. frondibus pinnatis: pinnis ſubrotundis crenatis baſi truncatis. *Hudſ. angl.* 453. *viride.*

Green

Green Spleen-wort.
Nat. of Britain.
Fl. June——September. H. ♃.

ebeneum. 7. A. fronde pinnata: pinnis lanceolatis subfalcatis serratis basi auriculatis, stipite lævissimo simplici.
American Spleen-wort.
Nat. of North America.
Cult. 1779, by John Fothergill, M.D.
Fl. September. H. ♃.

marinum. 8. A. frondibus pinnatis: pinnis obovatis serratis gibbis obtusis basi cuneatis. *Sp. pl.* 1540.
Sea Spleen-wort.
Nat. of Britain.
Fl. June——August. H. ♃.

Rutæ muraria. 9. A. frondibus alternatim decompositis: foliolis cuneiformibus crenulatis. *Sp. pl.* 1541.
White Maidenhair, Wall Rue, or Spleen-wort.
Nat. of Britain.
Fl. June——October. H. ♃.

Adiantum nigrum. 10. A. frondibus subtripinnatis: foliolis alternis: pinnis lanceolatis inciso-serratis. *Sp. pl.* 1541.
Black Maidenhair, or Spleen-wort.
Nat. of Britain.
Fl. April——September. H. ♃.

POLYPODIUM. *Gen. pl.* 1179.

Fructific. in punctis subrotundis, sparsis per discum frondis.

* *Fronde pinnatifida: lobis coadunatis.*

vulgare. 1. P. frondibus pinnatifidis: pinnis oblongis subserratis obtusis, radice squamata. *Sp. pl.* 1544. *Curtis lond.*

Common

Common Polypody.
Nat. of Britain.
Fl. June——September. H. ♃.

2. P. frondibus pinnatifidis: pinnis lanceolatis lacero-pinnatifidis ſerratis. *Sp. pl.* 1546. *cambricum.*
Welch Polypody.
Nat. of Britain.
Fl. June——September. H. ♃.

3. P. frondibus pinnatifidis lævibus: pinnis oblongis diſtantibus: infimis patulis; terminali maxima, fructificationibus ſerialibus. *Sp. pl.* 1546. *aureum.*
Golden Polypody.
Nat. of Jamaica.
Cult. before 1742, by Robert James Lord Petre.
Fl. March. S. ♃.

** *Fronde trifoliata: pedunculo foliolis tribus.*

4. P. frondibus ternatis ſinuato-lobatis: intermedia majore. *Sp. pl.* 1547. *trifoliatum.*
Three-leav'd Polypody.
Nat. of the Weſt Indies.
Introd. 1769, by Mr. George Houſtoun.
Fl. Moſt part of the Year. S. ♃.

*** *Fronde pinnata.*

5. P. frondibus pinnatis: pinnis lunulatis ciliato-ſerratis declinatis, ſtipitibus ſtrigoſis. *Sp. pl.* 1548. *Lonchitis.*
Rough Polypody, or Spleen-wort.
Nat. of Britain.
Fl. May——Auguſt. H. ♃.

6. P. frondibus pinnatis lanceolatis: foliolis ſubrotundis argute inciſis, ſtipite lævi. *Syſt. veget.* 937. *fontanum.*
Rock

Rock Polypody.
Nat. of England.
Fl. June——August. H. ♃.

**** *Fronde ſubbipinnata.*

Phegopteris. 7. P. frondibus ſub-bipinnatis: foliolis infimis reflexis: paribus pinnula quadrangulari coadunatis. *Sp. pl.* 1550.
Wood Polypody.
Nat. of Britain.
Fl. June and July. H. ♃.

criſtatum. 8. P. frondibus ſub-bipinnatis: foliolis ovato-oblongis: pinnis obtuſiuſculis apice acute ſerratis. *Sp. pl.* 1551.
Creſted Polypody.
Nat. of Britain.
Fl. June——Auguſt. H. ♃.

patens. 9. P. fronde bipinnatifida ſubtus villoſiuſcula: pinnis lineari-lanceolatis elongatis: pinnulis oblongis acutis integris: infimis longioribus. *Swartz prodr.* 133.
Filix non ramoſa minor, ſurculis crebris, pinnulis breviſſimis anguſtis. *Sloan. jam.* 1. *p.* 91. *t.* 52. *f.* 1.
Pubeſcent Polypody.
Nat. of Jamaica.
Introd. 1784, by Mr. Lindſay.
Fl. July——September. S. ♃.

Filix mas. 10. P. frondibus bipinnatis: pinnis obtuſis crenulatis, ſtipite paleaceo. *Sp. pl.* 1551.
Male Polypody.
Nat. of Britain.
Fl. June——Auguſt. H. ♃.

11. P.

11. P. frondibus bipinnatis glabris: pinnis obtusis argute serratis: superioribus ovatis; mediis oblongis; inferioribus lanceolatis pinnatifidis acutiusculis. *elongatum.*
Cut-leav'd Polypody.
Nat. of Madeira and the Azores. Mr. *Fr. Masson.*
Introd. 1779.
Fl. G. H. ♃.

12. P. frondibus bipinnatis: pinnulis lanceolatis pinnatifidis acutis. *Sp. pl.* 1551. *Filix femina.*
Female Polypody.
Nat. of Britain.
Fl. June——August. H. ♃.

13. P. frondibus bipinnatis: pinnis pinnatifidis integerrimis subtus undique polline tectis. *Syst. veget.* 937. *Thelypteris.*
Acrostichum Thelypteris. *Sp. pl.* 1528.
Marsh Polypody.
Nat. of Britain.
Fl. August. H. ♃.

14. P. frondibus bipinnatis: pinnis lunulatis ciliato-dentatis, stipite strigoso. *Sp. pl.* 1552. *aculeatum.*
Prickly Polypody.
Nat. of Britain.
Fl. June——August. H. ♃.

15. P. frondibus bipinnatis: foliolis pinnisque remotis lanceolatis: serraturis acuminatis. *Sp. pl.* 1552. *rhæticum.*
Stone Polypody.
Nat. of England.
Fl. June——August. H. ♃.

16. P. frondibus bipinnatis: pinnis basi sinuato-repandis, fructificationibus marginalibus. *Sp. pl.* 1552. *marginale.*

Marginal-flowering Polypody.
Nat. of Canada.
Introd. 1772, by Samuel Martin, M.D.
Fl. June——September. H. ♃.

fragile. 17. P. frondibus bipinnatis: foliolis remotis: pinnis subrotundis incisis. *Sp. pl.* 1553.
Brittle Polypody.
Nat. of Britain.
Fl. June——August. H. ♃.

***** *Fronde supradecomposita.*

axillare. 18. P. fronde tripinnata glabra: pinnis oblongis apice serratis adnatis paucifloris.
Slender Polypody.
Nat. of Madeira. Mr. *Francis Masson.*
Introd. 1779.
Fl. G. H. ♃.

umbrosum. 19. P. fronde tripinnata glabra: pinnis lanceolato-linearibus serratis adnatis multifloris.
Madeira Wood Polypody.
Nat. of Madeira. Mr. *Francis Masson.*
Introd. 1779.
Fl. G. H. ♃.

Dryopteris. 20. P. frondibus supradecompositis: foliolis ternis bipinnatis. *Sp. pl.* 1555.
Branched Polypody.
Nat. of Britain.
Fl. June——September. H. ♃.

æmulum. 21. P. fronde quadripinnatifida glabra: pinnis oblongo-linearibus incisis: pinnulis apice denticulatis.
Dwarf Madeira Polypody.
Nat. of Madeira. Mr. *Francis Masson.*

Introd.

Introd. 1779.
Fl. G. H. ♃.

22. P. fronde quinquepinnatifida ſubglabra membranacea, pinnulis acutis argute ſerratis, rachibus ramulorum marginatis. *Swartz prodr.* 134. *effuſum.*
Adiantum nigrum ramoſum maximum, foliis ſeu pinnulis tenuibus, longis, acutis, ſpinoſis. *Sloan. jam.* 1. *p.* 97. *t.* 57. *f.* 3. *inferior.*
Spreading Polypody.
Nat. of Jamaica.
Introd. 1769, by Mr. George Houſtoun.
Fl. November. S. ♃.

ADIANTUM. *Gen. pl.* 1180.

Fructific. in maculis terminalibus, ſub replicato margine frondis.

1. A. frondibus reniformibus ſimplicibus ſtipitatis multifloris. *Sp. pl.* 1556. *reniforme.*
Kidney-leav'd Maidenhair.
Nat. of Madeira.
Introd. 1778, by Mr. Francis Maſſon.
Fl. G. H. ♃.

2. A. frondibus pedatis: foliolis pinnatis: pinnis antice gibbis inciſis fructificantibus. *Sp. pl.* 1557. *pedatum.*
Canadian Maidenhair.
Nat. of Virginia and Canada.
Introd. before 1640, by John Tradeſcant, Jun. *Park. theat.* 1050. *f.* 3.
Fl. Auguſt and September. H. ♃.

Capillus veneris. 3. A. frondibus decompositis: foliolis alternis: pinnis cuneiformibus lobatis pedicellatis. *Sp. pl.* 1558.
True Maidenhair.
Nat. of Britain.
Fl. May——August. H. ♃.

villosum. 4. A. frondibus bipinnatis: pinnis rhombeis antice extusque fructificantibus, stipite villoso. *Sp. pl.* 1558.
Hairy-stalk'd Maidenhair.
Nat. of Jamaica.
Cult. 1775, by John Fothergill, M. D.
Fl. S. ♃.

fragrans. 5. A. frondibus bipinnatis glabris: pinnis lobatis: lobis obtusis.
Adiantum fragrans. *Linn. suppl.* 447.
Polypodium fragrans. *Linn. mant.* 307.
Sweet-scented Maidenhair.
Nat. of Madeira.
Introd. 1778, by Mr. Francis Masson.
Fl. G. H. ♃.

pteroides. 6. A. fronde supradecomposita: pinnis ovatis integris crenulatis, stipite lævi. *Linn. mant.* 130.
Heart-leav'd Maidenhair.
Nat. of the Cape of Good Hope.
Introd. 1775, by Mr. Francis Masson.
Fl. G. H. ♃.

TRICHO-

TRICHOMANES. *Gen. pl.* 1181.

Fructific. ſolitariæ, ſtylo ſetaceo terminatæ, margini ipſi frondis inſertæ.

1. T. frondibus ſupradecompoſitis tripartitis: foliolis alternis: pinnis alternis pinnatifidis. *Sp. pl.* 1562. *Jacqu. ic. collect.* 1. *p.* 121. *canariense.*
Hare's-foot Trichomanes, or Fern.
Nat. of Portugal, Madeira, and the Canaries.
Cult. 1741, by Archibald Duke of Argyle.
Fl. Moſt part of the Summer.

DICKSONIA. *L'Herit. ſert. angl.*

Fructific. margini averſo frondis ſubjectæ, reniformes, bivalves: valvula exterior ex ipſa ſubſtantia folii; interior membranacea.

1. D. frondibus ſupradecompoſitis villoſis: foliolis ſubintegris, caule arboreo. *L'Herit. ſert. angl. n.* 1. *tab.* 43. *arboreſcens.*
Tree Dickſonia.
Nat. of the Iſland of St. Helena.
Introd. 1786, by Mr. Anthony Hove.
Fl. Moſt part of the Winter. S. ♄.

2. D. frondibus ſupradecompoſitis glabris: foliolis ſerratis. *L'Herit. ſert. angl. n.* 2. *Culcita.*
Shining-leav'd Dickſonia.
Nat. of Madeira and the Azores. Mr. *Fr. Maſſon.*
Introd. 1779.
Fl. G. H. ♃.

PILULARIA. *Gen. pl.* 1183.

Flores *Masculi* ad latus frondis.
Fructif. *Feminea* ad radicem, globosa, quadrilocularis.

globulifera. 1. PILULARIA. *Sp. pl.* 1563.
Pepper-grass.
Nat. of Britain.
Fl. June——September. H. ♃.

ISOETES. *Gen. pl.* 1184.

Fl. *Masculi Anthera*, intra basin frondis.
Fl. *Feminei Capsula* bilocularis, intra basin frondis.

lacustris. 1. I. foliis subulatis semiteretibus recurvis. *Syst. veget.* 942.
Quill-wort.
Nat. of Wales and Scotland.
Fl. May——October. H. ♃.

MUSCI.

LYCOPODIUM. *Gen. pl.* 1185.

Anthera bivalvis, sessilis. *Calyptra* nulla.

1. L. foliis sparsis filamentosis, spicis teretibus pedunculatis geminis. *Sp. pl.* 1564. *clavatum.*
Common Club-moss.
Nat. of Britain.
Fl. July. H. ♃.

2. L. foliis sparsis integerrimis, spicis terminalibus foliosis. *Sp. pl.* 1565. *inundatum.*
Marsh Club-moss.
Nat. of Britain.
Fl. July. H. ♃.

3. L. foliis sparsis octofariis, caule dichotomo erecto fastigiato, floribus sparsis. *Sp. pl.* 1565. *Selago.*
Fir Club-moss.
Nat. of Britain.
Fl. August. H. ♃.

4. L. foliis sparsis decurrentibus, sarmentis repentibus, surculis erectis dichotomis. *Sp. pl.* 1566. *obscurum.*
Fan Club-moss.
Nat. of North America.
Introd. 1770, by Mr. William Young.
Fl. H. ♃.

complanatum. 5. L. foliis bifariis connatis: ſuperficialibus ſolitariis, ſpicis geminis pedunculatis. *Sp. pl.* 1567.

Arbor Vitæ Club-moſs.

Nat. of North America.

Introd. 1770, by Mr. William Young.

Fl. H. ♃.

helveticum. 6. L. foliis bifariis patulis: ſuperficialibus diſtichis, ſpicis geminis pedunculatis. *Sp. pl.* 1568. *Jacqu. auſtr.* 2. *p.* 57. *t.* 196.

Spreading Lycopodium.

Nat. of Switzerland and Madeira.

Introd. 1779, by Mr. Francis Maſſon.

Fl. G. H. ♃.

PALMÆ

P A L M Æ

FLABELLIFOLIÆ.

THRINAX. *Swartz prodr.*

Cal. 6-dentatus. *Cor.* nulla. *Stam.* 6. *Stigma* emarginatum. *Bacca* 1-sperma.

1. THRINAX. *Swartz prodr.* 57. *parviflora.*
Corypha palmacea, foliis flabelliformibus cum appendicula ad imum, petiolis tenuioribus flexilibus compressis. *Brown. jam.* 190.
Small Jamaica Fan-palm.
Nat. of Jamaica.
Introd. 1778, by William Wright, M. D.
Fl. S. ♄.

RHAPIS. *Linn. fil.*

HERMAPHR. *Cal.* 3-fidus. *Cor.* 3-fida. *Stam.* 6. *Pist.* 1.
MASC. *Cal. Cor. Stam.* ut in hermaphrodito.

1. R. frondibus palmatis plicatis: plicis marginibusque aculeato-denticulatis. *flabelliformis.*
Rhapis flabelliformis. *L'Herit. stirp. nov. tom.* 2. *tab.* 100.
Chamærops excelsa. *Thunb. japon.* 130.
Creeping-rooted Rhapis, or Ground-Ratan.
Nat. of China and Japan.
Introd. about 1774, by Mr. James Gordon.
Fl. August. S. ♄.

2. R.

arundinacea. 2. R. frondibus bipartitis: lobis acutis plicatis: plicis ſcabriuſculis.
Simple-leav'd Rhapis.
Nat. of Carolina. Mr. *John Cree.*
Introd. 1765.
Fl. September. G. H. ♄.

CHAMÆROPS. *Gen. pl.* 1219.

Hermaphrod. *Cal.* 3-partitus. *Cor.* 3-petala. *Stam.* 6. *Piſt.* 3. *Drupæ* 3, monoſpermæ.
Masc. dioici, ut in hermaphrodito.

humilis. 1. C. frondibus palmatis plicatis, ſtipitibus ſpinoſis. *Sp. pl.* 1657.
Dwarf Fan-palm.
Nat. of the South of Europe.
Cult. 1731. *Mill. dict. edit.* 1. Palma 2.
Fl. G. H. ♄.

CORYPHA. *Gen. pl.* 1221.

. . . *Cal.* . . . *Cor.* 3-petala, *Stam.* 6. *Piſt.* 1.
. . . *Drupa* 1-ſperma.

umbraculifera. 1. C. frondibus pinnato-palmatis plicatis filo interjectis. *Sp. pl.* 1657.
Great Fan-palm.
Nat. of the Eaſt Indies.
Cult. before 1742, by Robert James Lord Petre.
Fl. S. ♄.

PENNATI-

PENNATIFOLIÆ.

CYCAS. *Linn. mant.* 166.

MASC. *Amentum* ſtrobiliforme ſquamis ſubtus undique tectis polline.

FEM. *Spadix* enſiformis. *Germinibus* angulo immerſis, ſolitariis. *Styl.* 1. *Drupa* nucleo ligneo.

1. C. frondibus pinnatis: foliolis linearibus planis. *Sp. pl.* 1658. (excluſis ſynonymis Sebæ et Kæmpferi.) *circinalis.*
Broad-leav'd Cycas.
Nat. of the Eaſt Indies.
Introd. about 1763, by John Blackburne, Eſq.
Fl. S. ♄.

2. C. frondibus pinnatis: foliolis margine revoluto. *Thunb. japon.* 229. *revoluta.*
Arbor calappoides ſinenſis. *Rumph. amb.* 1. *p.* 92. *t.* 24.
Teſſio. *Kæmpf. amæn.* 897.
Narrow-leaved Cycas.
Nat. of China and Japan.
Introd. about 1758, by Mr. Richard Warner.
Fl. S. ♄.

COCOS. *Gen. pl.* 1223.

MASC. *Cal.* 3-partitus. *Cor.* 3-petala. *Stam.* 6.

FEM. *Cal.* 5-partitus. *Cor.* 3-petala. *Stigm.* 3. *Drupa* coriacea.

1. C. inermis, frondibus pinnatis: foliolis replicatis enſiformibus. *Syſt. veget.* 985. *nucifera.*
Cocoa-nut Tree.
Nat. of the Eaſt Indies.

Cult.

Cult. 1739, in Chelſea Garden. *Rand. chel.* Palma 11.
Fl. S. ♄.

aculeata. 2. C. aculeato-ſpinoſa, caudice fuſiformi, frondibus pinnatis, ſtipitibus ſpathiſque ſpinoſis. *Swartz prodr.* 151.
Palma pinnis et caudice ubique aculeatiſſimis, fructu majuſculo. *Brown. jam.* 344.
Palma tota ſpinoſa major, fructu pruniformi. *Sloan. hiſt.* 2. *p.* 119.
Great Macaw-tree.
Nat. of Jamaica.
Cult. 1731. *Mill. dict. edit.* 1. Palma 6.
Fl. S. ♄.

PHŒNIX. *Gen. pl.* 1224.

MASC. *Cal.* 3-partitus. *Cor.* 3-petala. *Stam.* 3.
FEM. *Cal.* 3-partitus. *Cor.* 3-petala. *Piſt.* 1. *Drupa* ovata.

dactylifera. 1. P. frondibus pinnatis: foliolis enſiformibus complicatis. *Sp. pl.* 1658.
Date Palm-tree.
Nat. of the Levant.
Cult. 1731. *Mill. dict. edit.* 1. Palma 1.
Fl. S. ♄.

ARECA. *Gen. pl.* 1225.

MASC. *Cal.* . . . *Cor.* 3-petala. *Stam.* 9.
FEM. *Cal.* . . . *Cor.* 3-petala. *Drupa* calyce imbricato.

oleracea. 1. A. foliolis integerrimis. *Syſt. veget.* 986.
Cabbage Tree.
Nat. of the Weſt Indies.
Introd. 1787, by Hinton Eaſt, Eſq.
Fl. S. ♄.

ELATE.

ELATE. *Gen. pl.* 1226.

MASC. *Cal.* 3-dentatus. *Cor.* 3-petala. *Antheræ* 6, sessiles.

FEM. *Cal.* 1-phyllus. *Cor.* 3-petala. *Pist.* 1. *Stigmata* 3. *Drupa* 1-sperma.

1. E. frondibus pinnatis: foliolis oppositis. *Sp. pl.* 1659. *sylvestris.*
Prickly-leav'd Elate.
Nat. of the East Indies.
Introd. about 1763, by John Blackburne, Esq.
Fl. S. ♄.

ZAMIA. *Gen. pl.* 1227.

MASC. *Amentum* strobiliforme, squamis subtus tectis polline.

FEM. *Amentum* strobiliforme, squamis utroque margine. *Bacca* solitaria.

1. Z. foliolis cuneiformibus rectis glaberrimis a medio ad apicem serratis subtus furfuraceis, stipite spinoso. *Linn. fil.* *furfuracea.*
Palma americana foliis Polygonati brevioribus leviter serratis et nonnihil spinosis, trunco crasso. *Pluk. alm.* 276. *t.* 103. *f.* 2. et 309. *f.* 5.
Palma americana crassis rigidisque foliis. *Herm. parad.* 210. *t.* 210.
Palmifolia femina. *Trew. ehret.* 5. *t.* 26.
Broad-leav'd Zamia.
Nat. of the West Indies.
Cult. 1691, in the Royal Garden at Hampton-court. *Pluk. loc. cit.*
Fl. July and August. S. ♄.

2. Z.

integrifolia. 2. Z. foliolis subintegerrimis obtusiusculis muticis rectis nitidis, stipite inermi. *Linn. fil.*

Zamia pumila. *Sp. pl.* 1659. (exclusis synonymis.)

Dwarf Zamia.

Nat. of East Florida.

Introd. 1768, by John Ellis, Esq.

Fl. July and August. S. ♄.

debilis. 3. Z. foliolis linearibus muticis apice serrulatis patenti-recurvis rachi canaliculata longioribus, stipite triquetro compresso inermi. *Linn. fil.*

Palma prunifera humilis non spinosa Insulæ Hispaniolæ, fructui jujubino similis, ossiculo triangulo. *Comm. hort.* 1. *p.* 111. *t.* 58.

Long-leav'd Zamia.

Nat. of the West Indies.

Introd. 1777, by Messrs. Kennedy and Lee.

Fl. July and August. S. ♄.

pungens. 4. Z. foliolis subulatis patentibus strictis rigidis mucronatis: margine exteriore baseos rotundato, stipite teretiusculo inermi. *Linn. fil.*

Palma sobolifera, ægyptia, foliis levioribus, fructu nigro. *Till. pis.* 129. *tab.* 45.

Needle Zamia.

Nat. of the Cape of Good Hope. Mr. *Fr. Masson.*

Introd. 1775.

Fl. G. H. ♄.

Cycadis. 5. Z. foliolis obliquis lineari-lanceolatis subulatis pilosis curvatis apice uni-vel trispinosis, stipite inermi. *Linn. fil.*

Zamia Cycadis. *Linn. suppl.* 443.

Cycas caffra. *Thunb. in nov. act. upsal.* 2. *p.* 283. *t.* 5.

α foliis apice unispinosis.

Entire narrow-leav'd Zamia.

β foliis

β foliis apice bi-vel tri-ſpinoſis.
Trifid narrow-leav'd Zamia.
Nat. of the Cape of Good Hope. Mr. *Fr. Maſſon.*
Introd. 1775.
Fl. G. H. ♃.

APPENDIX.

1. GINGKO. *Linn. mant.* 313. *Thunb. japon.* 358. *biloba.*
Maiden-hair Tree.
Nat. of Japan.
Cult. 1758, by Mr. James Gordon.
Fl. H. ♄.

ADDENDA.

ADDENDA.

Vol. I. *pag.* 25.

acinifolia. VERONICA floribus solitariis pedunculatis, foliis ovatis glabris crenatis, caule erecto subpiloso. *Sp. pl.* 19.

Basil-leav'd Speedwell.

Nat. of the South of Europe.

Introd. 1788, by Edmund Davall, Esq.

Fl. April and May. H. ⊙.

Pag. 31.

prismatica. VERBENA diandra, spicis laxis, calycibus alternis prismaticis truncatis aristatis, foliis ovatis obtusis. *Sp. pl.* 27. *Jacqu. ic. vol.* 2.

Germander-leav'd Vervain.

Nat. of the West Indies.

Introd. 1787, by Mr. Alexander Anderson.

Fl. May and June. S. ♂.

Pag. 33.

triphylla. VERBENA tetrandra, floribus paniculatis, foliis ternis, caule fruticoso. *L'Herit. stirp. nov. p.* 21. *tab.* 11.

Three-leav'd Vervain.

Nat. of Chili.

Introd. 1784, by John Sibthorp, M. D.

Fl. G. H. ♄.

Pag. 57.

crispa. IXIA foliis linearibus crispis, floribus alternis. *Thunb. Ixia, n.* 8. *Linn. suppl.* 91.

Curled-

Curled-leav'd Ixia.
Nat. of the Cape of Good Hope.
Introd. 1787, by Mr. Francis Maſſon.
Fl. G. H. ♃.

Pag. 59.

IXIA foliis enſiformibus reflexo-falcatis. *Thunb. Ixia, n.* 23. *Linn. ſuppl.* 92. *falcata.*
Sickled-leav'd Ixia.
Nat. of the Cape of Good Hope.
Introd. 1787, by Mr. Francis Maſſon.
Fl. G. H. ♃.

Pag. 65.

GLADIOLUS foliis enſiformibus glabris, ſcapo triquetro: angulis membranaceis, limbo corollæ patenti-reflexo. *viridis.*
Green-flower'd Corn-flag.
Nat. of the Cape of Good Hope. Mr. *Fr. Maſſon.*
Introd. 1788.
Fl. July. G. H. ♃.
OBS. Singularis corollis viridibus, externe ſtriis obſolete purpureis.

GLADIOLUS corollæ erectæ limbo campanulato, floribus ſecundis, foliis criſpis. *Thunb. Gladiolus, n.* 7. *Linn. ſuppl.* 94. *criſpus.*
Curl'd Corn-flag.
Nat. of the Cape of Good Hope.
Introd. 1787, by Mr. Francis Maſſon.
Fl. G. H. ♃.

GLADIOLUS polyſtachyus, corolla ringente, ſpathis lacero-ariſtatis, foliis enſiformibus glabris. *Thunb. Gladiolus, n.* 16. *bicolor.*

Two-colour'd Corn-flag.
Nat. of the Cape of Good Hope.
Introd. 1786, by Mr. Francis Masson.
Fl. March. G. H. ♃.

gramineus. Gladiolus polystachyus, scapo laxo, spicis capillaribus flexuosis, foliis ensiformibus glabris. *Thunb. Gladiolus, n.* 26.
Gladiolus gramineus. *Linn. suppl.* 95. (exclusis synonymis.)
Grass-leav'd Corn-flag.
Nat. of the Cape of Good Hope.
Introd. 1787, by Mr. Francis Masson.
Fl. Most part of the Year. G. H. ♃.

Pag. 68.

ciliata. Iris barbata, foliiis ensiformibus ciliatis. *Thunb. Iris, n.* 1. *Linn. suppl.* 98.
Fringed-leav'd Iris.
Nat. of the Cape of Good Hope.
Introd. 1787, by Mr. Francis Masson.
Fl. G. H. ♃.

tricuspis. Iris barbata, folio lineari longiore scapo subbifloro, petalis alternis trifidis. *Thunb. Iris, n.* 15.
Iris tricuspidata. *Linn. suppl.* 98.
Single-flower'd Iris.
Nat. of the Cape of Good Hope.
Introd. 1787, by Mr. Francis Masson.
Fl. June. G. H. ♃.

Pag. 74.

bituminosa. Iris imberbis, foliis linearibus spiralibus, scapo viscoso. *Thunb. Iris, n.* 42. *Linn. suppl.* 98.
Clammy Iris.

Nat.

Nat. of the Cape of Good Hope.
Introd. 1787, by Mr. Francis Maſſon.
Fl. G. H. ♃.

Pag. 80.

Cyperus culmo triquetro nudo, umbella ſimplici, ſpiculis linearibus confertiſſimis horizontalibus. *Sp. pl.* 69. *ſtrigoſus.*
Briſtled-ſpiked Cyperus.
Nat. of the Weſt Indies.
Introd. 1786, by the Rev. Samuel Goodenough, LL. D.
Fl. July and Auguſt. S. ♃.

Pag. 85.

CORNUCOPIÆ. *Gen. pl.* 72.

Involucr. 1-phyllum, infundibulif. crenatum, multiflorum. *Cal.* 2-valvis. *Cor.* 1-valvis.

1. C. ſpica mutica, cucullo crenato. *Syſt. veget.* 103. *cucullatum.*
Hooded Cornucopiæ.
Nat. of the Levant.
Introd. 1788, by John Sibthorp, M. D.
Fl. Auguſt. G. H. ☉.

Pag. 125.

Protea foliis bipinnatis filiformibus, pedunculis capitulis brevioribus, ſquamis calycinis ovatis baſi villoſis. *Thunb. Protea, n.* 5. *Syſt. veget.* 136. *ſphærocephala.*
Round-headed Protea.
Nat. of the Cape of Good Hope.
Introd. 1788, by Mr. Francis Maſſon.
Fl. G. H. ♄.

Pag. 126.

Protea foliis tridentatis glabris ſecundis, caule decumbente, *hypophylla.*

cumbente, capitulo terminali. *Thunb. Protea, n.* 16. *Syst. veget.* 137.
Trifid-leav'd Protea.
Nat. of the Cape of Good Hope.
Introd. 1787, by Mr. Francis Masson.
Fl. G. H. ♄.

nana. PROTEA foliis lineari-subulatis, capitulo terminali, calyce colorato. *Thunb. Protea, n.* 29. *Syst. veget.* 139.
Dwarf Protea.
Nat. of the Cape of Good Hope.
Introd. 1787, by Mr. Francis Masson.
Fl. G. H. ♄.

Pag. 129.

grandiflora. PROTEA foliis oblongis venosis capituloque hemisphærico glabris, caule arboreo. *Thunb. Protea, n.* 51. *Syst. veget.* 141.
Great-flower'd Protea.
Nat. of the Cape of Good Hope.
Introd. 1787, by Mr. Francis Masson.
Fl. G. H. ♄.

Pag. 147.

cordifolia. R. foliis perennantibus quaternis cordatis. *Syst. veget.* 152.
Heart-leav'd Madder.
Nat. of Siberia.
Introd. 1783, by Mr. John Bell.
Fl. July. H. ♃.

Pag. 149.

PENÆA. *Gen. pl.* 138.

Cal. 2-phyllus. *Cor.* campanulata. *Styl.* 4-angularis. *Capſ.* tetragona, 4-locularis, 8-ſperma.

1. P. foliis cordatis acuminatis. *Sp. pl.* 162. *mucronata.*
Heart-leav'd Penæa.
Nat. of the Cape of Good Hope.
Introd. 1787, by Mr. Francis Maſſon.
Fl. G. H. ♄.

2. P. foliis rhombeo-cuneiformibus carnoſis. *Sp. pl.* 162. *ſquamoſa.*
Scaly Penæa.
Nat. of the Cape of Good Hope.
Introd. 1787, by Mr. Francis Maſſon.
Fl. G. H. ♄.

Pag. 180.

CYNOGLOSSUM foliis ſpathulato-lanceolatis lucidis baſi trinerviis, bractea pedunculorum amplexicauli. *Syſt. veget.* 186. *virginicum.*
Virginian Hound's-tongue.
Nat. of Virginia.
Cult. 1759, by Mr. Ph. Miller. *Mill. dict. edit.* 7. *n.* 5.
Fl. July. H. ♂.

Pag. 209.

CONVOLVULUS foliis cordatis integris trilobiſque villoſis, calycibus lævibus, capſulis hirſutis, pedunculis ſubbifloris. *Sp. pl.* 219. *carolinus.*
Carolina Bindweed.
Nat. of Carolina.
Cult. 1732, by James Sherard, M. D. *Dill. elth.* 100. *t.* 84. *f.* 98.
Fl. July. H. ♃.

Pag. 221.

verticillata. CAMPANULA foliis floribusque verticillatis. *Linn. suppl.* 141.
Whorl'd-leav'd Bell-flower.
Nat. of Siberia.
Introd. 1783, by Mr. John Bell.
Fl. June. H. ♃.

Pag. 222.

thyrsoidea. CAMPANULA hispida, racemo ovato-oblongo terminali, caule simplicissimo, foliis lanceolato-linearibus. *Sp. pl.* 235.
Long spiked Bell-flower.
Nat. of Germany.
Introd. 1785, by William Pitcairn, M.D.
Fl. July. H. ♂.

Pag. 224.

fruticosa. CAMPANULA capsulis columnaribus quinquelocularibus, caule fruticoso, foliis lineari-subulatis, pedunculis longissimis. *Sp. pl.* 238.
Shrubby Bell-flower.
Nat. of the Cape of Good Hope.
Introd. 1787, by Mr. Francis Masson.
Fl. August. G. H. ♄.

Pag. 226.

hemisphærica. PHYTEUMA capitulo subrotundo, foliis linearibus subintegerrimis. *Syst. veget.* 211.
Linear-leav'd horn'd Rampion.
Nat. of Switzerland.
Introd. 1784, by William Pitcairn, M.D.
Fl. H. ♃.

pinnata. PHYTEUMA floribus sparsis, foliis pinnatis. *Sp. pl.* 242.

Winged-

Winged-leav'd Rampion.
Nat. of the Iſland of Candia.
Introd. 1788, by John Sibthorp, M. D.
Fl. G. H. ♃.

TRACHELIUM ramoſiſſimum diffuſum, ramis divaricatis recurvis, foliis ſubulatis. *Linn. ſuppl.* 143. *diffuſum.*
Shrubby Throat-wort.
Nat. of the Cape of Good Hope.
Introd. 1787, by Mr. Francis Maſſon.
Fl. Auguſt. G. H. ♄.

Pag. 240.
HYOSCYAMUS foliis caulinis petiolatis cordatis ſinuatis acutis: floralibus integerrimis, corollis ventricoſis. *Sp. pl.* 257. *reticulatus.*
Egyptian Henbane.
Nat. of Egypt.
Cult. 1731, by Mr. Ph. Miller. *Mill. dict. edit.* 1. *n.* 5.
Fl. July. H. ⊙.

Pag. 258.
CHIRONIA herbacea, foliis linearibus. *Sp. pl.* 272. *linoides.*
Flax-leav'd Chironia.
Nat. of the Cape of Good Hope.
Introd. 1787, by Mr. Francis Maſſon.
Fl. Auguſt. G. H. ♃.

Pag. 267.
RHAMNUS aculeis geminatis: altero recurvo, foliis ovato-oblongis. *Sp. pl.* 281. *Lotus.*
Barbary Rhamnus.
Nat. of Barbary.
Introd. 1784, by Monſ. Thouin.
Fl. G. H. ♄.

Pag. 269.

stipularis. PHYLICA foliis linearibus stipulatis, floribus quinquecornibus. *Linn. mant.* 208.

Stipuled Phylica.

Nat. of the Cape of Good Hope.

Introd. 1786, by Mr. Francis Masson.

Fl. G. H. ♄.

Pag. 271.

PITTOSPORUM. *Gærtn. sem.* 1. *p.* 286.

Cal. deciduus. *Pet.* 5, conniventia in tubum. *Caps.* 2-5-valvis, 2-5-locularis. *Sem.* tecta pulpâ.

coriaceum. 1. P. foliis obovatis obtusis glaberrimis coriaceis, capsulis bivalvibus.

Thick-leav'd Pittosporum.

Nat. of Madeira.

Introd. 1787, by James Webster, Esq.

Fl. May. G. H. ♄.

Pag. 275.

capensis. DIOSMA foliis linearibus triquetris subtus punctatis, *Syst. veget.* 239.

Hartogia capensis. *Sp. pl.* 288.

Cape Diosma.

Nat. of the Cape of Good Hope.

Introd. 1786, by Mr. Francis Masson.

Fl. G. H. ♄.

Pag.

Pag. 295.

ALLAMANDA. *Linn. mant.* 146.

Contorta. *Capſ.* lentiformis, erecta, echinata, 1-locularis, bivalvis, polyſperma.

1. ALLAMANDA. *Linn. mant.* 214. *cathartica.*
Willow-leav'd Allamanda.
Nat. of Guiana.
Introd. 1785, by Baron Hake.
Fl. S. ♄.

Pag. 326.

ERYNGIUM foliis radicalibus cordatis; caulinis palmatis auriculis retroflexis, paleis tricuſpidatis. *Sp. pl.* 337. *tricuſpidatum.*
Trifid Eryngo.
Nat. of Spain.
Introd. 1786, by Monſ. Vare.
Fl. September. H. ♂.

Pag. 338.

CONIUM ſeminibus ſubmuricatis, pedunculis ſulcatis, foliolis canaliculatis obtuſis. *Syſt. veget.* 278. *rigens.*
Fine-leav'd Hemlock.
Nat. of the Cape of Good Hope.
Introd. 1787, by Mr. Francis Maſſon.
Fl. June. G. H. ♄.

Pag. 366.

RHUS foliis pinnatis ſerratis, petioli extimis internodiis membranaceis. *Murray in commentat. gotting.* 6. (1784) *p.* 27. *tab.* 3. *ſemialatum.*
Service-leav'd Sumach.
Nat. of Macao. Mr. *David Nelſon.*

Introd.

Introd. 1780, by Sir Joſeph Banks, Bart.
Fl. S. ♄.

Pag. 388.

ſtrictum. LINUM calycibus ſubulatis, foliis lanceolatis ſtrictis mucronatis: margine ſcabris. *Sp. pl.* 400.
Upright Flax.
Nat. of the South of Europe.
Introd. 1786, by Monſ. Thouin.
Fl. May. H. ♂.

Pag. 423.

tatari-cum. ALLIUM caule planifolio umbellifero, foliis ſemicylindricis, ſtaminibus ſimplicibus, umbella plana. *Linn. ſuppl.* 196.
Tartarian Garlic.
Nat. of Siberia.
Introd. 1787, by Mr. Haneman.
Fl. H. ♃.

Vol. II. *pag.* 18.

faſcicula-ris. ERICA antheris ariſtatis, corollis groſſis, ſtylo incluſo, floribus faſciculatis, foliis pluribus linearibus truncatis. *Linn. ſuppl.* 219.
Cluſter-flower'd Heath.
Nat. of the Cape of Good Hope.
Introd. 1787, by Mr. Francis Maſſon.
Fl. G. H. ♄.

Pag. 20.

retorta. ERICA antheris ſubcriſtatis, corollis ovato-oblongis, ſtylo mediocri, foliis quaternis recurvis. *Linn. ſuppl.* 220.
Recurved-leav'd Heath.
Nat. of the Cape of Good Hope.

Introd.

Introd. 1787, by Mr. Francis Maſſon.
Fl. G. H. ♄.

Pag. 22.

ERICA antheris muticis incluſis, corollis infundibuliformibus quaternis, ſtylo ſubexſerto, foliis quaternis pubeſcentibus. *pyramidalis.*
Pyramidal Heath.
Nat. of the Cape of Good Hope. Mr. *Fr. Maſſon.*
Introd. 1787.
Fl. October. G. H. ♄.

Pag. 24.

ERICA antheris muticis exſertis, corollis campanulatis, ſtylo exſerto, foliis ternis aceroſis. *Syſt. veget.* 369. *umbellata.*
Umbel'd Heath.
Nat. of Portugal.
Introd. about 1782, by Meſſrs. Lee and Kennedy.
Fl. May and June. G. H. ♄.

Pag. 27.

3. GNIDIA foliis oppoſitis lanceolatis. *Sp. pl.* 512. *oppoſitifolia.*
Oppoſite-leav'd Gnidia.
Nat. of the Cape of Good Hope.
Introd. 1788, by Mr. Francis Maſſon.
Fl. G. H. ♄.

Pag. 35.

PAULLINIA foliis biternatis, petiolo intermedio marginato; reliquis nudis. *Syſt. veget.* 380. *barbadenſis.*
Barbadoes Paullinia.
Nat. of the Weſt Indies.
Introd. 1786, by Mr. Alexander Anderſon.
Fl. S. ♄.

Pag.

Pag. 50.

viminea. Cassia foliis bijugis ovato-oblongis acuminatis, glandula oblonga inter infima, ſpinis ſubpetiolaribus obſoletis tridentatis. *Sp. pl.* 537.
Twiggy Caſſia.
Nat. of the Weſt Indies.
Introd. 1786, by Mr. Alexander Anderſon.
Fl. S. ♄.

Pag. 58.

CADIA. *Forſk. deſcr.*

Cal. 5-fidus. *Petala* 5, æqualia, obcordata. *Legumen* polyſpermum.

purpurea. 1. Cadia. *Forſk. deſcr. p.* 90.
Panciatica purpurea. *Piccivoli hort. panciat.* 9. *cum fig.*
Purple-flower'd Cadia.
Nat. of Arabia.
Introd. 1775, by James Bruce, Eſq.
Fl. S. ♄.

Pag. 75.

COPAIFERA. *Gen. pl.* 542.

Cal. 0. *Petala* 4. *Legum.* ovatum. *Sem.* 1, arillo baccato.

officinalis. 1. Copaifera. *Sp. pl.* 557.
Balſam of Capevi.
Nat. of South America.
Introd. 1788, by Mr. Alexander Anderſon.
Fl. S. ♄.

Pag.

Pag. 92.

CUCUBALUS foliis obovatis carnosis. *Sp. pl.* 591. *fabarius.*
Thick-leav'd Campion.
Nat. of Sicily.
Cult. 1759, by Mr. Ph. Miller. *Mill. dict. edit.* 7. *n.* 7.
Fl. G. H. ♃.

Pag. 138.

EUPHORBIA foliis sparsis lanceolatis acutis lævibus, pedunculis dichotomis, capsulis muricatis. *mellifera.*
Honey-bearing Euphorbia.
Nat. of Madeira. Mr. *Francis Masson.*
Introd. 1784.
Fl. April and May. G. H. ♄.

Pag. 215.

POTENTILLA foliis radicalibus quinatis; caulinis ternatis, caule erectiusculo ramosissimo. *Linn. mant.* 76. *intermedia.*
Agrimony-leav'd Cinquefoil.
Nat. of Switzerland.
Introd. 1786, by Mr. Haneman.
Fl. H. ♃.

POTENTILLA foliis radicalibus quinatis cuneiformibus serratis; caulinis suboppositis, ramis filiformibus decumbentibus. *Sp. pl.* 713. *opaca.*
Scotch Cinquefoil.
Nat. of Scotland.
Fl. May. H. ♃.

POTENTILLA foliis radicalibus et caulinis infimis quinatis, caulibus dichotomis villosis basi procumbentibus. *Jacqu. ic. miscell.* 2. *p.* 349. *astracanica.*
Downy Cinquefoil.

Nat.

Nat. of Siberia.
Introd. 1787, by Mr. Haneman.
Fl. H. ♃.

Pag. 244.

puniceum. Delphinium labellis bipartitis pilosis, nectarii cornu recto, foliis multipartitis, bracteis calycinis nullis. *Linn. suppl.* 267.
Scarlet-flower'd Larkspur.
Nat. of Siberia.
Introd. 1785, by William Pitcairn, M.D.
Fl. July. H. ♃.

Pag. 306.

crispus. Leonurus foliis omnibus acute serratis, rugosissimis, margine inæqualiter reflexis, caulinis quinquelobis. *Syst. veget.* 538. *Murray in nov. comment. gotting.* 8. (1777.) *p.* 44. *tab.* 4.
Curled-leav'd Mother-wort.
Nat. of Siberia.
Introd. 1784, by William Pitcairn, M.D.
Fl. July. H. ♃.

Pag. 328.

arvense. Melampyrum spicis conicis laxis: bracteis dentato-setaceis coloratis. *Syst. veget.* 550.
Purple Cow-wheat.
Nat. of England.
Fl. June. H. ☉.

Pag. 329.

recutita. Pedicularis caule simplici, foliis pinnatifidis serratis, spica foliosa, calycibus coloratis, corollis obtusis. *Syst. veget.* 551. *Jacqu. austr.* 3. *p.* 33. *tab.* 258.
Jagged-leav'd Louse-wort.

Nat.

Nat. of Switzerland and Austria.
Introd. 1787, by William Pitcairn, M. D.
Fl. H. ♃.

PEDICULARIS caule simplici, spica foliosa, corollis galea obtusissima integra, calycibus quinquedentatis. *Linn. mant.* 86. *Jacqu. austr.* 2. *p.* 24. *t.* 139. *foliosa.*
Leafy Louse-wort.
Nat. of Switzerland and Austria.
Introd. 1786, by Edmund Davall, Esq.
Fl. H. ♃.

Pag. 341.

SCROPHULARIA foliis cordato-ovatis duplicate dentato-serratis, dentibus basin respicientibus, racemo composito aphyllo. *Murray in commentat. gotting.* 4. (1781.) *p.* 35. *tab.* 2. *altaica.*
White-flower'd Fig-wort.
Nat. of Siberia.
Introd. 1786, by Mr. Haneman.
Fl. May. H. ♃.

Pag. 373.

LEPIDIUM foliis pinnatis: foliolis lunatis: exterioribus elongatis, ramis mucronatis. *Syst. veget.* 586. *spinosum.*
Prickly Pepper-wort.
Nat. of the Levant.
Introd. 1787, by Mr. Haneman.
Fl. September. H. ☉.

Pag. 397.

CHEIRANTHUS foliis lyratis, siliquis apice tridentatis. *Sp. pl.* 926. *tricuspidatus.*
Trifid Stock.
Nat. of Barbary.

Cult.

Cult. 1759, by Mr. Ph. Miller. *Mill. dict. edit.* 7. *n.* 14.
Fl. July. G. H. ☉.

Farsetia. CHEIRANTHUS filiquis ovalibus compressis, foliis lineari-lanceolatis, caule fruticoso erecto. *Syst. veget.* 598.
Flat-podded Stock.
Nat. of the Levant.
Introd. 1788, by John Sibthorp, M. D.
Fl. G. H. ♄.

Pag. 398.
inodora. HESPERIS caule simplici erecto, foliis subhastatis dentatis, petalis obtusis. *Sp. pl.* 927. *Jacqu. austr.* 4. *p.* 25. *t.* 347.
Wild Roc t, or Dame's Violet.
Nat. of England.
Fl. May and June. H. ♃.

Vol. III. *pag.* 62.

SMITHIA.

Legumen articulis distinctis, monospermis, stylo connexis. *Stam.* divisa in 2 phalanges æquales.

sensitiva. 1. SMITHIA. TAB. 13.
Annual Smithia.
Nat. of the East Indies.
Introd. 1785, by John Gerard Kœnig, M. D.
Fl. October. S. ☉.
DESCR. *Caulis* decumbens, teres, lævis. *Rami* patentissimi. *Folia* alterna, abrupte pinnata, 4-10-juga: *foliola* obovato-oblonga, margine et subtus secundum costam setosa. *Petiolus* brevissimus; rachis setosa. *Stipulæ* binæ, persistentes, supra insertionem

Smithia sensitiva.

The material originally positioned here is too large for reproduction in this reissue. A PDF can be downloaded from the web address given on page [illegible] of this book, by clicking on 'Resources Available'.

ſertionem ſemilanceolatæ, integerrimæ, acuminatæ; infra inſertionem bifidæ: lacinia altera obtuſa, altera longiore, acuminata. *Racemi* axillares, 3-6-flori. *Pedunculus* petiolo longior, filiformis. *Pedicelli* calyce breviores. *Bracteæ* ſub ſingulis pedicellis ſolitariæ, ſtipulis ſimiles, ſed minus deorſum productæ; bracteæ ſub calyce binæ, calycem exteriorem ſimulantes, ovato-lanceolatæ, muricatæ. *Calyx* tuberculis ſetiferis muricatus. *Corolla* lutea.

Pag. 64.

HEDYSARUM foliis ſimplicibus, bracteis ſtrobilorum inflatis cordatis obtuſis. *Syſt. veget.* 673. *ſtrobiliferum.*

Beech-leav'd Hedyſarum.

Nat. of the Eaſt Indies.

Cult. 1787, by Aylmer Bourke Lambert, Eſq.

Fl. S. ♄.

Pag. 107.

HYPERICUM floribus trigynis, caule tereti, foliis ſeſſilibus lanceolatis baſi undulato-dentatis. *Linn. mant.* 106. *criſpum.*

Curled-leav'd St. John's-wort.

Nat. of Greece.

Introd. 1788, by John Sibthorp, M. D.

Fl. G. H. ♃.

Pag. 174.

GNAPHALIUM fruticoſum, foliis ſublanceolatis tomentoſis ſeſſilibus, corymbis alternis conglobatis, floribus globoſis. *Sp. pl.* 1194. *igneſcens.*

Red-flower'd Everlaſting.

Nat.

Cult. 1768, by Mr. Philip Miller. *Mill. dict. edit.* 8.

Fl. H. ♃.

Pag. 253.

PALLASIA. *L'Herit. ſtirp. nov.*

Recept. paleaceum. *Pappus* nullus. *Semina* verticalia, plana, marginato-ciliata. *Cal.* imbricatus.

halimifolia. 1. PALLASIA. *L'Herit. ſtirp. nov. tom.* 2. *tab.* 19.
Coreopſis limenſis. *Jacqu. ic. vol.* 2.
Downy Pallaſia.
Nat. of Peru.
Introd. 1786, by Monſ. Thouin.
Fl. July. S. ♄.

Pag. 277.

Lingua. OTHONNA foliis ovato-lanceolatis ſemiamplexicaulibus. *Linn. ſuppl.* 387.
Tongue-leav'd Rag-wort.
Nat. of the Cape of Good Hope.
Introd. 1787, by Mr. Francis Maſſon.
Fl. May. G. H. ♃.

Pag. 284.

ſurinamenſis. LOBELIA caule ſuffruticoſo, foliis oblongis glabris ſerratis, floribus axillaribus pedunculatis. *Sp. pl.* 1320.
Lobelia lævigata. *Linn. ſuppl.* 392.
Shrubby Lobelia.
Nat. of the Weſt Indies.
Introd. 1786, by Mr. Alexander Anderſon.
Fl. April. S. ♄.

Pag. 286.

pubeſcens. LOBELIA caulibus angulatis proſtratis foliiſque lanceolatis dentatis hirtis, pedunculis axillaribus unifloris.

Downy-

Downy-leav'd Lobelia.
Nat. of the Cape of Good Hope. Mr. *W. Paterſon.*
Introd. 1780, by the Counteſs of Strathmore.
Fl. May——Auguſt. G. H. ♃.

Pag. 303.

EPIDENDRUM caule tereti lævi, foliis enſiformibus, petalis lanceolatis glabris, labio recurvato latiore. *Sp. pl.* 1352. *enſifolium.*
Limodorum enſatum. *Thunb. japon.* 29. *Ic. Kæmpfer. t.* 3.
Sword-leav'd Epidendrum.
Nat. of China and Japan.
Cult. 1780, by John Fothergill, M. D.
Fl. S. ♃.

Pag. 380.

MOMORDICA pomis angulatis tuberculatis apice deciduo operculatis, foliis lobatis. *Sp. pl.* 1433. *operculata.*
Rough-fruited Momordica.
Nat. of the Weſt Indies.
Introd. 1787, by Mr. Alexander Anderſon.
Fl. June——September. S. ☉.

SYNONYMA ADDENDA.

Vol. I. *pag.* 10. Jasminum officinale. *Curtis magaz.* 31.
16. Ancistrum latebrosum. *Gærtn. sem.* 1. *p.* 164. *tab.* 32.
20. Veronica decussata. *Moench hort. weissenst.* 137.
29. Calceolaria pinnata. *Curtis magaz.* 41.
62. Gladiolus communis. *Curtis mag.* 86.
72. Iris spuria. *Curtis magaz.* 58.
Iris ochroleuca. *Curtis magaz.* 61.
79. Cyperus viscosus. *Swartz prodr.* 20.
96. Agrostis complanata.
Chloris petræa. *Swartz prodr.* 25.
122. Triticum unioloides.
Poa sicula. *Jacqu. ic. vol.* 2.
Briza cynosuroides. *Scop. insubr.* 2. *p.* 21. *tab.* 11.
159. Cornus alba. *Pallas ross.* 1. *p.* 51. *tab.* 34.
164. Elæagnus angustifolia. *Pallas ross.* 1. *p.* 10. *tab.* 4.
165. Elæagnus orientalis. *Pallas ross.* 1. *p.* 11. *tab.* 5.
194. Cortusa Matthioli. *Jacqu. ic. collect.* 1. *p.* 236.
202. Spigelia marilandica. *Curtis magaz.* 202.
227. Rondeletia hirta. *Swartz prodr.* 41.
228. Solandra grandiflora. *Swartz prodr.* 42.
229. Hamellia grandiflora.
Hamellia ventricosa. *Swartz prodr.* 46.

232. Loni-

Vol. I. *p.* 232. Lonicera tatarica. *Pallas roff.* 1. *p.* 55. *tab.* 36.

233. Lonicera cærulea. *Pallas roff.* 1. *p.* 58. *tab.* 37.

240. Hyofcyamus aureus. *Curtis magaz.* 87.

247. Solanum laciniatum.

Solanum aviculare. *Forft. fl. auftr.* 18.

250. Solanum fubinerme. *Swartz prodr.* 47.

256. Lycium japonicum.

Buchozia coprofmoides. *L'Herit. monogr.*

261. Ardifia excelfa.

Anguillaria bahamenfis. *Gærtn. fem.* 1. *p.* 372. *tab.* 77.

265. Rhamnus ellipticus. *Swartz prodr.* 50.

Ceanothus reclinatus. *L'Herit. fert. angl.*

277. Itea Cyrilla. *Swartz prodr.* 50.

291. Illecebrum divaricatum.

Illecebrum canarienfe. *Linn. fuppl.* 161.

295. Gardenia dumetorum. *Gærtn. fem.* 1. *p.* 140. *tab.* 28.

Gardenia aculeata.

Gardenia Randia. *Swartz prodr.* 52.

302. Cynanchum crifpiflorum. *Swartz prodr.* 52.

320. Ulmus pumila. *Pallas roff.* 1. *p.* 76. *tab.* 48.

328. Sanicula marilandica. *Jacqu. ic. vol.* 2.

376. Xylophylla latifolia.

Genefiphylla afplenifolia. *L'Herit. fert. angl. tab.* 39.

Xylophylla falcata. *Swartz prodr.* 28.

385. Statice finuata. *Curtis magaz.* 71.

401. Pitcairnia bromeliæfolia.

Hepetis anguftifolia. *Swartz prodr.* 56.

405. Hæmanthus fpiralis.

Amaryllis fpiralis. *L'Herit. fert. angl. tab.* 13.

Vol. I. *p.* 410. Narcissus odorus. *Curtis magaz.* 78.
433. Eucomis undulata.
Eucomis regia. *L'Herit. sert. angl. n.* 1.
444. Scilla campanulata.
Scilla hyacinthoides. *Jacqu. ic. collect.* 1. *p.* 61.
461. Lachenalia tricolor β.
Lachenalia tricolor. *Curtis magaz.* 82.
Lachenalia pendula.
Phormium bulbiferum. *Cyrill. neapol.* 1. *p.* 35. *tab.* 12.
474. Hemerocallis fulva. *Curtis magaz.* 64.

Vol. II. *p.* 5. Epilobium angustissimum. *Curtis magaz.* 76.
36. Sapindus rigida. *Gærtn. sem.* 1. *p.* 341. *tab.* 70.
Sapindus edulis.
Litchi chinensis. *Sonnerat it. ind.* 2. *p.* 230. *tab.* 129.
Scytalia chinensis. *Gærtn. sem.* 1. *p.* 197. *tab.* 42.
37. Haloragis Cercodia.
Cercodia erecta. *Gærtn. sem.* 1. *p.* 164. *tab.* 32.
65. Ledum latifolium. *Jacqu. ic. vol.* 2.
75. Styrax lævigatum.
Styrax glabrum. *Cavan. diss.* 6. *p.* 340. *tab.* 188. *f.* 1.
Styrax americana. *De Lamarck encycl.* 1. *p.* 82.
78. Saxifraga Cotyledon β.
Saxifraga Aizoon. *Jacqu. austr.* 5. *p.* 18. *tab.* 438.

129. Cuphea

Vol. II. *p.* 129, Cuphea viscosissima. *Gærtn. sem.* 1. *p.* 210. *tab.* 44.

149. Sempervivum arachnoideum. *Curtis magaz.* 68.

153. Cactus pendulus.
Rhipsalis Cassutha. *Gærtn. sem.* 1. *p.* 137. *tab.* 28.

159. Myrtus Gregii.
Greggia aromatica. *Gærtn. sem.* 1. *p.* 168. *tab.* 33.

182. Mesembryanthemum cordifolium. *Jacqu. ic. vol.* 2.

183. Mesembryanthemum limpidum.
Mesembryanthemum cuneifolium. *Jacqu. ic. vol.* 2.

184. Mesembryanthemum barbatum. *Curtis magaz.* 70.

193. Mesembryanthemum pinnatifidum. *Curtis magaz.* 67.

194. Mesembryanthemum pomeridianum. *Jacqu. ic. vol.* 2.

207. Rosa muscosa *Curtis magaz.* 69.

216. Potentilla grandiflora. *Curtis magaz.* 75.

231. Gordonia pubescens.
Franklinia alatamaha. *Marsh. arb. amer.* 49.

272. Helleborus lividus. *Curtis magaz.* 72.

279. Teucrium betonicum.
Teucrium betonicæfolium. *Jacqu. collect.* 1. *p.* 145. *tab.* 17. *f.* 2.

334. Antirrhinum triste. *Curtis magaz.* 74.

Vol. III. *p.* 10. Spartium junceum. *Curtis magaz.* 85.

11. Spartium decumbens.
Le Genet de Haller. *Reynier mem. pour l'hist. nat. de la Suisse* 1. *p.* 211. *cum fig*

Vol. III. *p.* 27. Ebenus pinnata. *L'Herit. ſtirp. nov. tom.* 2. *tab.* 38.

55. Colutea arboreſcens. *Curtis magaz.* 81.

63. Hedyſarum veſpertilionis. *L'Herit. ſtirp. nov. tom.* 2. *tab.* 43.

101. Monſonia ovata.
Monſonia emarginata. *L'Herit. geran. tab.* 41.

115. Sonchus fruticoſus. *L'Herit. ſtirp. nov. p.* 169. *tab.* 81.

119. Prenanthes alba. *L'Herit. ſtirp. nov. tom.* 2. *tab.* 55.

187. Tuſſilago alpina. *Curtis magaz.* 84.

245. Verbeſina gigantea.
Verbeſina pinnatifida. *Swartz prodr.* 114.

255. Gorteria fruticoſa.
Gorteria Aſteroides. *Jacqu. ic. vol.* 2. *collect.* 1. *p.* 88.

395. Montinia caryophyllacea.
Montinia fruticoſa. *Gærtn. ſem.* 1. *p.* 170. *tab.* 33.

NOVA

NOVA GENERA.

MONANDRIA MONOGYNIA.

POLLICHIA.

CAL. monophyllus, ſubcampanulatus, quinquedentatus.
COR. nulla.
STAM. *Filamentum* unicum, filiforme, longitudine calycis. *Anthera* ſubrotunda, didyma.
PIST. *Germen* ſuperum, fundo calycis immerſum, ovatum. *Stylus* filiformis, longitudine ſtaminis. *Stigma* bifidum.
PER. nullum, vel membrana tenuis.
SEM. unicum, fundo calycis incraſſati incluſum, medio ſquamæ receptaculi affixum.
REC. *Squamæ* ſub antheſi ſubrotundæ, ſucculentæ, ſingula in medio proprium florem ſuſtentans; dein carnoſæ, ſubpellucidæ, ovato-oblongæ, erecto conniventes, baccam lobatam ſuperne apertam ſimulantes.

DIANDRIA DIGYNIA.

CRYPSIS.

CAL. *Gluma* uniflora, bivalvis: *Valvulæ* oblongo-lanceolatæ, carinatæ, muticæ: *exterior* minor.
COR. *Gluma* bivalvis: *Valvulæ* lanceolato-oblongæ, muticæ: *interior* calyce longior; *exterior* calyce brevior.

STAM.

STAM. *Filamenta* duo, capillaria, corolla longiora. *Antheræ* oblongæ, cordatæ, incumbentes.

PIST. *Germen* ſuperum, oblongum. *Styli* duo, capillares, ſtaminibus breviores. *Stigmata* piloſa.

PER. nullum. *Corollâ* ſemen includens.

SEM. unicum, teretiuſculum.

TRIANDRIA MONOGYNIA.

ARISTEA.

CAL. *Spathæ* bivalves.

COR. *Petala* ſex, oblonga, ſubæqualia, patentia.

STAM. *Filamenta* tria, filiformia, petalis breviora. *Antheræ* oblongæ, erecto-incumbentes.

PIST. *Germen* inferum, trigonum. *Stylus* filiformis, filamentis longior, declinatus. *Stigma* infundibuliforme, hians, margine fimbriatum, ſubtrigonum.

PER. *Capſula* oblonga, trigona, trilocularis, trivalvis.

SEM. plurima.

DIGYNIA.

PEROTIS.

CAL. nullus.

COR. *Gluma* bivalvis: *Valvulæ* oblongæ, acutæ, ſubæquales, apice ariſtatæ.

STAM. *Filamenta* tria, capillaria. *Antheræ* oblongæ.

PIST. *Germen* ſuperum, oblongum. *Styli* duo, capillares, corolla breviores. *Stigmata* plumoſa, divaricata.

PER. nullum. *Corolla* ſemen includens.

SEM. unicum, lineari-oblongum.

TETRAN-

TETRANDRIA MONOGYNIA.

CURTISIA.

CAL. *Perianthium* monophyllum, quadripartitum: *laciniæ* ovatæ, acutæ.

COR. *Petala* quatuor, ovata, obtuſa, ſeſſilia, calyce longiora.

STAM. *Filamenta* quatuor, receptaculo inſerta, ſubulata, petalis breviora. *Antheræ* ovatæ.

PIST. *Germen* ſuperum, ovatum. *Stylus* ſubulatus, longitudine ſtaminum. *Stigma* quadri-vel quinque-fidum.

PER. *Drupa* ſubgloboſa, glabra.

SEM. *Nux* ſubrotunda, oſſea, quadri-vel quinque-locularis. *Nuclei* ſolitarii, oblongi.

PENTANDRIA MONOGYNIA.

ARDISIA. *Swartz prodr.* 48.

CAL. *Perianthium* pentaphyllum: *foliola* oblonga, perſiſtentia.

COR. monopetala. *Tubus* breviſſimus. *Limbus* quinquepartitus: *laciniæ* oblongæ, patulo-reflexæ.

STAM. *Filamenta* quinque, tubo corollæ inſerta, ſubulata, breviſſima. *Antheræ* lanceolatæ, erectæ, magnæ.

PIST. *Germen* ſuperum, globoſum. *Stylus* filiformis, ſtaminibus longior. *Stigma* ſimplex.

PER. *Drupa* globoſa.

SEM. unicum.

STRELIT-

STRELITZIA.

CAL. *Spatha universalis* terminalis, monophylla, canaliculata, acuminata, patenti-declinans, multiflora, basin florum involvens.

Spathæ partiales lanceolatæ, floribus breviores.

Perianthium proprium nullum.

COR. irregularis. *Petala* tria, lanceolata, acuta: *inferius* naviculare; *superiora* obtuse carinata.

Nectarium triphyllum: *Foliola duo inferiora* petalis paulo breviora, e lata basi subulata, margine undulata, complicata, includentia genitalia, versus apicem postice aucta appendice crasso, forma dimidiæ sagittæ; *Foliolum inferius* breve, ovatum, compressum, carinatum.

STAM. *Filamenta* quinque, receptaculo insidentia, filiformia: tria in altero foliolo nectarii, duo cum stylo in altero foliolo inclusa. *Antheræ* lineares, erectæ, filamentis fere longiores, inclusæ.

PIST. *Germen* inferum, oblongum, obtuse trigonum. *Stylus* filiformis, longitudine staminum. *Stigmata* tria, subulata, petalis altiora, erecta, initio florescentiæ conglutinata.

CAPS. subcoriacea, oblonga, obtusa, obsolete trigona, trilocularis, trivalvis.

SEM. numerosa, conceptaculo centrali duplici ordine adhærentia.

PLOCAMA.

CAL. *Perianthium* monophyllum, minimum, quinquedentatum, persistens.

COR.

COR. monopetala, campanulata, quinquepartita: *laciniæ* oblongæ.

STAM. *Filamenta* quinque, tubo inſerta, brevia. *Antheræ* lineares, incumbenti-erectæ.

PIST. *Germen* inferum, globoſum. *Stylus* filiformiſubclavatus, ſtaminibus longior. *Stigma* ſimplex, obtuſum.

PER. *Bacca* ſubgloboſa, trilocularis.

SEM. ſolitaria, lineari-oblonga.

HEXANDRIA MONOGYNIA.

AGAPANTHUS. Mauhlia *Dahl obſ. bot.* 25.

CAL. *Spatha* communis latere dehiſcens.

COR. monopetala, infundibuliformis, regularis. *Tubus* angulatus (quaſi e ſex unguibus compoſitus). *Limbus* ſexpartitus: *laciniæ* oblongæ, patentes.

STAM. *Filamenta* ſex, fauci tubi inſerta, corolla breviora, declinata. *Antheræ* reniformes, incumbentes.

PIST. *Germen* ſuperum, oblongum, trigonum. *Stylus* filiformis, longitudine ſtaminum, declinatus. *Stigma* ſimplex.

PER. *Capſula* oblonga, triquetra, trilocularis, trivalvis: *valvulæ* naviculares interne dehiſcentes ſecundum carinam centralem.

SEM. numeroſa, oblonga, compreſſa, membrana aucta.

CYRTAN-

CYRTANTHUS.

CAL. nullus.

COR. monopetala, clavata, curva, apice ſexfida: *laciniæ* ovato-oblongæ: tres interiores obtuſæ; exteriores terminatæ corniculo parvo.

STAM. *Filamenta* ſex, tubo adnata, filiformi-ſubulata, corolla paulo breviora. *Antheræ* oblongæ, erectæ.

PIST. *Germen* inferum, ovatum, obtuſe trigonum. *Stylus* filiformis, longitudine corollæ. *Stigma* trifidum.

LANARIA.

CAL. nullus.

COR. monopetala, ſubcampanulata, extus plumoſo-lanata. *Tubus* brevis. *Limbus* ſexpartitus: *laciniæ* lineari-lanceolatæ, ſubpatulæ.

STAM. *Filamenta* ſex, baſi laciniarum corollæ inſerta, filiformia, corolla breviora. *Antheræ* ovatæ, ſubincumbentes.

PIST. *Germen* inferum, turbinatum, extus lanatum. *Stylus* filiformis, erectus, longitudine ſtaminum. *Stigma* trifidum.

PER. *Capſula* ovata, trilocularis.

SEM. pauca.

POLYANDRIA MONOGYNIA.

APEIBA. *Aublet guian.* 537. *Swartz prodr.* 82. Sloanea *Loefl. it.* 312.

CAL. *Perianthium* pentaphyllum: *foliola* lineari-lanceolata, decidua.

COR.

Cor. *Petala* quinque, obovata, calyce breviora.

Stam. *Filamenta* numerosissima, receptaculo inserta, corolla breviora. *Antheræ* filamentis infra apicem longitudinaliter adnatæ.

Pist. *Germen* subglobosum, depressum. *Stylus* filiformis, staminibus longior. *Stigma* simplex, apertum.

Per. *Capsula* subglobosa, depressa, echinata, multilocularis, vix dehiscens.

Sem. plurima.

DIDYNAMIA ANGIOSPERMIA.

PENTSTEMON. *Mitchell in act. ac. nat. curios.* 8. *append. p.* 214.

Cal. *Perianthium* pentaphyllum, persistens: *foliola* lanceolata, subæqualia.

Cor. monopetala, bilabiata. *Tubus* calyce longior, basi supra gibbosiusculus, superne amplior, ibique subtus ventricosus. *Labium superius* erectum, bifidum: *laciniæ* ovatæ, obtusæ, labio inferiore breviores. *Labium inferius* tripartitum: *laciniæ* ovatæ, obtusæ, deflexæ, tubo breviores.

Stam. *Filamenta* quatuor, basi corollæ inserta, filiformia, apice divaricata, tubo breviora: duo inferiora longiora. *Antheræ* subrotundæ, distantes, inclusæ, bifidæ: *lobi* divaricati.

Rudimentum filamenti quinti inter superiora tubo insertum, longitudine staminum, filiforme, rectum, apice superne barbatum.

Pist. *Germen* superum, ovato-conicum. *Stylus* filiformis.

mis, longitudine tubi, apice deflexus. *Stigma* truncatum.

PER. *Capſula* ovata, acuta, compreſſa, bilocularis, bivalvis.

SEM. numeroſa, ſubgloboſa. *Conceptaculum* magnum.

DIADELPHIA DECANDRIA.

CYLISTA.

CAL. *Perianthium* monophyllum, quadripartitum, maximum, perſiſtens: *lacinia* ſuprema reflexa, apice bifida; reliquæ erectæ, oblongæ, acutæ.

COR. papilionacea, calyce paulo longior, perſiſtens.

Vexillum ſubrotundum, emarginatum, baſi utrinque lobulo auctum.

Alæ oblongæ, obtuſæ, vexillo breviores, baſi utrinque proceſſu auctæ.

Carina oblonga, apice et baſi fiſſa, alis longior.

STAM. *Filamenta* diadelpha (ſimplex et novemfidum), adſcendentia. *Antheræ* ſubrotundæ.

PIST. *Germen* ſuperum, ovatum, compreſſum. *Stylus* ſubulatus, adſcendens. *Stigma* ſubcapitatum.

PER. *Legumen* ovato-oblongum, compreſſum, uniloculare.

SEM. duo, ovalia.

SMITHIA.

CAL. *Perianthium* monophyllum, bilabiatum: *laciniæ* ovato-lanceolatæ, fere æquales.

COR

COR. papilionacea.

Vexillum obcordatum.

Alæ oblongæ, obtufæ, vexillo paulo breviores.

Carina lineari-oblonga, bafi fiffa, longitudine alarum.

STAM. *Filamenta* decem, connata in duas phalanges æquales. *Antheræ* oblongæ.

PIST. *Germen* bafi calycis coarctatum. *Stylus* capillaris, perfiftens. *Stigma* fimplex.

PER. *Legumen* calyce inclufum, 4-7-articulatum: articulis diftinctis, ftylo perfiftente connexis, orbiculatis, muricatis, monofpermis.

SEM. reniformia, compreffa, glabra.

MONOECIA TETRANDRIA.

EMPLEURUM.

* *Mafculi Flores.*

CAL. *Perianthium* monophyllum, campanulatum, quadridentatum, perfiftens.

COR. nulla.

STAM. *Filamenta* quatuor, fubulata, calyce longiora, patula. *Antheræ* oblongæ, fubtetragonæ, retufæ.

* *Feminei Flores* in eadem planta.

CAL. ut in mare.

COR. nulla.

PIST. *Germen* fuperum, oblongum, compreffum, terminatum proceffu foliaceo, erecto. *Stylus* nullus.

 Stigma

Stigma denticulo lateri germinis impositum, cylindraceum, deciduum.

PER. *Capsula* oblonga, compressa, processu foliaceo coronata, unilocularis, secundum marginem rectiorem dehiscens.

SEM. solitarium, oblongum, arillatum: *arillo* subcoriaceo, bivalvi.

OBS. *Rarissime duae capsulae ex uno eodemque calyce.*

INDEX

INDEX

OF

GENERIC NAMES.

A.

Aſperula,

C.

Chondrilla,

Eclipta,

H.

Mimulus,

Panicum,

Sanguinaria,

INDEX

OF

ENGLISH NAMES.

Biſtort,

C.

Crefs-

I.

Indian

Lychnidea,

M.

Mud-

Spinage,

T.

Tree-

Widow-

Monœcia Monadelphia, next to Sterculia, Vol. III. *p.* 378.

HERITIERA.

* *Masculi Flores* minores femineis.

CAL. *Perianthium* monophyllum, campanulatum, quinquedentatum.

COR. nulla.

STAM. in centro calycis, columnare, conico-ſubulatum, infra apicem cinctum *Antheris* (5-10) minutis, connatis in cylindrum.

* *Feminei Flores* in eadem panicula cum maſculis.

CAL. ut in maſculis.

COR. nulla.

STAM. *Filamenta* nulla. *Antheræ* decem, receptaculo ad baſin germinum inſertæ, duæ inter ſingula germina, didymæ, minutæ, forte ſteriles.

PIST. *Germina* quinque, ſemiovata, compreſſa, glabra. *Styli* conici, breves, ſub antheſi apice cohærentes. *Stigmata* clavata.

PER. *Drupæ exſuccæ* patentiſſimæ, ovales, ſupra planiuſculæ, ſubtus convexæ, carinato-alatæ, uniloculares.

SEM. ſolitaria, ſubgloboſa, magna.

OBS. *Character e ſpeciminibus ſiccis, collatis deſcriptionibus manuſcriptis Kœnigii, e vivis in loco natali factis.*

littoralis. 1. HERITIERA.
Samandura. *Linn. zeyl.* 433.
Nagam. *Rheed. mal.* 6. *p.* 37. *tab.* 21.
Looking-glaſs Plant.

Nat.

Nat. of the East India Islands: Zeylon, *J. G. Kœnig*, M. D. Pulo Condore. Mr. *Dav. Nelson.*

Introd. 1780, by Sir Joseph Banks, Bart.

Fl. S. ♄.

Obs. Atunus litorea *Rumph. amb.* 3. *p.* 95. *tab.* 63. foliis et fructu omnino convenit, sed flores diversissimos describit Rumphius.

THE END.

"Blending personal reflection with sharp political insight, *Taipei at Daybreak* offers a perceptive exploration of coming of age within Taiwanese activist movements. Set against the backdrop of Taiwan's pivotal Sunflower Movement — one of the world's most significant social movements of the 2010s — this candid and self-interrogating narrative sheds light on the experience of living on the fringes of global attention. Balancing the perspectives of Asian Americans and expatriates in Taiwan's political landscape, the book captures the tension of leaving America for Taiwan, and navigates themes of identity, activism, and belonging. A must-read for those seeking a deeper understanding of Taiwan's political struggles and the broader complexities of life on the margins."

Michelle Kuo, author of *Reading with Patrick*

"Part social commentary and part bildungsroman, Taipei at Daybreak dissects the anatomy of political movements and offers an incisive probe into the psyches of young misfits who find themselves drawn to activism. This debut is a meditation on the performativeness of protests and of life itself, told through the lens of a narrator full of angst and bravado while still holding himself at a cool distance until a shift overtakes him. Hioe's writing is direct and assured as it takes readers through scene-by-scene breakdowns of Taiwan's recent movements, sketching out urban scenes in vivid detail and raising illuminating questions about community, alienation, and the boxes — both literal and psychological — that trap us."

Karen Cheung, author of *The Impossible City: A Hong Kong Memoir*

"If you've ever wondered how to take over a government building — and why young people have regularly done so — this book is for you. *Taipei at Daybreak* is a vivid account of twenty-first century Taiwanese youth activism from the front lines, propelled by rage and ennui. From the perspective of a narrator who is perpetually both an insider and an outsider, Hioe digs deep into the roots and emotions of radical activism and the desire for a different world."

Wendy Cheng, author of *Island X: Taiwanese Student Migrants, Campus Spies, and Cold War Activism*

"You don't need to live in a foreign land to feel disconnected from the image of yourself you've constructed. We might speak the same language as our friends, yet feel like the only listener is the shadow lurking within. This is a book for those who, despite having so much, often feel like orphans of the world."

Johnson Yeung, Hong Kong activist

"If the story of a nation is told through literature and cinema, then Hioe offers plenty of references for those new to Taiwan: from *Orphan of Asia* to *Taipei People*; from Hou Hsiao-Hsien to Edward Yang. Experts on Taiwan will enjoy the vignettes of a Taipei life they know well: the (original) Eslite, Cafe Philo, run-ins with gangsters, anti-nuclear movements, Nylon Cheng, Lin Yi-hsiung, the sound of your neighbor through the walls of your six-ping-sized rooms."

Emily Y. Wu, co-founder of Ghost Island Media

Printed in the United States
By Bookmasters